I0815994

# Into the Void

**Outward Odyssey**
A People's History of Spaceflight

*Series editor*
Colin Burgess

# Into the Void

## Adventures of the Spacewalkers

**John Youskauskas and Melvin Croft**

**Foreword by Jerry Ross**

UNIVERSITY OF NEBRASKA PRESS • LINCOLN

The University of Nebraska Press is part of a land-grant institution with campuses and programs on the past, present, and future homelands of the Pawnee, Ponca, Otoe-Missouria, Omaha, Dakota, Lakota, Kaw, Cheyenne, and Arapaho Peoples, as well as those of the relocated Ho-Chunk, Sac and Fox, and Iowa Peoples.

For customers in the EU with safety/GPSR concerns, contact:
gpsr@mare-nostrum.co.uk
Mare Nostrum Group BV
Mauritskade 21D
1091 GC Amsterdam
The Netherlands

Library of Congress Cataloging-in-Publication Data
Names: Youskauskas, John, author. | Croft, Melvin, author. | Ross, Jerry Lynn, 1948– writer of foreword.
Title: Into the void: adventures of the spacewalkers / John Youskauskas and Melvin Croft; foreword by Jerry Ross.
Description: Lincoln: University of Nebraska Press, [2025] | Series: Outward odyssey: a people's history of spaceflight | Includes bibliographical references and index.
Identifiers: LCCN 2024030243
ISBN 9781496224125 (hardback)
ISBN 9781496243195 (epub)
ISBN 9781496243201 (pdf)
Subjects: LCSH: Extravehicular activity (Manned space flight) | Astronauts—Biography. | Outer space—Exploration.
Classification: LCC TL1096 .Y58 2025 | DDC 629.45/84—dc23/eng/20250210
LC record available at https://lccn.loc.gov/2024030243

Designed and set in Garamond Premier Pro.

To the people on the ground who have dedicated their lives
to one of the greatest of all human adventures.

# Contents

# Illustrations

# Foreword

Flying in space is an amazing experience, even for someone who has been fortunate to have done it several times. But *walking* in space is the ultimate high (pun intended). It is a surreal feeling to get into your space suit, depressurize the airlock, open the outside hatch, and emerge into the infinity of space. Thinking about those experiences now as I write this foreword some twenty years after my last space flight and space walks, I immediately smile and feel like I am right back where I was then and where I feel I still belong.

Performing a space walk is a very demanding physical undertaking, and almost all the physical effort is expended between the shoulders and the fingertips. And as physically tired as I was after each space walk, I came inside as fatigued mentally as I was physically. My mind was racing in overdrive while on the space walks thinking about what I was supposed to do, whether we were keeping up with the timeline, choosing what tool I should use, deciding how many times to turn the bolt and at what torque, making sure that my spacewalking buddy was okay, that my safety tethers were not getting snagged. And it was always so tempting as I was busy at work to sneak a peek at the earth, the space station, the universe. There are great views of Earth about two hundred miles below you and going by at five miles per second, and although you don't have much time to enjoy the view, you do want to take some mental pictures to carry with you for the rest of your life.

There you are, getting ready to go outside of your spacecraft to perform a space walk. You will be leaving the cozy environment of your spaceship to go out into the vacuum of space, with its hazardous radiation and micrometeorites and space junk flashing by and where the temperatures can be +/– 250 degrees F. The entire space suit weighs over 250 pounds, and that doesn't include your own weight or that of any tools or other equipment that you will be carrying. You could be spending ten or more hours in your space suit. The suit is pressurized to 4.3 psi with pure oxygen, which makes the suit very stiff. The mass, bulk, and stiffness of the pressurized space suit makes moving and working in the vacuum of space difficult. Things that are easy to do down here on Earth can be very taxing up there. Your visibility, mobility, dexterity, and tactility are severely reduced. Working in a space suit is very much like putting on a heavy, stiff overcoat, a pair of thick leather welder's gloves, and a welder's helmet and then trying to replace that hard-to-reach spark plug on your car. Except in space, you and your tools will float away if you don't properly restrain yourself at your worksite and tether the tools and other equipment. Additionally, EVA is a contact sport—bruises, blisters, lost fingernails, and shoulder surgeries can be expected. I still have places on my back and on the backs of my hands that have reminders of my space-suit-wearing days.

Ninety-five percent of the likelihood of a safe and successful EVA is determined before you launch into space. Because the space suits severely limit what a spacewalker can do, the spacewalking crewmembers have developed hardware design inputs for engineers to consider when designing EVA tasks. These inputs can help make the required spacewalking tasks more compatible with the crewmember's reduced capabilities when working in the space suit and in zero gravity. In addition to making sure the design is acceptable, spacewalkers also need to make sure that there are adequate translation aids and appropriate foot restraint locations, that the flight tools fit the flight hardware they will interface with, that there are no sharp edges or other hazards to the space-suited crewmembers, that the forces required are not too great, that loose pieces are not created by the required actions, and that adequate visual confirmations are provided to verify that the job has been performed properly. Additionally, to ensure that the equipment will function properly in the exotic and demanding environment of space, some hardware functionality will need to be verified in thermal vacuum chambers on Earth before the suit is launched into space.

At the same time the flight hardware is being developed and tested, the EVA operations team is actively working in our underwater training facilities, double checking that each separate assembly and maintenance task is feasible, developing translation aids and restraint locations to accomplish each task, and planning the choreographed sequence of actions each crewmember will perform during the six- to seven-hour space walk. Once the space walk timeline has been developed, the mission crew begins to train for the planned

EVAs. The crew will have many tasks to perform during their space walks, and although they will normally have 15 to 20 percent more time for tasks in orbit than they had during their training in the water tank, they still must keep things moving to get everything done in the time scheduled.

The success of the International Space Station would not have been possible without the meticulous preplanning that preceded its construction. This complex and challenging effort served to uncover—and correct—a host of problems before the ISS hardware was launched into space. That space walks were carried out with so few problems or surprises is a testament to the skill, professionalism, and dedication of the entire spacewalking team.

*Jerry Ross*

*NASA mission specialist, STS-61B, STS-27, STS-37, STS-55, STS-74, STS-88, and STS-110*

# Acknowledgments

The authors owe a huge debt to the staff at the University of Houston–Clear Lake Archives, who fulfilled numerous requests for documents during our early research.

We thank the following astronauts for spending hours talking to us about their experience in preparing and executing their space walks and answering many follow-up questions: Tom Akers, Ed Gibson, Linda Godwin, John Herrington, Rick Hieb, Jeff Hoffman, Jerry Linenger, Jack Lousma, Story Musgrave, Ken Reightler, Jerry Ross, Russell Schweickart, Winston Scott, Robert Stewart, Nicole Stott, Joe Tanner, and Kathy Thornton. Jerry Linenger also provided thoughtful follow-up and the "hole in the ice" concept for chapter 12. Astronaut Scott Parazynski gave permission to use material from lectures that he made to NASA employees on spacewalking several years after he retired from NASA. He also graciously answered many inquiries. Tom Jones provided valuable technical expertise on the shuttle EMU. Kate Igoe with the Smithsonian National Air and Space Museum also provided valuable assistance.

Additionally, the authors are grateful to NASA EVA trainers Christy Hansen, Kieth Johnson, Ed Rezac, and Tomas Gonzalez-Torres. Their insight on what it takes to prepare astronauts to perform a space walk has added a unique perspective to this account. Amy Ross shared numerous ideas about space suit development.

We are most thankful to Colin Burgess, the series editor for the popular University of Nebraska Press Outward Odyssey book series. Colin arranged permission for us to use Bert Vis's material from his cosmonaut interviews. His sage advice and editing made this book much better than the initial manuscript, and his willingness to answer question after question and offer advice has vastly improved this story of spacewalking.

We also thank the staff at the University of Nebraska Press. Senior editor Rob Taylor and his talented team guided us effortlessly through the complicated process of writing and organizing this book. We are most thankful to our copy editor, Richard Feit, for his keen eyes and thoughtful corrections.

Lastly, our wives, Katherine Croft and Heather Youskauskas, gave us their complete support, willing to keep themselves amused while we worked diligently for several years to accurately get this story written.

# Into the Void

# Introduction

Among the extremely small sliver of Earthlings who have left the planet for low Earth orbit and ventured out to the vicinity of our closest celestial neighbor—the moon—spacewalkers are an even more elite subgroup of humanity. It is estimated that in all of time, more than 108 billion human beings have inhabited the planet called Earth. Of the nearly 700 travelers who have left Earth to journey into the heavens, fewer than 300 have had the unique opportunity to exit the safety of their life-sustaining spacecraft and float in the vast emptiness of outer space.

Although no real "walking" ever occurred, the press coined the term *space walk* to describe to the world the technological achievements of Russia's Alexei Leonov when he first exited his spacecraft in 1965 and American Edward White when he followed suit three months later. NASA's official jargon used to describe such endeavors outside the spacecraft was *extravehicular activity*, better known by its three-lettered acronym EVA. The term applies to any activity undertaken exterior to the cabin, be it in orbit, during interplanetary coast, or on the surface of another celestial body either with or without an atmosphere. But aside from the fourteen EVAs conducted on the lunar surface in one-sixth gravity, every other EVA to date—the main subject of this book—has been undertaken in weightlessness.

Upon opening the hatch of their spacecraft, American and European astronauts, Russian cosmonauts, and now Chinese taikonauts find themselves in a bizarre world. Space is devoid of everything supportive to human life. Oxygen, air pressure, and water are present only in unmeasurable traces. Without any medium to propagate the tiny pressure disturbances that are sound waves, space is absolutely silent. While heat and cold exist in the extreme, gravity is seemingly absent, offset by the centripetal force created by the incredible speed of the spacecraft around the planet. As it and everything with it falls endlessly over a horizon that continuously curves away from it, the resulting orbit leaves the spacefarer floating without any resistance, like a swimmer in water with no viscosity. It is a world of nothingness.

Yet in this strange alien environment, a human's eyes and brain, genetically programmed over millions of years to operate in the gravity of a solid surfaced planet, can adapt quickly to operate effectively in raw space. Despite the overwhelming beauty, enormity, and uniqueness of the universe engulfing them when they first emerge from their spacecraft, spacewalkers are able to push aside the distractions, thanks to intense and ever more realistic training on the ground. There is no one perfect method to recreate a space walk on Earth, no "zero gravity room" in which to train. So the challenge for them is to take each of the individual, imperfect pieces of simulation and mentally blend them into one experience. Even then, the learning curve for a rookie spacewalker is steep once out of the hatch.

The first spacewalkers of the 1960s entered uncharted territory. As earthbound theory became spaceflight reality once they exited their spacecraft, they abruptly realized that their abbreviated training had inadequately prepared them to work effectively. With time, experience, and many lessons learned, the true art of walking and working in space was mastered. The Americans and Soviets, the only space players in town during the 1960s, both had their sights set on conquering the moon, and placing boot prints on the lunar surface required a space suit that could withstand the harsh extremes of the vacuum and temperatures of space, keep their occupants alive, and facilitate exploration. And that space suit and the knowledge of how to function inside it had to be fully fleshed out before a landing on the moon could be attempted.

The space walks performed by Leonov and White in 1965 were much less choreographed and structured than those of modern-day spacewalkers, and the amount and type of preparation for the early space walks pales in comparison to that of today. Not only did the early astronauts have difficulty maintaining control of their bodies outside the spacecraft during EVA; they also often became overheated by the constant exertion in their efforts to maintain stability. EVA has evolved into a complex, procedural, step-by-step process honed over six decades, not only to accomplish useful work after they climb out of the airlock but also to ensure the astronaut's safe return to the spacecraft's life-sustaining artificial atmosphere. Gemini astronauts Gene Cernan, Michael Collins, and Dick Gordon, who all struggled with their space walks and failed to complete their assigned objectives in the mid-1960s, likely would have marveled at the long and exacting space walks carried out today on the International Space Station (ISS).

In the aftermath of their lunar programs, both the Soviets and Americans shifted their focus to space stations in low Earth orbit. Mission objectives brought forth new justification for EVA—both the U.S. *Skylab* and Soviet *Salyut* space stations required external inspections and repairs. The U.S. space shuttle was conceived with simple EVA contingency plans that evolved into dramatic satellite captures and repairs—such as regular servicing of the Hubble Space Telescope—spectacularly iconic jet pack flights, and construction of the ISS. In a time when the limits of what could be accomplished by spacefarers working in space were being tested, nothing seemed impossible. Surprisingly, although improvements have been made in both the American and Russian suits, new EVA space suits haven't been utilized in nearly four decades. Thus, spacewalkers have had to adapt to the aging suits as new challenges were conceived.

A modern-day space walk is an all-encompassing exercise in intense focus and professionalism. Zipping along at an astonishing five miles per second, spacewalkers are all business, unlike fictional Hollywood portrayals of spacewalking astronauts, like Matt Kowalski in the popular sci-fi movie *Gravity*, cartoonishly gallivanting around the outside of the space shuttle as if on a joy ride. An EVA is considered a plum assignment for an astronaut, and with Mission Control, the crew, and the world watching your every move, the pressure to perform is palpable. Nobody wants to be the one who fails. It's like being an athlete on the spot in a world championship game for nearly seven hours straight.

Venturing outside a spacecraft, while breathtaking, is potentially deadly. Micrometeoroids and the space junk orbiting Earth that has accumulated over the past sixty-plus years have the potential to strike a spacewalker and compromise their space suit. These tiny projectiles have been striking the ISS for two decades now, leaving its exterior pockmarked with tiny metallic impact craters and sharp edges that threaten to tear a pressurized glove or suit. If not immediately fatal, even a slow leak of the suit's precious atmosphere could result in death. And yet against all these odds, we've never lost a spacewalker in space.

Fear is not a part of the spacewalker persona. The risk of being struck by a bullet-like piece of invisible space debris is far less visceral than the gut-wrenching experience of the launch they endured on their way to orbit. Spacewalkers are aware of the risks they take when they exit their spacecraft, but they mitigate those they can control and don't dwell on those they cannot. They have total confidence in the suit—a small personal spaceship reminiscent of a cocoon—and the engineers and technicians who built it, right down to the seamstresses who sewed their unique personal gloves. Tools and equipment have been rigorously tested on Earth to ensure they will work as planned in orbit. Normal and emergency procedures for the space walk have been planned and rehearsed dozens of times under the watchful eyes of professional and dedicated trainers. Earth-based environmental simulators such as giant pools of water, aircraft flying huge parabolic arcs that mimic weightlessness, and even vault-like vacuum chambers that suck ambient air pressure to zero all introduce real hazards well before the astronauts ever leave the ground. The training may not be perfect, but once the spacewalkers climb out of the spacecraft, they have come as close as possible to simulating working effectively in the inhospitable and unforgiving environment of space.

With the exception of a handful of free-flying spacewalkers, most astronauts who venture beyond the safe confines of their craft are well secured by lifelines: umbilicals provide life-sustaining essentials like oxygen and communication, and tethers secure the spacewalkers to the spacecraft itself. With such precautions in place, there's little chance of spacewalkers drifting off into space, lost forever to Earth orbit and the eventual—and inevitable—reentry into the atmosphere to burn up like a meteor. Today's space suits are self-contained, with a backpack life-support system, so astronauts now rely on simple steel tethers to ensure that they remain firmly attached to the spacecraft. Tethering is a regimen learned through months of practice on the ground, often under water in a large pool, until it's almost a primal instinct, and maxims like "Make before you break" (never disconnect one tether before connecting another one to your next location) are fully internalized. Spacewalking from a spacecraft with its own rocket thrusters allows the security of being able to thrust the spacecraft and scoop up a detached astronaut if necessary. But on a space station with limited means of propulsion, becoming untethered just feet away from the massive structure could be as deadly as being separated by miles.

Although they are trained professionals dedicated to the successful completion of their responsibilities, spacewalkers are human, too. Every one of them eventually takes the opportunity to sneak a peek, if only for a moment, at the unfathomable vista of their home planet, deep black space, their spacecraft, and time. Observing the spectacles of space through a relatively unobstructed space helmet is far more spellbinding than looking at the sights through the confining window of a spacecraft.

Picture yourself here; the shimmering blue globe of Earth is not perceived as "down" in a weightless world where your spacecraft or sprawling space station is the center of your existence, your home off the planet. The intense focus required to perform useful work in a hostile environment while besting a stiffened rubber bladder pressure vessel—your space suit—is only momentarily interrupted by the exquisite beauty of Earth far away. Yet still, in a pause, a brief glimpse, from 220 miles high—a familiar city, bay, river, or island, the thin iridescent blue band of our tenuous atmosphere, the rainbow of light prismed through it exploding into the blinding and intense white glare of an orbital sunrise. Or a bright glistening day turned to night in a matter of seconds when the sun quickly dips behind the earth. A glance at the deep inky blackness of space is a look into the infinity of time itself.

Between these two vastly different yet visually spectacular scenes—Earth and space—you hold on with one gloved hand to a metallic structure the size of an American football field, with all sorts of ungainly appendages sticking out at odd angles: enormous shiny gold solar arrays, silvery laboratory modules reminiscent of huge tin cans, the dull and subdued tan Russian portions of the station.

There is near constant radio chatter in your headset; the intra-vehicular (IVA) crewmember (NASA jargon for "the one who stays inside") and Mission Control are your best friends during an EVA, eager to help out where they can to ensure that you successfully and safely work through your assignments under intense time pressures. There is only so much air, water, and filtering capability in the backpack, so timelines are carefully monitored by those on the ground, and suit status is dutifully scrutinized.

Occasionally, when not in direct communication with Earth, all that is heard is a gentle breeze of air as you slowly inhale the four pounds per square inch of life-sustaining air of the space suit. The immensity of the Milky Way fills the night sky below, while flashes of lightning arc through distant gray cotton-ball-like clouds blanketing the earth for hundreds of miles. But these are fleeting moments, snapshots to be taken by the mind to savor later when there isn't so much riding on the task at hand.

The reactions to these moments are as unique as the spacefarers who have experienced them. Jerry Linenger found himself dangling on the end of a space crane in the pure blackness listening to rock 'n' roll music as he orbited the planet outside the dilapidated Russian space station *Mir*. Bob Stewart had the opportunity to turn away from space shuttle *Challenger*, and away from Earth, to find himself seemingly alone in the universe as he tested a personal rocket pack. Rusty Schweickart, perched on the top of the lunar module *Spider* in Earth orbit, was suddenly informed that he had free time to do whatever he pleased on his 1969 *Apollo 9* mission in Earth orbit, and those minutes became a life-changing experience for him. Awe, wonder, amazement, and even stark fear were ever-present emotions among the chosen few who have taken part in what some consider the most prized job in the astronaut corps.

This is a story about the surreal experience of floating and working in open space from the spacewalkers who were fortunate to have had the opportunity. It's also the story of the remarkable men and women who put their blood, sweat, and tears into preparing these spacewalkers and assuring that their EVAs could be safely carried out. So crawl into your space suit and leave everything behind as you float with them out into the greatest display of grandeur you can possibly comprehend. Your new workplace is serene and majestic yet potentially lethal. Push that fear aside and *focus*. Trust your training and get to work—the clock is running.

And whatever you do, *don't let go*.

# 1

# Skywalkers

For many are called, but few are chosen.

—Matthew 22:14

Two hundred and twenty miles above planet Earth, the pinnacle of multinational cooperation in space—the International Space Station—orbits the world sixteen times a day at over seventeen thousand miles per hour. As long as an American football field and a mass equivalent to one million pounds, the ISS has been continuously crewed by astronauts, cosmonauts, and the occasional space tourist from some twenty countries since the year 2000. In what has become routine in space operations, the long-term residents of the ISS, who typically spend six-month rotations aboard the station, periodically venture outside into the vacuum of space to perform maintenance, add new equipment, and repair balky gear. Extravehicular activity—EVA for short—has been perfected over six decades of ever-expanding capabilities, and today's "spacewalkers" benefit from generations of space pioneers who came before and assumed enormous risks to prove the feasibility of working in open space.

On the morning of 16 July 2013, two astronauts—Chris Cassidy of the United States, designated EV-1, and Luca Parmitano of Italy, designated EV-2—prepared for a space walk in support of Expedition 36, tasked with routing cables and making external electrical connections for a future Russian module. It was the second space walk of the six-month-long mission for the pair, and Parmitano led the way out of the Quest airlock hatch into the darkness outside. They moved hand over hand along the gold-colored rails to their worksite on the pressurized docking tunnel between two of the modules. As the sun burst above the crisp blue horizon and flooded the silver modules and golden solar arrays in pure white light, Parmitano struggled with a weightless, tumbling bird's nest of tethered tools, bags, and the coiled cable he was to install between the two countries' modules.

As he began plugging in the electrical cables and twisting their locking mechanisms, the unflappably calm-mannered Cassidy began stepping into a portable foot restraint that would secure him at his workstation nearby. Helmet-mounted cameras transmitted an astronaut's-eye view of their space-gloved hands going about their work, with tethers, safety hooks, and equipment in an endless, slow-motion, tumbling dance in front of the helmet's faceplate. Astronaut Shane Kimbrough served as the ground intravehicular (IV) communicator in Houston throughout the EVA, keeping the spacewalkers on time and advising them on procedures they hadn't been able to rehearse in the months since they left Earth.

The vital data from the two astronauts' space suits flowed from transmitters, through the space station's complex communication system, up to a data relay satellite orbiting twenty-two thousand miles above, and down to the consoles of the Mission Control Center (MCC) in Houston, where a team of engineers, headed by Karina Eversley, kept a watchful eye on their health. The lead EVA officer on console that day, Eversley had learned Russian in college at the University of Wisconsin and had lived in Russia on and off for two years working with early ISS crews. She was assigned her first flight as an EVA team member in the MCC in 1998. When Parmitano was first alerted to the failure of a carbon dioxide ($CO_2$) sensor in his suit, the team looked at the data and agreed that it was not critical and that he could continue. This was an oversight that would come back to haunt the team in the hour that followed.

The $CO_2$ sensors were known to have failed in the past. The component was susceptible to moisture, and after six hours or more of working outside, the perspiration of the astronaut would occasionally overwhelm the sensor and render it inoperative. But this failure occurred less than an hour into the space walk. As EVA office head Chris Hansen recalled:

> Because we were so used to seeing that $CO_2$ sensor fail, when it happened, nobody made the connection that the fact that it failed so early was a problem. We call that normalization of deviance. We got used to that sensor failing, and therefore even when it failed at this very unusual time, nobody connected the dots and realized that [it] might be a problem. In fact it wasn't just moisture building up in the suit from normal use. The suit was actually failing at this point and this was the first warning we got from the suit that something was going wrong and we missed that.

The pair had been working on their initial tasks outside for about forty-four minutes when Parmitano began to notice the feeling of moisture on his head. *Sweat? Should I say something?* Realizing that any hint of a problem to the ground might lead to an early termination of the EVA, with what he later described as a "superhuman effort, I force[d] myself to inform Houston." He radioed calmly to Kimbrough, "Shane, FYI I feel a lot of water on the back of my head, but I don't think it leaked from my [drink] bag." Cassidy asked if he had been working up a sweat as he struggled against his pressurized suit. "I am sweating, but it feels like a lot of water," he answered in his clipped Italian accent. The water was soaking into his Snoopy communication cap, but almost as quickly as he brought it up, he went back to work and routine technical talk between the two astronauts resumed.

A liquid in zero-g behaves nothing like it does on Earth with gravity. It doesn't pour or flow or splash on a floor. Water's surface tension causes the molecules to stick together, as on the edge of an overflowed glass where a small rim of liquid just barely hangs on before spilling over the side, only in all three dimensions. Left on its own, a small amount squeezed from a drink bag naturally forms a sphere, floating serenely in the spacecraft's cabin like a perfect crystal ball, unless disturbed by a playful astronaut who might blow on it or set it gyrating with a finger. But when the fluid does touch a surface, it spreads out—and sticks, be it to a wall or a washcloth or the head and face of an astronaut in a sealed space suit with no way to wipe it dry.

Parmitano continued to work, wedging his bulky, pressurized suit and its backpack in between three of the ISS's modules at their connecting point trying to reach with his gloved hand into the space between. "I have very little mobility," he reported, his faceplate pressed right up against one of the handrails. "If we were doing a skills run, this is where they'd want me to rescue you," Cassidy quipped, referring to the dozens of hours spent learning a broad range of general EVA tasks in Houston's enormous Neutral Buoyancy Laboratory (NBL). Several of these runs culminated in the retrieval of a simulated "incapacitated crewmember," a spacewalker playing dead in the water who then had to be stuffed back into the airlock by the other.

Eversley and her team had been considering the water Parmitano reported, and they had Kimbrough ask him if it was getting worse and if he could tell where it was coming from. He had no idea. Cassidy made his way toward his partner and could now see blobs of water sticking to the inside of his bubble helmet. He thumped on Luca's visor and could see the sticky, weightless puddles of water wiggle. Assuming it had to be coming from his drink bag, Luca drank it dry. When Kimbrough asked if the water was increasing, Cassidy reported, "Luca was able to drink a couple of water bubble blobs off the visor of his helmet, so it's less."

Given the tone of the conversation, Cassidy had figured that the water was a relatively minor issue and thought it might take Mission Control about fifteen minutes to come up with a solution. But now, just over an hour since the EVA began, Parmitano was sounding more concerned. The controllers told them to stand by for a few minutes while they discussed the issue, so the two spacewalkers took the time to take photographs of each other as they flipped through their cuff checklists to display various logos of their military services and schools, the glowing blue-and-white Earth presenting a stunning backdrop.

As they waited, Luca could feel the water, now thoroughly soaking his communication cap, breaching into the earcups of his headset. *This can't be the drink bag . . . The water keeps coming.* "I guess it must be sweat or urine then," surmised Cassidy. "How much can I sweat though?" Luca responded. With no indication from controllers of what the problem might be, he started back to work on the cables. Then he asked for a comm check from Cassidy. His transmissions were now becoming increasingly garbled, and the request itself made everyone in the control room perk up.

"The only other option, Chris, that I'm thinking is the LCVG." Parmitano was still troubleshooting the suit in his mind as he tried to continue his tasks, and having eliminated some other possibilities, he suggested that the Liquid Cooling and Ventilation Garment—the long underwear laced with tubing that circulated cooling water—might be the source of a leak.

Eversley's eyes shot right to the data on Luca's LCVG and immediately ruled out any problems with it, based on what she could see. Kimbrough passed her evaluation up to the crew but was more concerned with Parmitano's ability to communicate. "No, I can hear perfectly . . . but my head is really wet and I am feeling that it's increasing," Luca replied. Kimbrough, now standing and pacing behind his CAPCOM (capsule communicator) seat, turned and glared over the console immediately behind him to Eversley, looking for answers.

After a rapid-fire discussion on the internal MCC communication loop, the decision was made. "Based on what we heard . . . we think we're in a 'Terminate EVA' case for EV-2," Kimbrough radioed up. "So Luca, we'll have you head back to the airlock, and Chris, we'll get a plan for you to clean things up." Even still, the true severity of the problem with Parmitano's suit had not become totally clear. "Terminate EVA" was a detailed procedure to end the space walk while still allowing time to clean up equipment and secure unfinished work. "Abort EVA," a worst-case scenario, would entail stopping everything immediately and getting back inside.

It had taken the controllers of the space station's twenty-third U.S.-based EVA twenty-three minutes to realize that the safest course of action was to get Parmitano inside and out of the space suit. Those precious twenty-three minutes lost would become critical in what was about to happen. With Parmitano heading back to the airlock alone while Cassidy stayed behind to tidy up, Mission Control had just split up the spacewalkers at the worst possible time.

The two were face to face, and for the first time, Luca's confidence was shaken, now realizing that he was going to be on his own. "I just thought, hey the ground told me to go back. Chris—he is not concerned. I shouldn't be concerned and I can deal with it. But I just remembered . . . when I turned around I just had a feeling—*I really wish Chris could come with me.*"

The year 1965 ushered in humanity's first space walks: Alexei Leonov on his *Voskhod 2* flight—the first space walk in history—and NASA's Ed White on the *Gemini 4* mission just three months later. The United States and the Soviet Union were both hell-bent on winning the race to the moon, and NASA had developed the Gemini spacecraft, an interim program following Project Mercury—America's first manned space program—to help them gain many of the critical skills required to land on the moon. Gemini was intended to progress NASA's spaceflight knowledge in key areas—including EVA—while they were in the early stages of planning and building the Apollo program that would ultimately carry men to the lunar surface before the end of the 1960s decade.

Since those first steps outside the spacecraft by the Americans and Soviets during the Gemini and Voskhod programs, EVA has been a key element of every crewed space endeavor, excluding commercial space ventures. For the Americans, that includes the Apollo lunar program, *Skylab*, the space shuttle, and the ISS. Likewise, EVA was an integral element of the various Soviet and Russian space programs following Voskhod; Soyuz, the Salyut and Mir space station programs, and the ISS joint venture with the United States, Europe, Japan, and Canada. China launched their first space explorers, dubbed taikonauts, into space in 2003 and carried out its first space walk on the *Shenzhou 7* mission in 2008. China recently began assembly of its third space station, *Tiangong*, and has already conducted multiple EVAs from it. Through December 2024, 480 EVAs have been performed by 276 individual spacewalkers, most often two at a time.

During the Gemini and Voskhod programs, it had to be proven that humans could climb outside their spacecraft, survive there, and perform meaningful work. For the Americans it also provided a safety net if a lunar module returning from the surface of the moon had been unable to dock with the command module or if the docking tunnel connecting the two spacecraft could not have been opened. In such an event, the moonwalkers would have had to translate—space walk—from the lunar module to the command module for a safe return to Earth. The Soviets had developed similar spacecraft for their lunar landing program, a lunar orbiter (LOK) and a lunar lander (LK). Unlike the American command and lunar modules, the Soviet cosmonauts had no transfer tunnel linking their two vessels, and therefore the single cosmonaut bound for the lunar surface would have had to translate from the LOK to the LK via a space walk and, upon returning from the moon, translate back to the LOK from the LK—while lugging a precious cargo of moon rocks in tow. The world was united—at least for a short time—when Neil Armstrong and Edwin Aldrin settled down on the Sea of Tranquility on the lunar surface in July 1969 and carried out the first non-weightless EVA in history.

Once the United States had bested the Soviets, their objective quickly evolved into recovering as much scientific data about the moon as possible. Twelve astronauts conducted fourteen EVAs across the lunar surface, which netted 842 pounds of lunar samples bursting with information that would keep scientists on Earth busy for decades. Astronauts bounding over the stark lunar landscape on six missions spanning nearly three-and-a-half years mesmerized millions of people worldwide. Perhaps equally as astonishing, the Americans and Soviets had eked out only a paltry eight short space walks (excluding stand-up EVAs) in the Voskhod, Gemini, Apollo, and Soyuz programs before Armstrong and Aldrin first planted footprints on the moon.

The Salyut, Skylab, and Mir space station programs afforded longer-duration missions requiring astronauts and cosmonauts to work outside the space stations to carry out experiments, retrieve film, make scientific observations, and repair faulty equipment. Astronauts and cosmonauts executed 106 combined EVAs during the three space station programs between 1973 and 2000. Here, EVA finally became a routine function of living and working in space; critical repairs were made to the stations, experiments were conducted, and scientific data collected. But most importantly, confidence in the human ability to function safely in ever more complex space suits was gained.

The first space walk in the shuttle program was achieved on the STS-6 mission in April 1983 by two NASA astronauts, Story Musgrave and Don Peterson. Over the next fifteen years NASA honed their knowledge of EVA with over fifty shuttle space walks. Satellites were retrieved from orbit and brought home in the shuttle's payload bay for repair or were captured, repaired, and redeployed. In 1984 the Manned Maneuvering Unit (MMU), a self-contained untethered backpack with its own propulsion system that is worn by astronauts, was used to recover two communications satellites on STS-51A for return to Earth inside *Discovery*'s payload bay. Between 1993 and 2009 five servicing missions to repair and upgrade the Hubble Space Telescope required twenty-three audacious space walks. As a result of improvements made on these missions, the HST is still delivering cutting-edge scientific breakthroughs to this day.

As building the ISS turned from promise to reality, construction techniques were tested and practiced numerous times by spacewalking astronauts on shuttle missions looking to gain as much knowledge and experience as possible prior to tackling the assembly of the behemoth space station, which began in 1998. Spanning thirty missions over ten years, astronauts performed myriad tasks required to assemble the station, including the installation of trusses, attachment of electrical connections, deployment of solar arrays, and much more, beginning with a single module that eventually expanded to a monstrosity of sixteen interconnected pressurized modules and structures so big that it can easily be seen in the night sky from Earth. Spacewalkers—American, Russian, Canadian, Japanese, and European—were the construction workers that assembled the iconic orbiting facility, perhaps analogous to space-age versions of the fearless girder-striding "sky walkers" that built many towering skyscrapers in the early twentieth century.

Once the shuttle was retired in 2011, the capability of spacewalking astronauts to capture and repair satellites was temporarily lost. But EVA would continue to be a necessary skill for those astronauts and cosmonauts working on the space station. The ISS is an enormous facility, with thousands of mechanical and electrical components, many which can be accessed only from the outside during a space walk. Equipment fails, leaks occur, batteries become dated. Spacewalkers are often required to inspect and troubleshoot hardware, looking for the root cause of a new anomaly. Maintenance is required to keep the station operational, such as lubrication of the robotic Canadarm. Scientific experiments are often installed on the outside of the station and later retrieved. It's impossible to foresee all the different types of space walks that will be required during a long-duration mission on the ISS.

EVA capability has truly enriched humankind's understanding of our universe. Scientists have been able to reconstruct the origin of the moon and to gain a better grasp of our own planet's early beginning. The ability of EVA astronauts to repair satellites in orbit has contributed to scientists being smothered with data, which has led to remarkable discoveries throughout the universe. The HST has confirmed the age of the universe (fourteen billion years), proven that the expansion of the universe is not slowing down and that Pluto has five moons, provided irrefutable verification that black holes exist, and much more. Without EVA expertise, the ISS could not have been assembled, and research in the life and physical sciences, space medicine, meteorology, and astronomy in a weightless environment free of the cumbersome atmosphere would not have happened.

From the early days of space exploration, only a select few were chosen for a space walk, primarily due to the limited number of available EVA slots. Through the first eighteen years of spaceflight, less than thirty space walks (excluding lunar and stand-up EVAs) were conducted by the Americans and Russians prior to the first shuttle-based space walk. With multiple orbiters that could carry up to seven astronauts and with a significant increase in opportunities for EVA, the number of astronauts stepping into the void of space blossomed. Still, not every astronaut would be given the opportunity to walk in space.

Those chosen to perform a space walk on the Voskhod, Gemini, and early Soyuz missions depended on the crew position; the commander stayed inside, while the pilot—or flight engineer in the Soviet program—executed the EVA. However, in at least one instance, the pilot on a mission with a scheduled space walk was moved to another flight because he may not have been the best choice for an EVA. Deke Slayton writes in his memoir, *Deke!*, that although he wanted to assign Elliot See to the *Gemini 8* mission, he didn't believe that See was in good enough physical condition to conduct a space walk, and Deke assigned him instead to be the commander of *Gemini 9*. Likewise during the Apollo program, crew positions dictated which astronaut carried out EVAS on the lunar surface as well as deep space EVAS during the return to Earth. Throughout the Skylab missions—three flights all flown in less than a year—all nine crewmembers ventured outside the space station. The commander and flight engineer both performed space walks on the Soviet *Salyut*—six total—and *Mir* space stations (the Russians took over ownership of *Mir* following the breakup of the Soviet Union in 1991).

Except for the first four shuttle missions—test flights with only two pilots constituting the crews—space walks on the shuttle were also dictated by crew position. Naturally, on the first four missions, the pilots were required to have EVA expertise in case an emergency space walk was required. Although some of the pilots, for example Mark Kelly and Ken Bowersox, carried out space walks on missions to the ISS as a member of an Expedition crew, none of the pilots ever performed an EVA as a commander or pilot on a shuttle mission. Following the test flights, the shuttle was deemed operational, and mission specialists took over the responsibility of spacewalking; the commanders and pilots were not in the mix for a space walk.

Although spacewalkers were selected from among a pool of mission specialists, pilots could still play a role in EVA. The Upper Atmosphere Research Satellite was to be deployed on STS-48 using the robotic arm, and there were several scenarios that might have required an EVA. Pilot Ken Reightler was trained as EV-3 for that mission; Sam Gemar and James Buchli were trained as the two primary EVA crewmen (EV-1 and -2). Reightler participated in all the EVA ground training, and instead of referring to the potential EVAS as contingency, he called them "mission success EVAS," as they were not for emergency situations but were critical to the success of the mission if the satellite could not be deployed as planned. Reightler made all the pool runs with Gemar and Buchli wearing SCUBA gear. Initially, there was no plan to have him train in the space suit, but Reightler "begged for one short run, just to be able to understand what they could see or not and what the limitations would be," so he earned some pool time in the suit. He also trained for the EV-3 position as a pilot on his second mission, STS-60. Reightler recalled a few other pilots who trained as EV-3.

But not all mission specialists were assigned a space walk. The conundrum around the Astronaut Office at the Johnson Space Center was how mission specialists were selected for a space walk during the shuttle program. Most of the spacewalkers the authors interviewed had no idea why they, or anyone else, were chosen for a coveted space walk. "No idea why I was selected," veteran EVA astronaut Tom Akers quipped. "God's blessing I'm sure." Two-time spacewalker Linda Godwin explained, "We never knew the answer to those questions any more than we knew why we got on a specific mission." There was no formal grading done during EVA training, "but anyone would realize that you're being watched. I mean opinions are always being formed—yes." Jerry Ross, whose career spanned the entire thirty years of the space shuttle program, concluded, "That's one of the magical questions in life. I think it was probably a joint thing between George [Abbey] and John [Young]. . . . They always had a kind of a feeler system out there of people that they took inputs from, so you don't know exactly who else might have been providing information."

Ross can afford to laugh today about his assignment as an EVA crewmember. "I can be somewhat sarcastic and say well, I got it because they knew I couldn't swim!" he recalled. "Of course, swimming is part of the prerequisites of getting trained to be a spacewalker. And so I thought maybe they're trying to see if Ross is going to sink or swim out of this whole thing." He believes that his athletic ability, strong upper body, and mechanical aptitude contributed to his being chosen for spacewalking duties. Ross recalled that following his astronaut candidate training, he was given some choices on his first job-training assignment, and he leaned toward checkout and verification of shuttle software because he thought it might get him assigned to a mission earlier than other assignments. But the power brokers of the Astronaut Office had other plans for the rookie, and Ross went on to become one of NASA's premier spacewalkers over his historic seven spaceflights.

Ross recalls a somewhat formal process for selecting mission specialists for EVA during his time with NASA, and he admitted that not every astronaut is made for spacewalking. Size, strength, adaptation to the suit, and many other factors eliminate some astronauts from EVA consideration. "And we actually had a rating system that the more senior EVA astronauts in the office and the trainers would rate each crewmember as they went through their training, and we would come up with recommendations to the management of the astronaut office as to who was satisfactory to lead a space walk, who could be a good number two and who you probably ought to think about not assigning to space walks."

Scott Parazynski, selected as an astronaut twelve years after Ross, remembers that astronauts were evaluated during their space walks, which helped management make decisions later, such as whether to assign an astronaut as the lead EVA crewperson or whether a mission specialist shouldn't do a space walk. Like Ross, Parazynski stressed that spacewalking is "not a job meant for everybody."

The selection process for spacewalking astronauts during the shuttle program was not a straightforward one and varied throughout the program. Three-time spacewalker Kathy Thornton provided insight on why she believes she was selected for a space walk. She had trained for a contingency EVA on her first mission, and she quickly learned during training that the space suit had not been built with her short arm length in mind. Additionally, for someone of her small stature, the Display and Control Module mounted on the front of the suit restricted her reach. Consequently, she had to work extra hard to learn how to carry out the EVA tasks that she might be required to execute in orbit. She believes that her hard work, as well as her positive attitude—not making excuses and criticizing the suit—was recognized by the EVA training team. "I developed one-handed techniques where other people could use two hands because I couldn't get my other hand over there; I could not even come close to touching my fingers [from both hands] together in front of my chest. My work area had to be in front of my face because that was the only place I could get my two hands together."

There was another reason why she believes she was given the nod for EVA on her second mission, STS-49. The primary objectives of that mission were to rescue the Intelsat satellite and to practice techniques necessary for space station construction. Astronaut Dan Brandenstein was chief of the Astronaut Office when they were selecting EVA astronauts for the STS-49 mission, and according to Thornton, Brandenstein wanted a woman to perform a space walk on that flight. She believes that of the women who had done EVA training and were available, the trainers probably recommended her. Many more women would go on to perform numerous EVAs—as equal partners—alongside their male counterparts during shuttle missions and operation of the ISS. In 2018

Christina Koch and Jessica Meir carried out, from the ISS, the first all-female space walk in history.

Thornton acknowledged that "sometimes you're just in the right place at the right time." She doesn't believe she did anything special to get selected for a space walk beyond working as hard as she could to master the cumbersome space suit. "I think everything you do, somebody is watching and you sort of develop a reputation." On the other hand, she confessed that because her premier space walk on STS-49 didn't go as well as planned, some in the Astronaut Office were unhappy that she was assigned to the prestigious STS-61 Hubble servicing mission (for EVA). "I know there were—they told me." She kept her head down, continued to work hard, and contributed two nearly perfect space walks to make Hubble even better than originally intended.

Jeff Hoffman, with a total of four EVAs under his belt, was also one of four spacewalkers, along with Thornton, who repaired the Hubble telescope in 1993 on one of the most ambitious missions of the shuttle era. Due to the complexity of this repair, JSC decided to consider only astronauts with previous EVA experience, and there were only about a dozen active astronauts at that time that had spacewalking experience on their resume.

Several opportune events led to Hoffman's first space walk on the STS-51D mission in 1985, which qualified him for an EVA slot on STS-61. STS-51D carried only three mission specialists—Hoffman, Dave Griggs, and Rhea Seddon. Each mission required at least two astronauts to be prepared for a contingency EVA should an emergency arise. According to Hoffman, for 51D, the commander decided who would train for the contingency EVA slots, and Rhea Seddon was not enamored with the space suit due to her small size and therefore wasn't keen on doing a space walk. That left Hoffman and Griggs to serve as contingency EVA crewmen. Hoffman shared, with tongue in cheek, "The pilots were not allowed to do EVA. It's more important [for them] to bring the shuttle home safely—don't lose a pilot in space. I was expendable."

Secondly, Hoffman and Griggs had been assigned to—and had trained for—at least four missions before they ended up on STS-51D, and each of those missions had different types of EVAs and thus different training. Hoffman recalled, "So by the time we actually flew [51D], Dave and I had—I think it was almost fifty hours in the water, so we were better trained. At least we had more experience than your typical contingency EVA crew."

Once the 51D crew was in orbit, they deployed the Hughes Syncom-IV/

Leasat 3 communications satellite, but the booster failed to ignite and deliver the satellite into geosynchronous orbit as planned. NASA devised a scheme to attempt a repair—and it included a space walk. Hoffman recollected his EVA training: "We got good—clearly on 51D when they were determining if they should send us outside to fix the satellite, one of the first things they do is they go to the EVA trainers and say, 'What do you think of Jeff and Dave? They didn't train to do this job. Are you willing to send them outside?' I guess we got good grades, because they let us do it."

One un-flown mission specialist was assigned to a flight with two planned EVAs. There was one veteran spacewalker and two rookie astronauts assigned to the flight. Cognizant that assembly of the space station was looming just over the horizon and that EVA experience was direly needed, he approached his commander and asked if he would consider assigning one EVA to each of the two rookies (with the veteran performing both). The commander approved, and both rookies carried out highly successful space walks on that mission, gaining valuable EVA experience.

STS-37 commander Steve Nagel recalled that he had no control over his crew selection but was free to assign tasks to his crew as he saw fit, including space walks assignments. "It was whoever I wanted to pick." He strove to fairly distribute the workload, honoring crew preference when possible. But regarding who was chosen for a space walk, he conceded, "Oh, it's hard because everybody wants to do an EVA." He paired EVA veteran Jerry Ross with rookie Jay Apt, who successfully carried out an unscheduled contingency space walk to free a stuck antenna on the Compton Gamma Ray Observatory.

Like Hoffman and Griggs, Joe Tanner and Scott Parazynski—each with seven space walks to their credit—parlayed contingency EVA roles on STS-66 into their first planned space walks. Tanner recalled in 2020 that in 1993, he received four suited EVA training runs in the Weightless Environment Training Facility (WETF), which led him to being assigned, along with Parazynski, as contingency EVA crewmen on STS-66. During their nine months of mission training, they made a few additional runs in the pool. Tanner contends that both he and Parazynski performed well in their EVA training and were logical choices for the contingency EVA roles. Just as Linda Godwin asserted, Tanner believed that they were being scrutinized during their training, which undoubtedly played a role in their selection for a future space walk.

Tanner had a little inside help in his selection for a space walk on his final mission, STS-115. "I was in the management crew selection path by my last flight so there was no mystery in that selection. I sort of selected myself!" By the time John Herrington arrived at NASA as a new astronaut candidate (ASCAN) in August 1996, the ASCAN training had been revised to include rudimentary EVA training in the Neutral Buoyancy Laboratory. Constructed in the late 1990s, the NBL replaced the smaller WETF, providing a first-class neutral buoyancy training facility that can hold full-scale mockups of ISS modules and payloads. Herrington believes they were testing who might be suitable for space walks and who might not be the best candidates. "So, you fit in the suit and you've got the temperament, you're not claustrophobic, and you can work with tools—sure, why not?" Herrington went on to perform three highly successful space walks on the STS-113 mission.

Astronauts assigned as Expedition crew members on the ISS—approximately six-month stints—may be called on to perform a space walk anytime during their mission. Hence, EVA training is essential. When Nicole Mann, along with Japanese astronaut Koichi Wakata, ventured outside the ISS Quest airlock on 20 January 2023, she became the final member of Astronaut Class 21 to perform a space walk, making that class the only one in NASA's storied history to have all members earn the privilege to wear the coveted EVA patch.

Now that the space station construction has been completed, NASA has plans to return to the moon and other destinations beyond low Earth orbit. Spacewalking, at least in the near future, will continue to play an important role in the exploration of space. But in the early 1960s, humanity had just touched the realms of the heavens, and much uncertainty existed about the ability of humans to function in space, especially outside the spacecraft. Today's arduous process of selecting and training astronauts suitable for EVA has been forged by the lessons learned over nearly six decades of experience. Despite the unknowns, the pioneers—the first humans to leave the safety of their somewhat crude spacecraft in equally crude space suits—were up to the task. Soon, the art of EVA would be mastered. But it was not going to be easy.

# 2

# A Steep Learning Curve

For the things we have to learn before we can do them,
we learn by doing them.

—Aristotle

On 22 February 1965 *Cosmos 57*, an uncrewed, highly modified version of the Vostok-type spacecraft that had carried Yuri Gagarin into orbit and home less than four years earlier, orbited Earth. Inside, a fully functional Berkut space suit was pressurized in accordance with procedures. Extending from the side of the shiny, ball-shaped spacecraft was an inflated fabric tube about the size of a small shower stall, just barely large enough to hold a single space-suited man. Fitted up against the ship's entry hatch, its sidewalls were pumped full of air to form a semirigid chamber that would allow a cosmonaut to leave the spacecraft without bleeding the entire cabin of its atmosphere.

Days before its launch into space, engineers had discovered a potential problem with the airlock's outer "sortie" hatch, a concern that in a vacuum it might not close completely when commanded by the ground. An emergency meeting with chief designer Sergei Korolev was called, and even though there was some dissent over changes to procedures so close to launch, the decision was made that an alternative technique could be used during the flight to ensure that the hatch would close.

As the test in orbit was being carried out, two simultaneous commands were sent from separate tracking and control stations for the motor to close the hatch. Suddenly, all communications with the spacecraft were lost. Unseen by the Russian engineers on the ground, the spacecraft silently exploded in the vacuum of space, spraying shards of the metallic ship, its soft airlock, and whatever remained of the test space suit out into a thousand different orbits of their own.

It was later determined that when the dual commands to close the hatch were sent, they somehow interfered with each other, causing the ship's braking rocket to fire. With the off-center mass of the airlock still attached, the spacecraft tumbled, resulting in the detonation of the self-destruct mechanism. Intended for use only should an out-of-control ship be careening toward a populated area, the explosion wiped out any chance of fully testing the gear to be used on the world's first space walk by a man.

A very unhealthy Sergei Korolev sat dejected in Alexei Leonov's hotel room just hours after the explosion. Korolev explained to Leonov and his commander, Pavel Belyayev, that all the data from the test flight had been lost. The Voskhod spacecraft was an interim design to conduct a short series of missions before the more advanced Soyuz was built, and the one remaining capsule was the one in which they were scheduled to fly. If the decision was made to outfit this one for another test, Leonov's historic first space walk could be delayed by a year or more while a new ship was built.

Most of the systems required to carry out the space walk had been verified before the errant signals were sent. The suit was airtight, and the airlock extended properly and was pressurized with air just as it would be for the crewed flight. The only functions not able to be tested were the separation of the disposable chamber by means of pyrotechnic charges and the security of the main hatch during reentry into Earth's atmosphere.

As Leonov later wrote in his book *Two Sides of the Moon*, the chief designer said to them, "I cannot tell you what you should do. . . . There is no doubt that there are risks. It is your decision." After a moment of consideration, Korolev added coyly that he believed that the Americans were gearing up for what was going to be at least a stand-up EVA—one in which the astronaut would not fully exit the open spacecraft—in the spring of 1965. That was all it took. Leonov and Belyayev had no intention of having an astronaut's name ahead of theirs in any history books. With an aura of invincibility, they confidently announced to the imposing Korolev, "*My gotovy.*" "We're ready."

In early 1963 a civilian-attired Alexei Leonov stood alongside several of his fellow cosmonauts, including the now world-famous Gagarin, humanity's first space voyager, on the factory floor of OKB-1, the Soviet's primary space design bureau. Among a row of large, silvery, spherical spacecraft in various stages of construction were the two Voskhod spacecraft. Protruding from each of the advanced capsules were identical plastic-looking tubes, which none of them recognized for what they were. Sergei Korolev approached the group as they studied the odd contraptions and confidently announced that they would be learning to "swim in open space," just as a sailor must learn to swim in the ocean if crewing a boat. He regaled them with visions of working in space to build space stations and repairing equipment.

Then his eyes focused on the young Leonov. Referring to him as "my little eagle," he instructed the rookie cosmonaut to get suited up, climb into the spacecraft, and demonstrate that he could successfully maneuver himself out of the capsule through the crude airlock mock-up. As quickly as the chief designer had arrived, he spun on his heel and walked away, leaving Leonov stunned. It had been over a year since he began training as a cosmonaut, and he had no idea he had earned Korolev's trust. Gagarin slapped Leonov on the shoulder, shaking him from his brief trance. "Now it is you who have been selected," he said with a smile.

It was nearly fifty years before Leonov would learn what attracted the attention of Korolev to him among all the other cosmonaut candidates, as he shared with NASA's public affairs commentator Rob Navius. Found among the minutes of a meeting Korolev attended to discuss the proposed EVA were his contemporary observations of Leonov. "I would like to outline the most important feature of Mr. Leonov, and this is his wit, his ability to quickly act, his ability to guess correctly, his ability to accept technical materials," he opined, among many other personal accolades. "And I think that this person strongly deserves our attention."

"That's what Mr. Korolev told about me, which I never knew," Leonov recalled with a smile. "And these words determined my fate; determined my destiny." What went unsaid at the time but must have been on everyone's mind was that the Americans, at the urging of President John Kennedy, had announced the goal of putting astronauts on the moon by 1970. Being able to work outside of a spacecraft wasn't just part of some science-fiction-like dream like building cities in space. For the young cosmonauts, it was not a stretch to envision themselves exploring the surface of another planet.

In addition to the specialized airlock design for the *Voskhod 2* mission, a protective space suit was designed and manufactured that would allow Leonov to survive outside of his spacecraft in this unimaginably harsh environment. The Berkut suit consisted of four layers of high-strength nylon and two redundant rubber pressure bladders, all covered by an off-white multilayer thermal insulation.

Strapped onto the suit was a hard backpack containing three oxygen bottles that fed through external hoses into the helmet then throughout the space suit itself. An additional hose and tether line would connect Leonov to the Voskhod, providing radio communications, telemetry on his health back to the ground, and a backup supply of oxygen from the ship. The helmet, with its clear faceplate, also contained an inner sun visor that could be raised and lowered over the eyes, similar to those found on typical flight helmets.

The suit was designed to operate at one of two pressures as selected by the cosmonaut. Originally intended to function at 270 hPa, or about 0.27 atmospheres, it was also capable of switching to 400 hPa, a pressure designers used exclusively to protect the crew from decompression sickness in the event of an emergency during launch or landing. The decision to utilize the suits in a rescue mode would come to have serious implications during the proposed EVA and lead to a decades-long misinterpretation of the events that unfolded.

The fully suited cosmonaut would strap on his backpack, open the inner hatch of the Voskhod, and enter the inflated airlock, and the commander would close the hatch behind him. The air in the chamber would be slowly vented to space; once released, the outer hatch could be opened at the extended end of the tube, and Leonov could exit into the unknown void of low Earth orbit. After evaluating his suit and his ability to work outside, the entire process would be reversed—Leonov would reenter the enclosure feet first, the outer hatch would be closed, and the airlock would be filled with air again. Then the inner hatch would be opened, and the first spacewalker would return to the safety of his spacecraft.

It seemed a relatively easy operation to carry out in concept, but the engineers and cosmonauts found it exceedingly difficult to simulate on the ground and in the brief seconds of weightlessness provided by a Tu-104 zero-gravity aircraft. These challenges, due largely to working in uncomfortable, hot equipment as the jet climbed and dived, were highly demanding of the cosmonauts. But the preparations for this historic feat were necessarily grueling, as the unknowns of man's ability to survive outside of a spacecraft were as numerous as those with Yuri Gagarin's single orbit of the planet.

In a time of deep secrecy surrounding the Soviet space program, no public announcement was to be made of Leonov's mission to conduct the first space walk. While surrounded by great fanfare from those who prepared the two cosmonauts for launch on 18 March 1965, few in the world outside of the snow-covered Baikonur launch site, located near the town of Tyuratam in Kazakhstan, were aware of the historical moment that was about to unfold. Leonov rarely discussed his work with his wife, including the fact that he would be undertaking such a dangerous and potentially fatal task. He wrote years later that he didn't want her to worry, but it's likely that the secrecy surrounding the mission prevented him from discussing it with her. If the space walk was successful, the Soviet propaganda machine would be churning out the news to the world with great pride. If not, Leonov's attempted feat would likely not have been acknowledged for years to come.

Orbiting high above the planet, Leonov was allowed no time to acclimate to his new environment before pressing ahead with the space walk. His pioneering EVA *was* the mission, and the conventional wisdom of the time was to complete the task before something else failed and he was prevented from carrying it out. But the new weightless world he now found himself in played havoc with his senses. Even though he felt like he was upside-down for much of his first orbit, Leonov prepared his suit, backpack, and umbilical for the challenge ahead. Once in the airlock, he rested while his suit flushed his body of nitrogen to prevent what divers refer to as "the bends," a painful experience resulting from the excess gas rapidly dissolving out of the bloodstream if air pressure is reduced too rapidly.

As the Volga airlock was bled of air, his suit inflated into a nearly unbendable rigidity. With all his checks completed, Belyayev, now alone in the spacecraft, commanded the motorized outer hatch to open, swinging inward into the crowded tube. Leonov craned his neck back against the stiffened suit as far as he could and was mesmerized by what he saw. *Voskhod 2* was just crossing the terminator—the sharp division between the sunlit planet below and the darkness of night opposite—and he could see the western coast of Africa slowly slide into view across the open porthole above him.

The cosmonaut was still safely enclosed in the airlock, and if anything went wrong with his suit, the hatch could be quickly closed and the tube repressurized to save him. But the moment finally came when ground controllers gave him the order to exit. "Diamond Two [Leonov's call sign], it is time for you to start your mission." Leonov raised himself up and out of the airlock, grasping a ring of handholds around the hatch, leaving behind his relatively comfortable home in space—into the nothingness.

Leonov was now truly in space, surrounded by a panorama of the brilliant blue-and-white Earth and the deep, endless black of infinity beyond. "*Zemlya kruglaya!*" he exclaimed through the crackling radio. "The earth is round!" It was a seminal moment in humanity's existence. Never before had a man been so high and travelling so fast in such an unforgiving environment yet had been able to survive and function in a meaningful manner. It was one thing to do so in the relative safety of a strong-hulled spacecraft. It was quite another to be outside, surrounded by the cruel vacuum that would mercilessly kill one in an instant were it not for his protective space suit. The implications for the future of space exploration may have been beyond the average Soviet citizen at the time, but the surreal images of one of their countrymen daring the unknowns of space high above would fill them with nationalistic pride. Some accounts of the space walk suggest that Leonov's fellow Soviets were able to watch the event live on television, but in the secretive days of the space race that had begun with the 1957 launch of the Sputnik satellite, this would have been virtually impossible. State officials would never allow a live broadcast, not knowing if the first space walk might end in success or failure. There were two black-and-white television cameras on the airlock chamber that transmitted live images to engineers on the ground and a color film camera Leonov attached to the lip of the hatchway that captured his every move, but the revelation to the Soviet people would wait until he was safely back inside.

Leonov dangled from the end of the airlock tube for a few moments, activated one of the cameras, and took in the awe-inspiring view around him. The bold red letters *CCCP* were clearly visible on his white helmet, and the lower part of his face not covered by the visor began to feel the intense heat of the sun. He gently pushed off the spacecraft and floated away against a backdrop of the Mediterranean. He gave a quick wave to the camera as his snaking umbilical twisted his bloated, white-suited body in a slow pirouette. "Man has entered open space!" exclaimed Belyayev over the radio. The statement was so stunning to the young cosmonaut dangling in the ether outside of *Voskhod 2* that it took him a moment to realize that it was *him* to whom his commander was referring.

Before him lay an astonishing three-dimensional map of the world—the entirety of the Black Sea, and with a slight glance to his left, Greece and the unmistakable boot of Italy. Half a decade later, Leonov recalled, "What remain[s] etched in my memory was the extraordinary silence . . . my heart beating." The form of his rigidly inflated suit caused Leonov's legs to remain nearly straight and close together. He could move his arms about but not nearly as much as had been planned. A still camera on Leonov's suit, intended to photograph the exterior of the spacecraft, was rendered useless, as he couldn't reach down far enough to touch the activation switch.

After a few moments of gently moving about, Leonov came back in contact with his ship and pushed himself off into a wild tumble in multiple axes at once. It would seem to an observer that he was completely comfortable and confident rolling and flipping like a circus gymnast in the weightlessness of a black sky and piercing white sunlight. But the world's first spacewalker was about to endure a life-or-death sequence of events.

Leonov was allowed only eight minutes outside his spacecraft, keeping him in daylight and in radio communication with far-flung ground controllers during the space walk. But over that brief time, his seemingly effortless motions in the void were anything but. He was sweating profusely from exerting force against the pressurized suit, and his hands and feet were no longer able to reach into the gloves and boots of the overstretched garment. It was time to get back into the airlock, but to do so would require Leonov to coil up the long umbilical in order to stuff himself and all his gear back into the inflated enclosure. What would happen next has become an inextricable mix of history and legend.

Leonov could barely get the long umbilical under control as he fought against the suit. He was running out of time to get back inside, limited by the oxygen in his backpack and the lack of exterior lighting once the spaceship passed into the darkness of the night side of Earth. The stiffness of the suit and the fact that his hands were becoming useless in the outstretched gloves led Leonov to make a crucial decision—he had to lower the pressure of his suit to reduce its rigidity if he was going to make it back into the airlock.

It has been suggested over the years, perhaps due in part to the difficulties of the Russian language, that Leonov improvised and simply opened a valve on the suit and bled air out into space, an extremely hazardous undertaking. But in reality, the suit was originally designed to be operated at 270 hPa, and Leonov had even anticipated having to select the lower pressure as a result of his experience training in the zero-g simulating aircraft. Leonov had been clearly ordered to report on everything he was doing, but in this case, he made an exception. "I broke the rules and didn't report to mission control to avoid spreading panic, and raise a whole host of questions," he recalled in an interview with the Fédération Aéronautique Internationale (FAI). "After all, nobody could have helped me in that situation."

With the pressure now regulated down, Leonov's suit was far more flexible and substantially easier to move about in. Even so, for over fifty years, he claimed that he was not able to enter the airlock feet-first as planned, although he was not clear why this was an issue. Leonov described diving head-first into the airlock and then contorting himself in the pressurized suit back to a feet-first position within the tight confines of the tube in order to close the outer hatch. He was most keen to retrieve the color film camera and not have it float off into space, carrying with it the spectacular imagery he hoped it had captured.

In a report written by Leonov immediately following the mission but only made public following his death in 2019, Leonov describes holding the camera in his right hand and pulling himself into the chamber with his left. This sequence of events is confirmed by video footage that, too, was only released in recent years but that clearly shows Leonov coming back into the airlock feet-first, holding the camera and reentering as he reported. Even if he had come into the airlock head-first, there was no need to flip around inside to close the outer hatch, as it was operated electrically by Belyayev inside the cabin.

It is possible that if he did come in head-first and the video footage failed to capture it, Leonov would have to flip to a feet-first position in order to reenter the cabin and his seat. But this would have been much easier to accomplish once the airlock was repressurized and his suit had lost its rigidity than if he had to fight against a pressurized suit still surrounded by vacuum. Whatever the reasons behind the disparities of his return to the safety of *Voskhod 2*, Alexei Leonov had accomplished what no man had before and had brought Sergei Korolev's grand visions of heroic cosmonauts working in open space one step closer to reality.

In the post-flight report to the FAI, Leonov reported several conclusions from his brief experience of extravehicular activity. "Leaving the spacecraft into outer space is quite possible and is no longer mysterious to a man," he wrote. "A man in a special space suit with a self-contained life support system may not only exist in space but also perform certain aimed and coordinated operations; [and] it is possible to carry out some physical work, to make scientific observations in outer space."

Leonov was chosen personally by Korolev to be the first to attempt an extremely dangerous excursion into the unknown, but he never looked back with any boastful claims about beating the Americans. "As far as I was concerned I was there to prove to my fellow man what human beings were capable of," he later wrote. Leonov's star would continue to rise, and it became known only after the fall of the Soviet Union that he was assumed to become the commander of his country's first lunar landing. The secretive failures of the N-1 moon rocket would never allow him that opportunity, yet he returned to space in 1975 as the Soyuz commander of the joint Apollo-Soyuz Test Project.

After the historic success of the premier extravehicular activity on *Voskhod 2*, the Soviets planned to move forward aggressively with their lunar program, the first piece of which was the larger, more advanced Soyuz ("Union") spacecraft and another space spectacular intended to stun the world. While the Soviets usurped the Americans with the first space walk in history, NASA was not sitting by idly.

Gemini pilot Ed White struggled inside his gleaming white space suit to firmly anchor himself into the seat of the *Gemini 4* spacecraft following his twenty-two-minute space walk. White's ballooned suit, adorned with a large American flag on his left shoulder, was not cooperating. Command pilot Jim McDivitt was vigorously pulling on White's suit wherever he could get a handhold to keep him from floating back outside the spacecraft. Working together, the two astronauts had not been able to get the hatch closed to the point where White could begin to engage the closing latch, much less secure the door for reentry. The machinery had acted up in a similar fashion during training in the vacuum chamber; fortunately, White and McDivitt had learned the nuts and bolts of the latch during training and were confident they could make it work—if it didn't break first. If the hatch could not be properly closed and securely fastened, death would be imminent upon reentry.

McDivitt believes that White's suit was a little stiffer at the end of the space walk than when the EVA began—his arms and legs were not bending as earlier—which made it far more difficult to climb all the way back inside the spacecraft. During zero-g training, White would wedge himself down in his seat, grab hold of a canvas handle attached to the hatch, and pull downward to close it. Then he would use his right hand to ratchet the hatch down and secure it. White's chest pack was also impeding his mobility, and the whole effort was taxing the muscles in his legs. White remembered post-flight, "I got back down and started to wedge myself down, and I got two fat cramps at the bottom of my thighs in both legs, where the muscles started to ball up a little."

This method of closing and securing the hatch had always worked in the zero-g airplane, but it wasn't coming together in orbit. Not surprisingly, White's training kicked in. "But once I got my hands underneath the instrument panel, I was back pretty well in familiar grounds—the work that we'd done five dozen times in the zero-g airplane, and I knew the technique pretty well."

Those minutes spent trying to engage the latch must have been exasperating for the two astronauts, as McDivitt disclosed: "We broke the [handle] attachment about three or four times on the zero-g airplane. Every time they kept telling us it wasn't made out of the right kind of stuff, and the stuff we were going to have in the spacecraft would be the right material." White described his predicament: "I was pulling down with my legs as hard as I could. I was pulling on the handle. I remember one time you [McDivitt] said, 'Hey, don't pull on that handle so hard! You're going to break it!'" White confessed, "I was heaving on the handle as I was pulling it down each time. It felt like to me that the handle was giving. But I didn't give a darn! If it broke, it was going to break." Manipulating the small, sharp components of the metallic latch with the fingers of White's pressurized gloves presented another potential problem; it could have resulted in a puncture, a slow and certain death for White in the vacuum of space.

"As soon as I had gotten up there to operate the gain lever, I couldn't operate the canvas handle anymore," White said. "I couldn't apply any torque or pull there. It was the most interesting moment of the flight, but I think it was probably the most . . . dramatic moment of my life—about those 30 seconds we spent right there. The dogs started latching. I could feel them going in, and then I could feel them come over dead-center." McDivitt yelled out, "Success!" By the time the hatch had been secured, White was nearly exhausted, his face covered in sweat, which rendered him nearly blind, but the Americans had inched closer to catching the Soviets.

McDivitt and White were scheduled to open the hatch again after pressurizing the spacecraft following the EVA to discard the unneeded EVA equipment, but even though they were confident they could manage the hatch locking mechanism, they decided to not risk it and elected to forego another hatch opening.

This was not the first time on the mission that the hatch had nearly foiled them. Following depressurization of the crew cabin in preparation for White's space walk, they couldn't get the hatch to open. White and McDivitt considered canceling the EVA, but given their extensive training with the hatch-locking mechanism, they were reasonably confident that if they could get the door opened, they could also close and secure it. Still, there was a shred of doubt in McDivitt's mind; during training, except for an exercise in the vacuum chamber wearing his pressurized space suit, he was normally wearing his sports clothes while assembling and disassembling the latching mechanism.

White explained in the *Gemini 4* flight crew debriefing, "Let me say one thing about the decision to go ahead and open the [h]atch. If we hadn't done so much work together with this hatch and run through just about every problem that we could possibly have had, I would have decided to leave the hatch closed and skip with EVA when we first started having trouble with it. We had encountered just about every conceivable problem that we could possibly have with the hatch." McDivitt was adamant; had they not been thoroughly trained on the hatch and the vagaries of getting it closed and secured, the EVA would have been canceled. They decided to not inform Mission Control of their decision, and after fiddling with the latching mechanism for a few more minutes, the hatch suddenly popped open; it was time to do a space walk.

With the exception of the hatch problems, White's space walk went spectacularly well. He proved that NASA could send an astronaut outside the protective comforts of the spacecraft and survive in a space suit in the vacuum of space. Coming just two and a half months after Leonov's premier space walk, NASA was buoyed by the success of White's initial foray into EVA, which to the casual observer he made look and sound easy. But White was immensely capable physically, and his tasks were relatively simple and unstructured, which led to a misplaced complacency; NASA would soon find that they had much to learn about working in open space. Those seemingly benign problems with the hatch portended more extensive shortcomings of NASA's understanding of EVA during the following three space walks carried out over the next year. Not until Edwin Aldrin's spacewalk on the *Gemini 12* mission—and many lessons learned—did NASA pull off a successful space walk that gave them confidence that they could perform meaningful work outside the spacecraft.

The Soviet Union had seemingly bested the Americans in the early beginnings of spaceflight, and because Apollo would not be ready to fly until the latter years of the 1960s, Project Gemini was developed as a bridge program to gain the skills necessary to land humans on the moon—rendezvous and docking of two spacecraft, long duration spaceflight, and EVA. Aside from near disaster on the first docking of two spacecraft on the *Gemini 8* mission—the result of a thruster stuck in the open position that caused the spacecraft to tumble out of control—the first three objectives were accomplished with few unexpected problems. Spacewalking, however, would not be as easy to master. Ten Gemini missions flew in just a little over a year and a half, and the program was moving at such a breakneck pace that equipment and procedures weren't always being improved from mission to mission based upon the crewmember's feedback. Although EVA proved to be a much bigger challenge than NASA anticipated, the Gemini series provided the ideal proving ground for NASA to solve many of the unknowns of spacewalking. During the Gemini program, NASA caught up with and eventually surpassed the Soviet's space program.

While spaceflight is inherently dangerous, depressurizing a spacecraft and climbing outside adds even more risk, including the possibility of losing a spacewalking astronaut. There is some probability, albeit a small one, that an astronaut floating outside the spacecraft might be struck by a micrometeoroid (a particle smaller than a grain of sand) or a larger piece of space junk, although there was little human-generated debris orbiting Earth during the Gemini missions. David Shayler, in *Gemini 4: An Astronaut Steps into the Void*, reported that Ed White was struck by a small micrometeor[oid]; fortunately, it did not penetrate the inner layers of his suit.

All the Gemini spacewalkers were attached to the vehicle with an umbilical during their space walks, so it would have been unlikely for one of them to become detached from the spacecraft and float away, a potentially fatal outcome for a spacewalking astronaut. However, if a firing thruster had stuck undetected during a space walk—as on *Gemini 8*—it would likely have sent the spacecraft tumbling out of control, causing the umbilical to wrap around the spacecraft, entangling the hapless spacewalker and leading to an agonizing, fatal end.

Other hazards threatened the Gemini spacewalkers. They had to maintain a comfortable and safe distance from the thrusters scattered around the outside of the spacecraft for attitude control. Typically, the command pilot had to maneuver the spacecraft to maintain its proper attitude due to the spacewalking astronaut's tugs on the umbilical, which caused the spacecraft to gyrate. *Gemini 9* spacewalker Gene Cernan remembered that the thrusters received his upmost attention: "Anytime that Tom [Commander Stafford] was firing the thrusters or . . . the automatic mode was firing the thrusters and I was holding the spacecraft either handrail or hatch, I could feel the thrusters firing . . . pow, pow, pow!" Stafford and Cernan had to communicate regularly to ensure a thruster was not fired while he was close enough for the suit to be damaged by thruster exhaust.

Cernan also recalled a potentially dangerous and sharply pointed spring-loaded antenna that came out with a bang when it was deployed. "The antenna startled me. I knew it was going to come out, but it didn't dawn on me at that time to look for it. . . . Had I been on the other side of the spacecraft or had my leg or back near this antenna, I'm sure it would have gone right through my suit without any difficulty at all." He was also concerned about the spring-loaded handrails that were not deployed until the spacecraft reached orbit; if one struck him, it may have injured him, or perhaps worse.

Mike Collins, in *Carrying the Fire*, acknowledged that an EVA astronaut could die during a space walk, which prompted him to write an internal NASA memorandum seeking advice on what should be done if a Gemini astronaut perished or became incapacitated during an EVA. Likely, the command pilot would have been unable to maneuver him back inside the spacecraft. "However," Collins wrote, "it is a harsh ground rule and one which means that, for a successful re-entry, the left-seat astro must disconnect his buddy's hoses and lines, close the hatch, and leave him in orbit. There are obviously many implications here, and I would like to hear any ideas you have on the subject." His request went largely ignored.

Once the first Gemini mission—*Gemini 3*—launched on 25 March 1965, the pace of spaceflights progressed at a rapid pace, exceeding that of the Mercury program—on average a flight about every sixty days. The astronauts who flew these missions were already well trained; some were veterans of previous flights. But once assigned to a mission—typically, a little over six months before they were scheduled to fly—they began training in earnest to accomplish the objectives for their flight. Six months may seem like an eternity, but there was a plethora of training to be done in all facets of the mission, not just for EVA. Not surprisingly, the hectic launch schedule precluded much sharing of lessons learned between the Gemini spacewalkers; there simply wasn't enough time in the day to squeeze in more meetings when there was more important training to be done.

For the astronaut selected to perform a Gemini space walk—the junior pilot in all cases—there was a regimented training process that varied little until neutral buoyancy training in a boys' school swimming pool was used for the final Gemini mission. From *Gemini 4* up through *Gemini 11*, preparation for EVA was divided between training in Earth's gravity field, known as one-g training, and weightless exercises in the KC-135 aircraft. One-g training included simulations in the altitude chamber, on the Air Bearing Floor, and in a body harnesses and slings. Edwin Aldrin was the only Gemini spacewalker who worked in the pool prior to flying his mission.

The EVA one-g training was very effective and much more convenient than the zero-g or other simulations. The *Summary of Gemini Extravehicular Activity* report outlined the need: "Objectives of one-g training were to familiarize the flight crews with the procedures, the hardware, and the spacecraft stowage related to EVA and to develop a coordinated work effort between the crewmembers." A detailed checklist was developed for the space walk, and the astronauts walked through those procedures step by step using mockups built as close to the configuration of the actual flight vehicle as possible; they often wore space suits and full flight equipment during these training exercises.

The command pilot trained just like the pilot during the one-g simulations so that he understood every task assigned to the spacewalker in order to help conserve his energy and also judge progress and assess the work pace of the spacewalker. During training stints in the altitude chamber if training in the spacecraft, they used the actual flight article. They also practiced donning and checking out the backpack-like life support equipment worn during space walks; the Extravehicular Life Support System (ELSS), the Extravehicular Support Package (ESP), and the Astronaut Maneuvering Unit (AMU) "jet pack," depending on which one was required for the space walk.

Much of their EVA training, including use of the Hand-Held Maneuvering Unit (HHMU), was carried out on the Air Bearing Floor, which is a very smooth, level, polished surface where, like a giant air hockey table, test articles float on a cushion of compressed air. *Gemini 10* spacewalker Michael Collins referred to it as the "slippery table." The floor eliminates most of the friction with the test article, but not all of it, allowing limited but valuable weightless training. But there is another downside to the table—astronauts can't maneuver around on the two-dimensional table as if they were floating in space. There are six different ways in which an object in three-dimensional space can move, known as the six degrees of freedom: forward-backward, left-right, up-down, yaw, roll, and pitch. These can be further divided into three directional and three angular movements.

Unfortunately, the Air Bearing Floor can be used to simulate only three degrees of freedom at a time; two directional and one angular. But with careful planning, it can be used to simulate all six degrees of freedom in multiple exercises. On the Air Bearing Floor, in the standing position, astronauts could move forward, backward, and to the left or right. They could also pivot (yaw) left or right while remaining in a comfortable position. But to simulate other degrees of freedom, they would have to lay on their backs or sides.

On their backs—relative to their perspective—they could translate up (toward the head) or down (toward the feet), left or right, or spin horizontally to the floor while looking up at the ceiling, which is called roll. Lying on their sides, they could translate up or down as when on the back, forward and backward as when standing, but had the ability to spin parallel to the floor—known as pitch. Each body position allowed for movement about only three axes and had no ability to simulate the complex coupling of inertial movement in multiple axes simultaneously, as would be the case in the zero-gravity vacuum of space.

Due to the hectic training schedule caused by the fast pace of the Gemini missions, the EVA astronauts were limited to just a few hours on the floor. Ed White spent twelve hours on the table, Dave Scott just over twenty (his *Gemini 8* space walk was canceled following the harrowing incident with the stuck thruster). Dick Gordon spent a little over thirteen hours on the floor, and Edwin Aldrin spent twenty. Gene Cernan spent a total of 140 hours in various training regimes in preparation for the first test flight of the AMU. In addition to the standard training procedures, he practiced in a fixed-base simulator at Edwards Air Force Base in California. A motion-based simulator was developed at Ling-Temco-Vought (LTV), where Cernan and other astronauts refined the intricacies of rendezvous using the AMU, developing procedures much different than those proposed by the contractor. The astronauts occasionally trained using a body harness and slings designed to simulate weightlessness. The device could be used to simulate four degrees of freedom and was generally used in conjunction with a spacecraft mockup.

Zero-gravity training in an air force KC-135 aircraft featured prominently in the curriculum, whereas today, with years of EVA lessons learned, it is no longer a significant part of EVA training. The aircraft offered a volume approximately seven by ten by sixty feet and often incorporated a mockup to approximate the tasks they were going to practice. The exercises included methods of hatch ingress and egress, umbilical management, translation outside of the spacecraft, entry into the adapter equipment section at the rear of the spacecraft, the donning of chest packs, tether attachment, and experiments operation. Typically, they wore their training space suit for these exercises, but on flights just prior to their missions, the prime pilots usually wore their flight-rated space suits to gain confidence in their operation.

The prime and backup crews participated in approximately five training sessions, averaging about forty gut-wrenching parabolas on each flight. Numerous problems with equipment were discovered, and corrections were made prior to flight. However, film footage taken during these flights clearly shows that the Gemini astronauts often struggled to carry out their tasks, a far cry from astronauts carrying out space walks on the ISS today. Michael Collins can be seen bouncing around the inside of the aircraft during each weightless parabola, often tumbling out of control before blindly grabbing a handhold and then returning to the task at hand. Isaac Newton's third law of motion (for every action, there is an equal and opposite reaction) was clearly in effect. Every time Collins grabbed something, it pushed back and sent him cartwheeling in the opposite direction. Several times he ended up in a situation where he couldn't do anything, kind of like a turtle on its back. He tumbled and spun while trying to dampen out the motions caused by his attempts to carry out his tasks. Through sheer willpower he eventually accomplished his objectives.

Today many customers pay over $5,000 to experience weightlessness in commercial zero-g aircraft, flying about fifteen parabolas that last approximately twenty-five seconds each. Some become quite sick, which is why the aircraft is often referred to as the Vomit Comet, but the limited number of parabolas flown usually spares the client from the dreaded barf bag. But when training for Gemini, with its forty or fifty parabolas per flight, its effort to cram in as much training as possible, and the overheating from poorly ventilated suits, any anticipatory joy often quickly turned to dread. Lots of participants on those flights vomited, including astronauts. Collins recalled his love-hate relationship with his suit. He loved it because it kept him alive; he hated it because it was uncomfortable and cumbersome; he often experienced bouts of claustrophobia during training. Fortunately, by flight day the suit had been tailored to fit comfortably.

The amount of training the Gemini EVA astronauts received pales in comparison to that of later programs. The moon was beckoning as Apollo matured, and the engineers planning that program needed to know sooner rather than later if the objectives of Gemini could be met and mastered. Thus, spreading out the Gemini training over a longer timeframe was not an option. It didn't help that EVA presented a steep learning curve that NASA initially failed to understand, so even if they had found the time for additional training, they didn't know *how* to properly train the astronauts. It took a small company under contract with NASA's Langley Research Center to point them in the direction of underwater "neutral buoyancy" training, which revolutionized the art of preparing for EVA.

Ed White had a spectacular adventure during his space walk on the *Gemini 4* mission. "I feel like a million dollars!" he exclaimed shortly after climbing out through the hatch into the vast emptiness of space. He reveled in the vivid blues of the Gulf of Mexico and Caribbean Sea as he tumbled around; his gleaming white suit formed to a seated position, lending a slight cowboy-like pose to this new American hero. He was also fascinated with the lights of Perth, Australia: "What a view! By golly, you can see the black sky up above. Boy! Oh, boy, oh, boy!" When McDivitt ordered him to climb back inside the spacecraft, he was downright disappointed. "Coming in. Listen, you could almost not drag me in, but I'm coming. . . . This is the saddest moment of my life." Post-mission, he reported that he felt completely natural in the vacuum of space, with no disorientation whatsoever.

Unfortunately, White's space walk was not a harbinger of what was to occur on the next three Gemini space walks. The simple tasks he had been assigned, coupled with being very strong and in outstanding physical condition, misled NASA into believing they had mastered EVA. Gene Cernan, Mike Collins, and Dick Gordon soon learned that performing meaningful work in a weightless environment was extremely difficult and that their training, especially the zero-g flights, had not prepared them for the tasks they had been assigned. In fact, NASA had become convinced that their training was more than sufficient. Neither Cernan, nor Collins, nor Gordon successfully completed all their assigned tasks on their weightless umbilical space walks, although Collins and Gordon performed much simpler successful standup EVAs—standing up in their seats—with minor but manageable problems.

Cernan had been tasked with evaluating the ELSS for the first time on the *Gemini 9-A* mission, but his major objective was to test the air force's Astronaut Maneuvering Unit, a self-contained backpack propelled by hydrogen peroxide. The AMU was similar in concept to the Manned Maneuvering Unit flown by some of the shuttle astronauts in 1984, the big difference being that the MMU was flown untethered, whereas the AMU would have been attached to the Gemini spacecraft with a tether at all times.

Opening the hatch, Cernan jettisoned some equipment, deployed a handrail, retrieved an experiment package, mounted a camera, took some photographs, and attached a docking bar, although it was unexpectantly difficult to keep his body inside the spacecraft. The experiment package that he had handed inside to Stafford seemed intent on floating back out of the open hatch. Cernan contorted his body into a position from which he could trap it with his feet. Cernan's movement while standing in the spacecraft kept torquing the capsule, requiring Stafford to correct its attitude using the thrusters.

He exited the spacecraft hatch and evaluated a Velcro hand pad, which provided some stability, but even using both hands attached to the Velcro, his feet kept rising up over his head, leading to loss of body control. Plus, the ELSS kept riding up his chest, which made it almost impossible to grab anything with both hands. Cernan then translated to the nose of the vehicle, but as with the Velcro pad experiment, he could not keep himself in a stable position, even using both hands to hang onto the spacecraft. After slowly pushing himself to the end of the umbilical several times, he returned to the cabin and rested before heading back to the adaptor section where the AMU awaited. It was clear the tasks that seemed doable in the zero-g aircraft were much more difficult to accomplish in orbit.

Cernan then encountered the next worrisome problem of the EVA: closing the hatch to within about two inches to keep the cockpit thermally comfortable was much more physically challenging than he had anticipated. With some difficulty, he then carefully translated to the adapter section at the rear of the capsule.

Once there, all he had to do was perform a series of seemingly simple tasks required to bring the AMU to life, climb into it, strap it on, and make the test flight. Surprising to Cernan, every task that he attempted was much more difficult than in the zero-g aircraft. Once he had secured his feet into stirrups designed to stabilize his body, he tried to attach a small hook, something that he never had any trouble with during training. It was a two-handed operation, but when he attempted it in orbit, his feet slipped out of the stirrups and he rolled left, which flung him over the top of the spacecraft. Stafford was well aware that something was happening back in the adaptor section of the spacecraft as he could hear a strange banging noise—and it had to be Cernan. Cernan worked on attaching the hook—all the while fighting the bulky ELSS mounted on his chest—but he'd done it in training, so he was confident he could do it in orbit. Eventually, he begrudgingly abandoned the hook but lamented after the flight, "I came within a cotton-picking gnat's eyebrow of getting that thing on."

He discovered that he could maintain his body position by bending his ankles up rigidly and firmly locking them into the stirrups, but this required working his hips and knees and thighs and every muscle in his legs he could recruit for the task, which meant he was generating a lot of force and hence workload. During this time, Stafford began to notice that Cernan's respiratory rate had begun to increase rapidly. During training for these tasks, his rate would range from thirty to thirty-five breaths per minute, which is a high workload. Cernan informed Stafford that he was becoming fatigued, which was no surprise to Stafford, who had timed his respiratory rate to be at least forty.

To complicate matters, Cernan suddenly began to feel extreme warmth on his backside. It felt as if he had backed into a hot campfire or furnace—almost blistering hot. He adjusted his suit cooling loop to high flow and waited for the sun to go down. Post-flight, it was determined that the outer layers of his space suit had torn, allowing the harsh heat from the sun's rays to scorch portions of his back.

Cernan continued to slowly make progress in preparing the AMU to fly, but he was exerting far too much effort and energy, which caused his visor to fog up, eventually to the point where the fog disrupted his vision. He had practiced these procedures religiously—albeit much faster than he could in space—and didn't need to see to complete them. The fogging continued to worsen, prompting him to admit that even if he accomplished all the necessary tasks to activate the AMU, where could he go from there? He couldn't see outside his helmet. Both astronauts came to the realization that it was time to terminate the EVA, and when Cernan finally climbed back into the safe confines of the vehicle, his helmet was still largely fogged over.

The umbilical space walks that Michael Collins and Dick Gordon carried out on their *Gemini 10* and *11* missions were similar; they both translated over to a previously launched Agena target vehicle where they planned to carry out their assigned tasks. But the similarities ended there. Collins had to propel himself over to an undocked Agena, whereas for Gordon, the Gemini spacecraft was docked to the Agena. Collins was tasked with retrieving several experiment packages—one from the Agena—replace it with another package, and evaluate the HHMU. Gordon was also slated to test the HHMU and to attach a hundred-foot tether between the spacecraft and the target vehicle. The two spacecraft were going to undock on the thirty-first revolution of the mission, and then the Gemini spacecraft would have backed off to the end of the tether for a station-keeping exercise. Their assigned tasks were different, but they had trained the same way, primarily using the Air Bearing Floor and zero-g aircraft, just like Cernan had trained for his "space walk from hell," as he referred to it once back on the ground.

Collins climbed out of the hatch on the *Gemini 10* mission, and after taking care of some chores, it was time to make his way over to the undocked Agena spacecraft and get to work. He pushed off gently from the spacecraft, trying to impart a low velocity trajectory; slamming into the target vehicle at high velocity was not a good idea.

As he neared the Agena, Collins grabbed at it with his bulky gloves but couldn't hold on. Careening away from the target vehicle, he quickly began taking the slack out of his umbilical and used his HHMU to return to the Gemini spacecraft, all the while imparting forces on the spacecraft that Young had to dampen out using the Gemini thrusters, thereby using up precious fuel. His post-flight descriptions of his difficulty in controlling his body movements are very reminiscent of his challenging zero-g training flights. Collins found that the HHMU wasn't extremely accurate, but with a bit of luck, he was able to position it near his center of gravity, squeeze off a few short spurts of thrust, and approach close enough to the Agena that he could snag the leading edge of its docking cone. But the smooth, blunt cone was not a very good handhold for his hard pressurized gloves. "I was able to get my hand down inside a recess between the main body of the Agena and the docking collar where there were some wires," Collins recalled, "and then I was able to get the experiment package off."

After he had recovered the experiment package, he quickly ran into another problem. Part of the Agena nose had come loose, and a piece of it was hanging in his way, posing a grave concern of becoming tangled up in it. Commander John Young was also concerned about the mess and realized that he had used too much fuel maintaining the attitude of the spacecraft. He promptly directed Collins to come back inside. Collins discarded the now useless new experiment package and utilized the umbilical to pull himself—with just a slight nudge—back to the spacecraft. There was no further evaluation of the HHMU as planned. Collins recalled that despite all the difficulty he had encountered on his space walk, he didn't want to come back inside.

Dick Gordon's EVA on *Gemini 11* didn't fare any better than Collins had attempted; he was over heated well before he climbed out of the spacecraft for his umbilical EVA. Gordon and command pilot Pete Conrad had been allotted three hours for EVA preparation time, but they had trained so well that they accomplished it in about forty-five minutes. The ELSS had been hooked up, and the only way they could keep Gordon cool was by running the ELSS, which caused problems with the cabin atmosphere. Hence, they decided to disconnect him from the ELSS and reconnect him to the suit loop of the vehicle. Complicating matters further, Gordon encountered problems attaching his EVA visor. What should have taken mere minutes took well over half an hour, and because he was not connected to the ELSS, he had no cooling and became overheated—all before he had even left the spacecraft. Once the hatch had been opened, he tackled several tasks while standing within the spacecraft, but he kept floating up and out of the hatch, requiring Conrad to keep yanking him back inside, which caused more exertion and sweat.

Once he intentionally exited the spacecraft—already tired, hot, and with an accelerated heart rate—Gordon attached a camera, which also took more effort than he had expected. The pesky problems had piled up and taken their toll on him, and by the time he had accomplished this last task, he was severely overheated, fatigued, and out of breath. Still, he pressed on.

Gordon pushed off the spacecraft for the Agena target vehicle tether location. He had already decided to use a thruster located about midpoint as an assist to push off again, but he miscued and ended up missing the Agena. Conrad pulled him back to the spacecraft using the umbilical, which Conrad remembered was like reeling in a fish. Gordon's second attempt was better executed, and he was able to grab a handhold located on the Agena. He had intended to maneuver his legs inside the docking cone to wedge himself securely, which would free up both his hands, just like he had done in training in the zero-g aircraft. Unfortunately, it didn't work in orbit, and he had to use his left hand to hang onto a handhold and then attach the tether with his right glove, a two-handed task. He was unable to maintain his position and work on attaching the tether at the same time. Post-flight, Gordon explained, "And this was a monumental task as far as I was concerned." Conrad ordered Gordon to rest for several minutes a number of times as it was obvious that his crewmate was running out of steam. Finally, with great effort, Gordon was able to attach the tether. By this time he had overtaxed the cooling capability of the ELSS. He still had one more task to perform at this location—attach a docking bar mirror—but after an unsuccessful attempt to remove the mirror cover, exhausted and with stinging sweat in his right eye, he gave up and returned to the hatch. The plan was for Gordon to translate back to the adapter to check out equipment and test foot restraints, but he was fatigued, and it was clear that performing tasks required much more effort and energy than he had in reserve. So after about a half hour into the planned 107-minute EVA, the space walk was ended. Following the mission, Gordon protested that attaching the tether took about half an hour—a task that he could do in less than thirty seconds in the zero-g aircraft.

The end of the Gemini program was one flight away in September 1966, and NASA had yet to demonstrate that their astronauts could conquer spacewalking. Some of the Gemini pilots referred to the "dominant effect of small forces," which in a weightless environment became significant and were in aggregate responsible for their spacewalking difficulties. These dubious forces were not well understood and led to the design of tethers and restraints, one of the keys to eventually mastering spacewalking.

Some of the astronauts suspected that there were mysterious external body forces that caused a spacewalker to float up and away from Earth. Cernan described his loss of control in orbit as "tail over tea kettle. I always tended to roll back over to the right and over the top of the spacecraft." Later missions following *Gemini 9-A* investigated these alleged external forces, but no preferred direction of movement was noted. It was concluded that there were a variety of explanations for the inability of astronauts to stabilize themselves during EVA.

Once a part of the space suit was moved, it tended to return to a neutral position. The spacewalkers were also unintentionally imparting small inadvertent forces. The motion of the spacecraft may have affected the astronaut's movement. It was also possible that the astronauts were never able to achieve a condition of no movement, which made it appear as if external forces were acting upon them.

Collins was quite blunt about NASA's understanding of EVA in his oral history interview in October 1997:

> Well, the fundamental problem was, I guess, stupidity. We had not, in our designs, really thought through what happens to objects that bang together in weightlessness. . . . In space, where there's no anchoring, you not only rotate this way and call that yaw, you rotate in pitch, and you rotate in roll. So if I touch that table, I go off in some totally three-dimensional random direction. So as soon as I start doing that, I've got to stop, and so I grab, and then I go swinging back the other way with greater force, and I need greater force to stop that. Very soon you're just out of control.

He pointed out that Aldrin's successful space walk on *Gemini 12* had been better thought out and planned; there were handholds and workstations designed specifically for EVA. "I was going over to this Agena, propelling myself with this dorky little gun, which in itself was a very difficult thing because, again, going back to the pitch, roll, yaw, if you did not have a gun aligned absolutely through your center of your body, if you were off a little bit, then when you squirted the gun, instead of going that way, you would not only go that way, but you would start twisting and turning."

Rusty Schweickart, *Apollo 9* lunar module pilot, never trained on the Air Bearing Floor with the HHMU, but he did work with Ed White while he was training for the *Gemini 4* mission, and he shared Collins's assessment of the gun. "We ended up with a gas bearing floor that Ed White would lay sideways in his suit and squirt around and try to use that hand-held gun which was pretty ridiculous. No way that was going to work (laughing), but it was the only possibility we had for him maneuvering on EVA."

Collins had to attempt to grab hold of equipment that had not been designed to be grabbed with big hulking gloves, all while wearing a bulky space suit that's not very elastic. "I mean, it was not clean analytical engineering. I mean, it was more acrobatics and a guy on a trapeze and stuff that you don't think about in the space program. But as I say, the fundamental thing was stupidity."

For Gordon, the tethering exercise was just plain hard. His legs floated up and out of the docking cone where he had tried to wedge them, and he was trying to hold onto a handrail with one hand and do the tethering with the other. Conrad summed up Gordon's near exhaustion rather succinctly post-flight, inventing a word to describe Gordon's condition: "In two- or three-minute periods, you sat there and blurped." Gordon felt that he had to work four times harder on his EVA than in the zero-g aircraft. Cernan had warned him of this; but Gordon's training had tricked him into believing that he was prepared.

Dick Gordon acknowledged that there wasn't time to share information about the difficulties incurred on the three space walks (nine, ten, and eleven): "We were on two-month launch centers. It was almost like just passing each other to and from the pad. And there wasn't a great deal of opportunity to assimilate the information from previous flights. So, I don't think we did a very good job of transmitting to each other the things that were happening." The first time he learned about some of the problems that Cernan had encountered on his EVA was when he read Cernan's autobiography—published in 1999!

Sam Mattingly and John B. Charles chronicle the evolution of using underwater neutral buoyancy to simulate weightlessness in *A Personal History of Underwater Neutral Buoyancy Simulation*. The concept dates back to 1963–64, when a small company named Environmental Research Associates (ERA) was working with NASA Langley Research Center to investigate airlocks for future space station concepts. Mattingly and Harry Loats, also with ERA, believed that training underwater would be an effective way to simulate weightlessness; unfortunately, they had no space suits to test the concept. But they did have access to a full-scale plastic airlock model that Langley had constructed for a demonstration to their management. Matttingly and Loats borrowed an out-of-service navy flight pressure suit, and in June 1964, Otto Trout from Langley placed the model in the swimming pool at the Langley Air Force Base Officers' Club. Mattingly donned the borrowed pressure suit, and using whatever they could find to make the suit neutrally buoyant—lead shoes, body-worn weights, hand-held dumbbells—he was able to approximate a neutrally buoyant position in the pool. He succeeded in negotiating himself through the airlock, but he was unable to simulate all six degrees of freedom.

Building on this nominal success, Mattingly was granted access to an indoor pool owned by the McDonogh School, a nearby boy's military institute in Owings Mills, Maryland. The weighting process was improved by employing welding shot in bags tied to the suit to better approximate neutral buoyancy. Soon Mattingly was able to perform a complete test of airlock ingress and egress procedures and was astute enough to have the simulations filmed to later help convince management to support additional tests. Certified SCUBA divers were used to assist in the pool. Simulations in zero-g aircraft were also carried out and compared with the underwater simulations, which strongly suggested that the underwater work was an effective way to simulate weightlessness.

However, those in charge at the Manned Spacecraft Center (now Johnson Space Center) in Houston—even after Cernan's failed *Gemini 9-A* EVA—were still convinced that they would be able to train astronauts for EVA utilizing zero-g aircraft and the Air Bearing Floor. NASA engineer Kenneth Kleinknecht stated that "up until that time, underwater training was sort of below the dignity of an astronaut."

Mattingly and Loats continued to progress their work and invited NASA representatives to view their demonstrations. Eventually, a NASA official requested that they carry out a simulation of the upcoming *Gemini 10* space walk and provided a Gemini adapter section mockup for the test. The demonstration pointed out that an astronaut *could* accomplish one of the EVA tasks planned for the mission—but would need three hands to do so. Collins's experience on his *Gemini 10* flight confirmed the underwater results. Mercury astronaut Scott Carpenter also performed an underwater simulation in the McDonogh pool and found that neutral buoyancy tasks clearly took more time than anyone realized, and he helped convince the MSC managers that ERA's work was worth pursuing. Gene Cernan also participated in a *Gemini 9-A* post-flight demonstration and confirmed that the simulations were representative of his space walk.

Dick Gordon didn't have time to squeeze in underwater simulation training before his *Gemini 11* EVA, but Charles Matthews, the program manager of NASA's Project Gemini, recommended to Bob Gilruth that Edwin Aldrin participate in the underwater training for his *Gemini 12* mission. Aldrin's first simulation was very helpful, resulting in modification of the EVA plan, including the elimination of the complicated AMU from the space walk. Next, a preflight evaluation of the revised *Gemini 12* space walk was conducted and repeated in seven sessions over four weeks just before the mission. Aldrin conducted five of the sessions and his backup, Gene Cernan, did two passes, all in the McDonogh School swimming pool. Aldrin also did a post–*Gemini 12* underwater simulation to further assess the value of underwater training and gave it an enthusiastic thumbs up.

Aldrin's successful space walk on *Gemini 12* was the last box that needed to be checked before transitioning to the Apollo program. Using waist tethers and foot restraints, Aldrin was able to successfully—with minimal effort—turn bolts, make and break electrical and fluid connectors, cut cables, hook rings, and strip patches of Velcro while he worked at astronaut-friendly workstations. Two-minute rest periods were sprinkled in throughout the two-hour and nine-minute-long space walk. Six hours prior to his umbilical EVA, he performed a stand-up EVA to allow him to acclimate to the foreign environment of space before heading outside. His heart and respiration rates were monitored during the EVA so he could be slowed down if he was exerting too much energy. Film footage of Aldrin during his umbilical space walk clearly show that he was in complete control at all times. However, he was anchored and tethered to the spacecraft during the entire EVA instead of free-floating like Collins and Gordon on their difficult spacewalks.

Post-mission, Aldrin surmised that conquering EVA was more difficult than the other objectives Gemini was charged with mastering: rendezvous, docking, and long-duration flight. These could all be simulated rather easily on the ground, whereas astronauts needed to gain practical EVA experience in orbit to learn the proper techniques.

Aldrin proved that meaningful work could be carried out in the weightless environment of space if planned properly and that the underwater simulations were a good proxy for spacewalking. The ramifications were huge; if a task could be accomplished in the pool, there was a high probability that it could be done in outer space. Underwater training became a mainstay for future EVA training, and over the years, it has been polished to near perfection, allowing astronauts to carry out weightless space walks during the shuttle and ISS programs for well over seven amazing and productive hours at a time. Without the groundbreaking work of Matttingly and Loats in the little swimming pool at the McDonogh School, neutral buoyancy training may have come along too late to contribute to the race to the moon. But at the end of 1966, the stars were aligned, and with the confidence brought by Gemini's success, NASA was ready to fly to the moon. On the other side of the world, the Soviets still had much to prove before they too could carry out a lunar mission.

The Soyuz spacecraft consisted of two pressurized modules: a bell-shaped descent module containing crew seats, control panels, and the heat shield and parachutes needed for return to Earth, and an elongated spherical orbital module mounted on top. A hatch between the two components could be closed, and the orbital module would serve as an airlock. An unpressurized service module extended beneath these two, housing propulsion systems, fuel, and two deployable solar panels. Gone was the claustrophobic inflatable airlock of Voskhod; the Soyuz orbital module could accommodate two fully suited spacewalkers at a time. In late 1964 Korolev called for a new mission concept for the maiden crewed mission of the Soyuz program—the launch and docking of *two* spacecraft in orbit.

However, after the successful space walk by Leonov, an even more ambitious task was assigned to the crews of *Soyuz 1* and *2*—an EVA transfer of two cosmonauts from one of the docked vehicles to the other. This would be necessary on the upcoming lunar landing missions, as the proposed command ship and lunar lander had no hatches or transfer tunnel between them. It was an audacious plan for the first crewed flight of a new spacecraft, given that *two* examples of the design would be orbited at the same time, after only one fully successful automated flight on 7 February 1967.

To carry out the new objective, a new EVA suit was ordered. The Yastreb suit, which leveraged many new features from the suit for the Soviet reconnaissance aircraft of the same name, featured a closed-loop system that circulated air through an evaporating heat exchanger to cool the body and a lithium hydroxide cartridge to scrub exhaled carbon dioxide from the system. A backpack similar in appearance to the one Leonov used was quickly developed.

An electrical umbilical was the suit systems' only connection to the spacecraft, providing power, telemetry of data, and communications for the crewmember. The Yastreb was the Soviet's first dedicated EVA suit, intended only to be donned in space and used outside of the spacecraft, rather than a dual-role emergency pressure suit. It had adjustable sections in the arms and legs to prevent a recurrence of Leonov's inability to reach fully into his ballooned gloves and boots, increased arm mobility, an emergency oxygen bottle, and an external sun visor on the helmet. Extensive testing in zero-g aircraft had revealed that it was easier for crewmembers to pass through the open Soyuz hatchways with the life support system packs mounted on the front of the suit, between the legs, rather than on the back.

The suits and airpacks were ready for flight in April of 1967, when the flight of *Soyuz 1*, piloted by a lone cosmonaut, Vladmir Komarov, lifted off from Baikanour. Aleksei Yeliseyev and Yevgeny Khrunov had trained extensively for their daring mission to space walk from their *Soyuz 2* spacecraft, commanded by Valery Bykovsky, over to *Soyuz 1*, and return to Earth with Komarov.

But almost immediately after reaching orbit in the revolutionary new Soviet spacecraft, Komarov began experiencing serious malfunctions with his ship. The launch of *Soyuz 2* was called off, leaving Komarov to struggle through his problems alone in orbit. After overcoming many of them, the daring cosmonaut became the first space explorer to lose his life during a mission when both his main and backup parachutes failed to deploy properly, tangling around each other as he plummeted to Earth.

The details of the doomed *Soyuz 1* mission and the secretive attempt to launch *Soyuz 2* for the docking and EVA were not formally acknowledged by the Russians until after the fall of the Soviet Union more than twenty years later.

Aleksei Yeliseyev, standing in the bitter cold near the Site 1 launch pad on 15 January 1969, was craving a cigarette. He and his crewmates, Evgeny Khrunov and commander Boris Volynov, were bundled up in heavy winter coats over their thin flight suits, each topped with a traditional Russian *ushanka*, the thick fur hat that Westerners typically associated with their Cold War adversaries. The crew had arrived at the launch pad early and were out of sight behind a nearby building until the dignitaries arrived to be filmed seeing them off.

The day prior, *Soyuz 4*, piloted by Vladimir Shatalov, was launched safely into orbit and would soon pass overhead, allowing a precisely timed launch leading *Soyuz 5* to a rendezvous with Shatalov's craft. The crews of the docked spacecraft would then perform the EVA, as Yeliseyev and Khrunov had trained for two years prior, before the *Soyuz 1* tragedy. General Leonid Goreglyad was parked nearby in a compact Volga sedan, and Yeliseyev approached him asking for a smoke. The general worried what the higher-ups might think if they saw that and suggested the young cosmonaut sit in the car and have his cigarette in hiding.

The crew of *Soyuz 5* had arrived at Tyuratam nearly two weeks before. Following the disastrous *Soyuz 1* mission and months of corrective work on the spacecraft design, the next attempted docking of *Soyuz 3* with the unmanned *Soyuz 2* was unsuccessful. While both crews awaited their respective launches, General Nikolai Kamanin, the head of cosmonaut training, informed them that the United States had announced a 28 February launch date for *Apollo 9*, which was to test all the components of a lunar mission, including a test flight of the lunar lander and an EVA transfer demonstration from the lander to the command module. But while it had been suggested that the *Soyuz 4/5* mission was set to upstage the Americans, post–Soviet Union history revealed that the plan had been in the works as part of the Russian lunar program for years.

In fact, as the final preparations were being made for the dual launch of the Soyuz spacecraft, a nearby pad was being readied for an unmanned launch for a flight around the moon. The massive N-1 rocket intended to fly the first Soviet cosmonauts to the moon was in its final stages of processing for its maiden test flight.

For the two commanders—Vladimir Shatalov aboard *Soyuz 4* and Boris Volynov on *Soyuz 5*—the possibility of both having to return to Earth alone was never far from their minds. "We wore no spacesuits—no EVA suits," Volynov explained to author Bert Vis in a May 2001 interview. All the cosmonauts would ride to orbit in shirtsleeves, with only the two spacewalkers donning their Yastreb suits for the transfer. With the commanders sealed off in their descent modules while Yeliseyev and Khrunov floated between the two depressurized orbital modules, neither commander could be of any help if anything went wrong. "Shatalov and I were in the same position," Volynov recalled. "In case there would be an emergency situation, [they] had no chance to survive. It could turn out that the [EVA] hatch could not be sealed."

Yeliseyev had not even burned through half his borrowed cigarette in the stuffy little Volga when the general, keeping watch in the brutal chill, rapped on the window. "*Pora*," he ordered. "It's time."

The docking of *Soyuz 4* and *Soyuz 5* was flawlessly performed on 16 January, one day after the latter launched into an orbital race to catch up with its counterpart. According to Soviet media, the pair now formed "the first experimental station in space," a moniker that would confuse Western press and space agencies as to the true plans for the future of the Soviet space program.

Immediately following the docking, Yeliseyev and Khrunov aboard *Soyuz 5* began their preparations for the EVA transfer, floating up into the habitation module and unpacking their space suits. Volynov soon joined his crewmates, setting up a movie camera to record the activities. "I also helped to suit up, and with the life support system," he recalled. "The backpacks with the oxygen supplies, ventilation, communications, and so on. Switching them on, checking them out . . . many, many important details."

Distorted images of the brightly lit compartment and the cosmonauts were reflected in the shiny gold visors of their helmets, a pricey yet effective means of reflecting the most amount of sunlight while outside the spacecraft. Just before snapping each man's helmet over their head, Volynov planted a customary kiss on each cheek of his crewmates; he would not see them again until after they had all returned to Earth.

The commander returned to the safety of the descent module, closing the hatch between him and the spacewalkers. After slightly lowering the pressure in the habitation module to check their suits' integrity, Yeliseyev and Khrunov gave him the okay to vent all the air out of the compartment and command the outer EVA hatch to open. As Khrunov began his exit from the spacecraft, another film camera recorded the beginning of the space walk, but when Yeliseyev retrieved the camera for stowage in the habitation module, where the commander would later retrieve it, he had difficulty securing it and had to leave it bouncing gently around inside the module.

With Khrunov leading the way across the mated ships to the now open hatch of *Soyuz 4*, Yeliseyev joined him in open space as the pair came into communication contact with ground controllers. A ghostly black-and-white television image appeared on the screens in the control room, and the first thing engineers noticed was the wayward film camera, with all its beautiful color imagery of the space walk, tumbling slowly out of the hatch, lost in the blinding orbital sunlight.

Yeliseyev found the transfer between the two spacecraft pleasantly effortless, simply moving hand over hand across the rails mounted on the exterior of the ships. He found a "sensation of limitless space and freedom" as he floated, holding on with barely a fingertip at times. He looked to his left at the spacecraft he had left the planet in and thought of his commander sitting alone inside. To his right, Khrunov was already "standing" in the hatchway of the ship that would return them to Earth, holding onto his cable. The blue horizon of home seemed "far, far away" as the clouds and land masses slid slowly by. Yeliseyev felt so ebulliently confident that he took a moment to stop and try to memorize all he was seeing and feeling. Though still secured between the ships by his electrical cable and a tether clipped to the handrail, he recalled that in a brief, overwhelming moment of peace and calm, "I slowly unclenched my fingers to try to float without holding onto the handrail."

Once he reached the opposite habitation module, Khrunov tucked himself against the far wall of the spherical compartment, allowing Yeliseyev to float freely through the hatch. With the outer hatch closed and the diving-bell-like airlock flushed with air, *Soyuz 4* commander Shatalov opened the inner hatch and welcomed his two new crewmates. The three bounced around wildly as they enjoyed being reunited, and the spacewalkers even surprised Shatalov with newspapers and letters from home that they had secreted inside their space suits during the transfer.

Yeliseyev and Khrunov's space walk had lasted just thirty-seven minutes and had proven out a major objective of the Soviet's plan to reach the moon, even if the likelihood of beating the Americans to that celestial goal had slipped their grasp. But with the Iron Curtain solidly blocking news of the many details of their lunar ambitions, Western observers didn't quite know what to make of the spectacular accomplishment.

The declaration by Soviet media that the cojoined *Soyuz 4/5* formed the "first orbital station" led to some contradictory speculation in the free press of the world. As author Dominic Phelan explained in *Cold War Space Sleuths*, the BBC told their audience, "These launches indicate that the Russians are moving on an entirely different track from the Americans, establishing platforms in Earth orbit." The exception was the *Daily Express*, which broke ranks and ran the headline "Moon Race! Russia out to Beat U.S. Russia Is Going Flat out to Get a Man on the Moon before America." Nearly every other newspaper echoed the BBC's opinion.

In reality, the Soviets at the time did indeed intend to land on the moon utilizing a one-man lander and an even more advanced space suit. But a month after the mission, on 21 February 1969, the first N-1 moon rocket destroyed itself just over a minute after launch. This was followed by three consecutive failures of the launch vehicle—more powerful than the U.S. Saturn V—dashing all hopes of a Soviet cosmonaut reaching the moon before the United States.

*Apollo 9* lifted off on 3 March 1969 and accomplished a successful test of the Apollo space suit during EVA in Earth orbit. The next Soviet EVA wouldn't come for more than eight years, after the United States had completed not only fourteen historic EVAs on the lunar surface but three more on the trans-Earth coast back home and ten space walks from the orbiting *Skylab* space station. For nearly a decade, working in the nothingness of open space was dominated by the Americans.

# 3

# We Choose to Go to the Moon

You have a zipper sir, and it's a stupid zipper!

—Astronaut Story Musgrave

The splashdown of *Gemini 12* in the Atlantic Ocean on 15 November 1966 signaled the end of an era and the beginning of Apollo, which was charged with running the last laps of the race to the moon. Project Gemini's many contributions had put the United States decisively ahead of the Soviets in the quest for the first lunar landing. The maiden launch of an Apollo spacecraft had been planned for 21 February 1967, but tragically, America's space program suffered a crippling blow on 27 January. Commander Virgil Grissom, pilot Roger Chaffee, and Ed White, America's first spacewalker, were killed when a fire broke out inside their Apollo Block I spacecraft during a test on Pad 34 at Cape Kennedy. An investigation led to numerous improvements in the new Block II spacecraft, and eighteen months later, Apollo was ready to resume its path to the moon.

The first flight, *Apollo 7*, successfully ran the command module through its paces in low Earth orbit in October 1968, leading to *Apollo 8* making ten orbits around the moon before the end of the year. Next up was *Apollo 9*, an Earth orbital mission, with a throng of objectives necessary to land men on the moon, including the first Apollo-era space walk and test of the lunar space suit. Lunar module pilot (LMP) Russell (Rusty) Schweickart, along with commander Jim McDivitt and command module pilot (CMP) Dave Scott, were charged with testing the A7L space suit and its Portable Life Support System (PLSS), the suit that astronauts would wear while exploring the surface of the moon. The plan was for Schweickart to exit the lunar module through the front hatch, tethered by a nylon rope, then use handrails to maneuver over to the docked command module hatch and climb partially into the cabin to demonstrate an effective method of rescuing astronauts returning from the moon if they were unable to transfer through the docking tunnel.

Initially, the mission ran smoothly, but on day three of the flight, an event played out that could have ended the mission and that highlighted another potential danger to a spacewalking astronaut. Schweickart experienced a bout of what is now called Space Adaptation Syndrome (SAS); he felt extremely nauseous and began to vomit. Doing so inside a space suit helmet while outside in the vacuum of space is a potential death sentence. There would be no way for the astronaut to clear the obstruction from his throat and face; suffocation was a real possibility unless the incapacitated spacewalker could be quickly rescued. Like motion sickness on Earth, SAS is caused by upsets in the vestibular and visual systems due to the microgravity environment; symptoms commonly include fatigue, loss of appetite, nausea, vomiting, and headaches. Despite the setback, Schweickart was encouraged; he soon began to feel better. But when he heaved a second time a bit later, McDivitt canceled the EVA planned for the next day. Perhaps the entire vision of landing a man on the moon before the end of 1969 was resting on Schweickart's shoulders; he shared that if he had not been able to test the space suit, it might have pushed the lunar landing into 1970. He confided that his malady was one of the low points of his life.

The next day McDivitt and Schweickart proceeded as if they were going to perform the EVA; Schweickart would don the suit, helmet, and PLSS inside the pressurized cabin, but they would not depressurize the spacecraft for Schweickart to climb outside. They'd proceed as if they had performed the space walk and then pick up the checklist at that time and continue. Today it is known that astronauts who succumb to SAS eventually overcome it, but in 1968, very little was known about the illness. Remarkably, by the next day, barely twelve hours later, Schweickart felt better and was confident that he was up for the EVA. "I mean, it was pretty clear that at that point the motion sickness was rapidly becoming history. I was more aware of the fact that I was past that episode than Jim [McDivitt] was." McDivitt didn't tip his hand on his assessment of Schweickart's condition as they were working through the checklist, moving closer and closer to the simulated cabin depressurization. "So when Jim looked at me and said, 'It looks like you're feeling better,'" Schweickart was encouraged. He and McDivitt had become close friends through the long months of training, and it was immediately clear that they were both thinking the same thoughts. Less than thirty minutes before the simulated space walk was scheduled to occur, McDivitt informed Schweickart, and the ground, that they were proceeding with the space walk.

Schweickart climbed out of the lunar module hatch and stood on the LM porch gazing out into the vast void of space. Encouraged by McDivitt to test the handrails, he began to translate to the command module over the top of the lunar module. The EVA lessons learned during the Gemini program were paying huge dividends. The planned two-hour space walk was shortened, Schweickart explained, due to a change in the timeline. Once the EVA had been canceled on day three, the ground decided to let the crew sleep in an extra hour, and when the EVA was reinstated, that lost hour placed them behind schedule. "There was nothing other than the timeline that we were up against that prevented me from going over to the command module." Schweickart had recently seen a photograph taken by CMP Dave Scott—one that he had likely seen before but had forgotten about—that showed him on top of the lunar module a mere six to eight feet from the command module. At that time, the camera being operated by Scott, who was standing in the open hatch of the command module, malfunctioned, which led McDivitt to eventually terminate the EVA. But Schweickart was confident that he could have easily made it to the command module hatch as planned. Following some free time outside—about five precious minutes—while Scott attempted unsuccessfully to repair the camera, Schweickart climbed back inside the lunar module forty-six minutes after the start of his space walk. He reflected on the canceled EVA and subsequent rise from the ashes: "And so from low to high. As you can imagine, that EVA was more than just a normal EVA. I mean, that was just in twelve hours going from as low as I've ever been to about as high as I've ever been." Following the successful flight of *Apollo 9*, only one lunar dress rehearsal mission was flown, with no EVA, before *Apollo 11* landed on the moon in July 1969.

In retrospect, McDivitt's decision to send Schweickart outside barely twelve hours after having been sick was a bold one. During the space shuttle era, NASA learned that about three-fourths of astronauts become sick in space. Vomiting is not unusual when suffering from SAS, and it is not uncommon for the sickness to last several days. Today it's doubtful that NASA would allow an astronaut to perform a space walk twelve hours after becoming sick. But in 1969 there was no database to even hint if the sickness would pass in a matter of hours or if it would continue to haunt Schweickart for the entire mission. Regardless, McDivitt's decision may have preserved NASA's quest of the moon before the end of the decade.

When asked if NASA would have flown an additional flight to test the space suit had he not been able carry out the space walk, Schweickart responded adamantly:

> I don't think we would have risked going ahead with the [Apollo] 10–11 sequence if on 9 we had not actually been able to go outside. It was pretty critical. The suit hadn't been tested. The only time the suit was really tested before Neil and Buzz ran around on the surface was the Apollo 9 flight; 10 didn't do it—it would have been extremely difficult to add that to the Apollo 10 mission per se; I think we would have had an extra flight in there. Whether we would have made the end of the decade; that's an interesting question. We were flying at a pretty rapid rate. But certainly, in our minds, when we canceled the EVA we were well aware of the potential implications of that without any question. We could have missed the Kennedy goal. More than just canceling an EVA, the night before the EVA the real possibility was going through our minds that we might have to cancel the rest of the mission; we might have to come home early. That was really a big shadow hanging over us the night before, particularly hanging on me. Yeah it was no question that it was an important and rather momentous decision we made to cancel that EVA the night before. That was a big deal!

Training in the pool for the Apollo missions added another challenge. The fit of any space suit ranges from being comfortable to tolerable to painful depending on the astronaut. Even a soft Gemini or Apollo suit became very hard once pressurized. Additionally, the suit can be made neutrally buoyant in the water, but the astronaut inside is still subjected to the laws of gravity and slowly sinks to the low part of the suit. Inverted ops—upside down—are particularly challenging in the pool. Schweickart reflected on his training:

> When the suit is upside down in the water, suspended there in the middle of the tank, you're crammed into the top of the suit and the blood is rushing to your head. You're not really in weightlessness—your suit is not rising or falling or rotating if you're lucky, but you're crammed into the side of the suit if you're lying on your side or into the helmet if you're upside down or whatever. In fact, it [the pool] is a much harder

environment [than space]. We'd come out of the suit with some of that early testing with blood running down our shoulders.

Not only can training be difficult and unpleasant; it can also be dangerous. Schweickart believes that most people understand that going outside the spacecraft into the vacuum of space is risky, "but they don't have any notion that the testing on the ground is actually far more dangerous." Schweickart reasoned, "I mean in flight one thing that is 100 percent certain is the vacuum is not going to fail. But the device that creates the vacuum on the ground has many horrifying possible failures connected with it. The [sudden] re-pressurization failure of that big vacuum chamber—that would just destroy the whole building let alone little old me in a space suit. . . . That's a pretty scary environment!"

Space suit technician Jim Leblanc nearly lost his life in a vacuum chamber in December 1966 testing an early version of a lunar space suit. Leblanc was standing, holding onto a raised platform, as the pressure was sucked out of the chamber. He reported that all was going well just before his pressurization hose disconnected, which almost instantaneously placed his body in a near-vacuum environment, causing him to black out. He rolled backward, striking the floor of the chamber hard with his back. Rapid re-pressurization of the chamber was initiated, and it took a coworker from an adjacent partially pressurized anteroom, wearing an oxygen mask, only twenty-five seconds to reach him. The saliva on his tongue had begun to bubble as he lost consciousness, but he was able to stand up shortly after the incident. Afterward, his ears ached, but otherwise he suffered no injuries or long-term effects.

There seemed to be no lack of potential safety hazards when walking in space. Schweickart was apprehensive about a mission rule on his *Apollo 9* space walk that is standard operating procedure today on the ISS—for good reason: spacewalking astronauts are required to be attached to the space station with a safety tether. Schweickart shared that he and many astronauts believed that the tethers on his mission were a threat to safety and were unnecessary. Long tethers are prone to twist around anything or hang up on the smallest of bumps or protuberances on the spacecraft, distracting the spacewalker from what they should be focused on—safety hazards. It also takes considerable time to manage the unruly fiend. "Even if you for some stupid reason let go of the spacecraft and push away from it," Schweickart explained, "no big deal.

You take the spacecraft and fly over to the guy and pick him up. The lunar and command module could have easily done a couple squirts and picked you up. It wasn't a big deal." He contends that the tether was required because it made the managers at headquarters feel better.

Schweickart was chosen in astronaut group three in 1963, the first group selected that was not required to be test pilots. He'd dreamed of flying airplanes as a child, earned his bachelor of science degree in aeronautical engineering in 1956, joined the U.S. Air Force and Massachusetts Air National Guard, where he learned to fly fighter aircraft, and was awarded a master of science degree in aeronautics and astronautics in 1963 from the Massachusetts Institute of Technology. In addition to his piloting skills, he brought strong scientific credentials to the astronaut office. Considered a free spirit by many, he also had a keen interest in the arts, which may have contributed to his ability to add perspective to his space walk beyond the simple curt comments from most of the pilot astronauts.

When Dave Scott's camera malfunctioned during Schweickart's space walk, McDivitt gave him five minutes to attempt to repair it and then informed Schweickart to sit tight. He briefly considered going over the checklist in his mind but instead decided to contemplate where he was and enjoy the scenery. Freeing one hand from the rail, he swung around and gazed at the illuminated Earth below him with the sun and blackness of space in the opposite direction. He described the view as spectacular, and the suit was completely silent when no one was talking. He was content to float inside his suit and soak in the beauty of space and marvel at his home planet from the unobstructed view inside the helmet—far superior to looking out of any spacecraft window. Schweickart describes on his website his mindset of those precious five minutes he had outside the spacecraft with no official duties:

> OK I'm just going to be a human being here and look at what's happening—how did I get here? Humanity has reached this point where we're moving out from the Earth; you know I'm a small part of that but that's what's going on. How does that happen in history? What does it mean? When I say how did I get here—who am I? Am I me, or am I us? It's very clear that you are there as a representative of humankind. This is humanity moving out, and you're just the representative on that frontier.

Landing on the moon changed the EVA game for the near future. The fourteen moonwalks by twelve astronauts during the lunar surface program benefitted from the gravitational field of the moon, despite its being a weak one. The exploits of the twelve astronauts who walked on the moon have been chronicled in many books, including Colin Burgess's *Footprints in the Dust* and Andrew Chaiken's *A Man on the Moon*. Each moonwalker brought their own unique talents and perspectives to the program, but the thrill of walking on the desolate lunar surface affected each of them in diverse ways.

Neil Armstrong, *Apollo 11* commander: "It [the moon] probably affects different people in different ways. It is spectacular, and I think everyone is touched by it when they have the experience."

Buzz Aldrin, *Apollo 11* LMP, from Supriya Jain's collection "40 Best Buzz Aldrin Quotes": "It certainly didn't make me feel lonely, except to realize that we were as far away as people had ever been. Once we were on the surface of the moon we could look back and see the Earth, a little blue dot in the sky. We are a very small part of the solar system and the whole universe. The sky was black as could be, and the [lunar] horizon was so well defined as it curved many miles away from us into space."

The moonwalkers employed most of the techniques used in the Gemini program to train for their lunar surface EVAs; one-g training, the pool, and zero-g flights remained important parts of the training regimen, whereas the Air Bearing Floor was less used. New techniques were also invented. The Reduced Gravity Walking Simulator was employed to approximate the moon's weaker gravitational pull, roughly one sixth that of Earth. On this odd-looking contraption of cables, straps, and an overhead crane, the astronauts were suspended at an angle to a wall where they practiced walking, jumping, and assessing how much energy was required to navigate in the reduced gravity environment.

There were many hours of numerous one-g mission walk-throughs where the astronauts practiced a myriad of tasks that would be required of them on the lunar surface, including deployment of scientific experiments and sample collection. Indoor and outdoor moonscapes were employed to practice many of the tasks they were assigned, such as the use of tools to collect rock samples and gadgets to scoop up lunar regolith. The relentless summer heat of Houston presented a challenge; the PLSS was not intended to work in atmospheric conditions and therefore the astronauts had to be artificially cooled when training outside, which didn't always work very well.

Pete Conrad, *Apollo 12* commander: "Oooh, is that soft and queasy. Hey, that's neat. I don't sink in too far. I'll try a little . . . [Letting go of the ladder and stepping out of the LM shadow]. Boy, that sun is bright. That's just like somebody shining a spotlight."

Alan Bean, *Apollo 12* LMP, from *Painting Apollo: First Artist on Another World*: "I felt a long, long way, 239,000 miles at least, from most of the people and places I love. It seemed unreal . . . impossible . . . but my space gear, and the familiar voices from earth told me otherwise."

Geology field training became an integral part of their preparation, much more so for the last four missions to land on the moon when geologic exploration became a higher priority. The astronauts participated in custom designed field excursions to carefully selected geologic field sites—led by top notch geologists—to teach them how to see the moon and what to bring home so the experts on the ground could reconstruct the history of Earth's closest neighbor and add geological insights of our own planet.

The Apollo suit presented its own challenges on the lunar surface. *Apollo 16* lunar module pilot Charlie Duke explained that despite all the hours training in the suit, he had to learn how to operate it once on the surface in the reduced lunar gravity. Bending over at the waist took much effort, and bending the knees was equally as difficult. Following each of his lunar EVAs, once safely back inside the lunar module, he was exhausted. The gloves were particularly tiring; Duke compared manipulating the pressurized gloves to squeezing a rubber ball for seven or eight hours. Often the flight surgeon had to slow him down, forcing him to rest while on the lunar surface; his heartbeat exceeded 140 beats a minute numerous times. He often cramped. Fingernails took a beating in the suit from rubbing up against the inside of his gloves. "Some of the fingernails . . . were . . . sort of black and blue from blood bruises," he recalled, "and those kinds of things we had that were as a result of the Apollo pressure suit. It kept you alive, though."

Many of the dangers that plagued previous weightless spacewalkers were still in play. Perhaps the biggest concern during the lunar surface program was a rupture of the suit or damage to it or the PLSS from astronauts falling on the lunar surface. Prior to launch, the moonwalkers became intimately familiar with their suits during the myriad one-g and weightless simulations, and once they acclimated themselves on the lunar surface, they quickly gained a level of self-confidence that worried some on the ground. They jumped, hopped, stumbled, slid, and often fell hard to the ground. Granted, in the reduced gravity of the moon, a tumble is much less dangerous than on Earth; but better to not take that unnecessary risk.

According to Story Musgrave, the zipper was a definite weak point of the suit. Fortunately, the zipper never failed on the lunar surface, but it was a huge motivation for the development of a suit with a hard upper torso for the shuttle era program, which had no zippers.

Alan Shepard, *Apollo 14* commander: "Now planet Earth is only four times as large as the moon, so you can really still put your thumb and your forefinger around it at that distance. So it makes it look beautiful; it makes it look lonely; it makes it look fragile. You think to yourself, just imagine that millions of people are living on that planet and don't realize how fragile it is . . . That was an overwhelming feeling in seeing the beauty of the planet on the one hand but the fragility of it on the other."

Ed Mitchell, *Apollo 14* LMP: "Wow, I'm on a pristine surface where humans have never been. I'm where humans may never be again, at least for a long, long time. Wow, there's an Earth up there. Look at that little planet. It's [the moon] so much smaller than ours, you can see the curvature of the horizon from the surface. . . . Yes, with one-sixth gravity, you can bounce like a trampoline even in all this heavy regalia. Yes, there's a lot of awesomeness, there's Wow!, there's exhilaration. Well, there's the sense of being the first to ever be at this place. That's awesome."

*Apollo 17* lunar module pilot Jack Schmitt felt extremely comfortable with the suit simply because he had "a long experience in using the suits, dealing with them, of knowing the people that make them, how they were made." Schmitt and his fellow moonwalkers tumbled to the lunar surface numerous times, but the suits performed magnificently for over eighty hours of lunar exploration.

Fortunately, none of the moonwalkers experienced any serious injuries on the lunar surface; still, they encountered minor problems. NASA flight surgeon Rick Scheuring noted in 2018 (in *Muscularskeletal Injures in US Astronauts*) that the moonwalkers reported nine injuries that occurred during their EVAs. Seven of them were nagging injuries to the hands and wrists, for example, a laceration incurred when the suit wrist ring cut into the skin. There was also a shoulder injury and general muscle fatigue. Abrasions were suffered when the suit sleeve rubbed against the skin. It's not surprising that most of these incidents occurred on the last three missions, each with three lengthy EVAs.

One of the moonwalkers conceded that performing a fourth EVA would have been challenging due to his sore hands.

Another obstacle to additional forays onto the lunar surface was dust; in addition to finding its way into wrist and hose locks, it abraded the suit—especially the gloves—caused seal failures, and seemed to cover everything it came in contact with. Grains of lunar dust are very small and angular with jagged edges, giving them an affinity to stick to the fabric of the suit. Additionally, the surface of each grain has a positive electrostatic charge due to ultraviolet rays from the sun, which also caused it to adhere to the suits. The astronauts routinely attempted to remove it with brushes, especially prior to climbing back inside the lunar module, but that only ground the dust deeper into the fabric of the suit. Even Velcro was rendered useless from the clogging dust.

Dave Scott, *Apollo 15* commander, from *Two Sides of the Moon*: "The low sun angle of the early lunar morning laid long shadows over the spectacular scenery spread before me. It was like an exhibition of the exquisite images by the great photographer Ansel Adams. There was no color, but great contrast between the brightly illuminated surface and the black shadow of the mountain slopes and craters where no sunlight fell. Oh, boy, what a view!"

Jim Irwin, *Apollo 15* LMP, from *To Rule the Night: The Discovery Voyage of Astronaut Jim Irwin*: "Dave, that reminds me of a favorite biblical passage in the Psalms: 'I will lift up mine eyes unto the hills, from whence cometh my help.' But of course we get quite a bit from Houston too."

Every year millions of visitors pass through several museums scattered around the United States that are dedicated to preserving the legacy of aerospace history. Among the plethora of aircraft and spacecraft dating back to the beginning of humankind's quest to fly with the Wright Flyer, are several gray-hued Apollo space suits. More than fifty years after traversing the flat plains and highlands, including the Sea of Tranquility, the Hadley Apennine Mountains, and the valley of Taurus-Littrow, these formally pure-white A7-L's and A7-LB model suits are still embedded with the ancient lunar regolith of the moon.

With their customized fitting, particularly of the A7-LB, a dedicated space historian can look at a photo and tell you who the faceless astronaut behind the gold-plated visor is without even reading a caption. Amazingly, those identifying shapes are still there today. Dave Scott's suit looks like Dave Scott in the photos from *Apollo 15*; Gene Cernan's lunar space suit was displayed for decades in the Smithsonian's Air and Space Museum in Washington DC in his familiar John Wayne pose beside an engineering model of the lunar rover.

The earlier models of the suits worn on *Apollo 11* through *Apollo 14* appeared less customized in shape than those of the later J-series lunar landings. In fact, after *Apollo 12*'s photos came back with moonwalkers Alan Bean and Pete Conrad appearing nearly indistinguishable, red stripes were added to the commander's arms and legs to differentiate the astronauts on all subsequent lunar flights. What many museum visitors might overlook, though, is that while most of the Apollo suits' exteriors look like they were worn while rolling in the ash of some terrestrial volcano, the crew patches and flags affixed to them appear unusually clean.

After their missions to the moon, every Apollo astronaut was gifted the fireproof beta cloth patches from their space suits, with the notable exception of *Apollo 11*. It was decided between NASA and the crew that given the historic significance of being the first to land on and explore another heavenly body, these suits should remain intact.

John Young, *Apollo 16* commander: "Okay, Houston; I'm standing (actually, he's kneeling) out on the porch. I've got the ETB [Equipment Transfer Bag] in one hand, and we're just sort of looking around here. My golly, what a view!"

Charlie Duke, *Apollo 16* LMP: "We were three-and-a-half/four miles to the south. We were several hundred feet above the valley floor. And you could look—from this advantage, you could look out all the way across the Cayley Plains and the valley that we had landed in. You could see, in the distance, [Smokey] Mountain and North Ray Crater. And . . . right out in the middle there was our little lunar module that was [colored] Mylar . . . orange. And then looking off to the northwest—as far as the eye could see-was just the rolling terrain of lunar surface, you know, shades of gray. It was really an impressive sight."

When not on public display, most of the Apollo suits, along with some three hundred significant experimental designs, Mercury, and Gemini program suits, are now kept in climate-controlled storage at the Udvar-Hazy Center of the Smithsonian Institution at Dulles, Virginia. While they were designed to withstand the extreme environment of deep space, they were not engineered to withstand the test of time, which has taken its toll on the fabrics, plastics, rubber, and dissimilar metals of which they were constructed. In fact, a few lunar-worn suits, not yet recognized as the historic treasures they are today, were later downgraded to training status and worn by other astronauts. This too made the preservation and restoration of them even more difficult.

Neil Armstrong's A7L underwent a thorough analysis and preservation process in 2018, which allowed it to be publicly displayed the following year for the fiftieth anniversary of *Apollo 11*'s enduring accomplishment. The suit that protected the first man to have the soft soil of an alien world under his feet represents not only what a country can accomplish when put to the task by a charismatic president but what the human race can accomplish when an urgency, be it real or contrived by geopolitics, presents itself. The image of a lone, white-suited astronaut standing on the starkly sunlit lunar surface against a pure black sky immediately became a source not only of national pride, but of a global sense that we, Earth's inhabitants, were becoming a species not forever moored to our home's shores.

The space suits in the Smithsonian's collection serve two forward-looking purposes other than presenting history to the public. Several of the companies that produced the designs have long been out of that business, engineers have grown old or passed on, and original construction and testing documentation has been lost. By examining the suits firsthand, today's generation of space suit designers are able to see the solutions that their predecessors developed for the many challenges of protecting astronauts in space. This "knowledge re-capture" is a vital component in building even more robust suits engineered to last for years in the harsh environments of space, the surface of the moon, and even Mars.

The second purpose these old suits serve is to inspire. For a child who may have never thought about spaceflight let alone the challenges of surviving in the unforgiving wilderness of outer space, seeing an Apollo space suit is something akin to coming face to face with a superhero. Towering above the millions of youngsters who see them every year, the worn, gray-stained, tattered suits look like something that has *been somewhere*. As the kids stare at their convex reflections in the now dulled gold visor of the lunar extravehicular visor assembly, they listen to the museum docents tell them of the larger-than-life men who wore them as they walked, bounced, and even drove a car across the alien landscape of our nearest celestial neighbor, decades before any of them were born.

They may find themselves wondering, "What was the person inside that suit of armor seeing? What were their experiences? Could that be me one day?"

Many employees working for NASA or the commercial space companies of today can trace their inspiration back to one event witnessed, one visit to a museum, or one fortuitous meeting with an engineer, scientist, or astronaut. Perhaps someday, one of these young boys or girls will be the first person to set foot on a Martian moon or the rusty red dunes and craters of Mars itself. From there, the entire solar system will still lie ahead, with endless possibility for new discoveries.

Gene Cernan, *Apollo 17* commander: "If I could take everyone in this world . . . and stand them alongside me on the surface . . . and look back home, where you see no religious borders, no cultural borders, no color borders, no whatever, you just see an Earth with the bright blues of the oceans and the land, and we all live here together, I've got to believe the world [would know] no political differences, I got to believe the world would truly be a better place to live."

Jack Schmitt, *Apollo 17* LMP: "It wasn't until the flight plan called for me to go some seventy-five meters away from the lunar module in three different points around it and take a panorama at those three points to document the site before we had really screwed it up, that's the first time I had a chance to see this magnificent valley that we were in, a valley deeper than the Grand Canyon, seven thousand feet, six- to seven-thousand-foot mountains on either side, thirty-five miles long, and about four miles wide where we had landed. The slopes of the valley walls were brilliantly illuminated by this little sun. . . . The sun itself was brighter than any sun that I had ever seen."

Apollo accomplished many objectives. It achieved its primary goal set by John Kennedy of "landing a man on the moon and returning him safely to Earth" nearly six years after an assassin's bullets took his life in November 1963. America gained the respect of millions of people around the world by accomplishing an audacious goal challenged by a young president looking to make his mark on the twentieth century. Perhaps most importantly, Apollo brought back a treasure trove of scientific knowledge about the moon and how it was formed. It also taught us about the Earth. Rocks older than three billion years are rare on the nearly five-billion-year-old Earth, whereas most of the rocks brought back from the moon are older than three billion years. Jack Schmitt illuminated this point: "Earth history beyond three billion years ago is very difficult for us to look at here on Earth. . . . On the moon, we just start looking at history at three billion years and go back from there. So, it's a pitted and dusty window into our own past. No question about that."

Live video of astronauts gallivanting across the dusty surface of the moon mesmerized hundreds of millions of people on Earth. But the CMPs, rarely showcased during these missions, also played a pivotal role in the exploration of the moon. While their crewmates were on the lunar surface, they were busy making their own observations and scientific investigations from lunar orbit, adding an important contribution to unravelling the complexity of the moon. Just as the scientific investigations were ramped up for the final three forays to the lunar surface, the CMPs also had a beefed-up scientific program. And while their role dictated that they remain in lunar orbit, their efforts required them to perform their own space walks shortly after leaving the moon, the only deep space weightless EVAs ever done in the history of spaceflight.

# 4

# From the Moon to the Earth

In the deep space of the sea I have found my moon.

—Jacques Cousteau

As NASA transitioned its Apollo program to the more ambitious J-series missions beginning with *Apollo 15*, unique requirements came into focus that would result in a most unusual opportunity for a select few astronauts. These missions featured audacious plans to explore the moon's surface to greater distances with a lunar roving vehicle for up to three days. They would also debut an extensive suite of instruments contained within the scientific instrument module (SIM) bay of the Apollo service module. Operating alone in lunar orbit for the three days that his crewmates spent on the surface, the command module pilot was to conduct unprecedented observations utilizing powerful cameras, some of which were adapted from the secretive U-2 spy plane program.

In an age before high-resolution transmittable images, large rolls of exposed film would have to be removed from these cameras and brought into the command module, the only heat-shielded part of the vehicle that would return its precious cargo of men and moonrocks to Earth. That necessitated an EVA, but not just any ordinary EVA. After the rendezvous and docking with his returning crewmates and the time-critical rocket burn to send them back toward Earth, the CMP's space walk would occur during what was known as "trans-Earth coast"—the voyage home.

As such, the CMPs of the final three Apollo missions are a most exclusive fraternity. Hundreds have now conducted spacewalks; only twelve have conducted EVAs on the lunar surface. But Al Worden of *Apollo 15*, Ken Mattingly of *Apollo 16*, and Ron Evans of *Apollo 17* are the only astronauts to have experienced a space walk while not in Earth orbit. Every other spacewalker had had their incredible view bisected by a panoramic, curved horizon dividing the bright blue of the planet from the inky black of space. For these three astronauts, there would be no vast expanse of familiar continents passing by *down there* (or *up there*, depending on one's orientation), only infinite blackness in every direction, a receding moon behind them and a distant, blue-and-white orb ahead that they were imperceptibly falling toward.

Apollo had no airlock, so for the duration of the EVA, the entire crew was in vacuum, protected by their pressure suits. The only way out was the large main hatch that had been sealed tight since launch day. In the opened hatchway, the spacewalker would be assisted by the lunar module pilot, essentially performing a "stand-up EVA" as Dave Scott had on *Apollo 9*. Film and television cameras were mounted to the open hatch, and the LMP assisted with the CMP's umbilical hose and in getting the film cassettes back inside.

In a way, it was a small consolation for the CMP, who tended to the mother ship in lunar orbit while his two crewmates got the plum assignment of exploring the ancient, dusty craters of the moon's surface. For decades following their missions, they were endlessly asked about their time spent flying solo around the moon, cut off from all of humanity for half of each orbit. "Were you lonely up there in the command module all by yourself? Did you feel separated from Earth while on the *dark side of moon*?" They all unanimously dismissed such suggestions; they were simply too busy conducting their own lunar observations and housekeeping of the ship to allow any such feelings to creep into their minds, let alone admit it.

But that would all change as these three astronauts stepped from their spaceships out into the blackness of a world between worlds, where the human emotion of being *gone* penetrated the very soul.

*Apollo 16* commander John Young was *not* happy with having to open his command module's hatch on the way back from the moon. In his mind, it was a risk to the mission and his crew that he'd rather not take. As his spacewalking CMP, Ken Mattingly, related to NASA's Rebecca Wright in 2001:

> Like other things, I had planned this thing with nauseating detail, and John didn't look at it at all. . . . So I knew we were going to put our suits on and we were going to go dump trash—throw it outside. John got more and more nervous about, "You turn that and you're going to vent the cabin to a vacuum." Right . . . then, "Do you know what you are doing?" "Yes, I think so." John was really, really nervous. It didn't match his philosophy of if you don't have to do it, don't, which no one can argue with that

philosophy, especially when you're that far from home. So I really didn't know if John was going to let me do it or not, but he finally [said] okay.

The commanders of each of the three J-series missions were all spaceflight veterans, with enormous responsibility for the success of their missions and the safety of their crews. Dave Scott, John Young, and Gene Cernan all had extensive EVA experience by the time they were returning from the moon, and they oversaw all-rookie crews, who were by definition doing all of this for the first time. Cernan said to Ron Evans, "Okay, Babe. When you get out there, just take it nice and slow and easy. You got all day long. Feel yourself around, and it's nice and easy to get around. Just don't let your body start moving too fast down there."

After returning from the lunar surface, the commander and lunar module pilot brought back *a lot* of dust and moon pebbles embedded within the beta cloth fabric of their space suits, storage bags, and clothing. The microscopically razor-sharp edges of lunar regolith played havoc with some of their gear, most concerning of which were the zippers and the metal ring connections of the pressure suits' gloves and helmet. Mission Control had them lubricate the rings as best they could prior to donning the suits, and a leak check was performed before completely depressurizing the cabin. "I tell you," John Young said, "it's pretty good for these old suits to be holding air with all this moon dust in them, Houston."

Once the suits were deemed airtight, it was time to open the vent valve on the CM hatch, releasing the remaining air within the ship to space. Each crew ran through the final checklist items to be done before their venture could begin.

CAPCOM: *Apollo 16*, you're GO for cabin depress.

YOUNG: Okay . . . GO for depress. GN2 valve handle, pull.

MATTINGLY: Okay, GN2 is to PULL.

YOUNG: Gage minimum, leave in vent position.

MATTINGLY: Okay, the gage is minimum, and it is in vent.

YOUNG: Verify helmet and gloves locked.

MATTINGLY: Ok, I got two gloves that are locked. You checked my helmet.

YOUNG: Confirm GO for depress from Houston. We got that.

MATTINGLY: All set.

As the artificial atmosphere rushed through the small valve with an audible *wooooosh*, it took with it a lot of the errant lunar dust, rocks, and stray hardware that had been floating in the nether reaches of the spacecraft for days.

YOUNG: Don't—don't let that screw go in there, Ken.

MATTINGLY: Oh, I tried.

In their post-mission debrief, Mattingly noted that despite their best efforts to clean up the cabin before depressurization, once the small hole was opened in the side of their spacecraft, everything was going to flow through it with the river of air heading for the exit. The screw that he watched enter the open valve "worried me for the next hour and a half because I didn't know whether it came out the other side." Young and Mattingly agreed that any small pebbles and rocks wouldn't jam the valve—they would just be crushed—but something like a screw might prevent the valve from closing, leaving them unable to repressurize the cabin.

Once the remaining wisps of air were gone from the spacecraft and the crews' dirty space suits were reassuringly inflated, it was time to open the hatch. In a choreographed exercise that the CMP and LMP had rehearsed endlessly in the KC-135 zero-g aircraft, the two maneuvered through a snaking maze of twisted umbilical hoses to work through the procedure. Even with the center couch temporarily removed and stowed, it was a tight fit. The spacewalker had the additional bulk of the CDR's red-striped emergency oxygen pack strapped behind his helmet, which itself was wrapped in a soft cover with an integrated hard shell, housing several movable shades and the iconic mirrored gold faceplate. This lunar EVA visor assembly, or LEVA, encased the astronaut's head and eyes, protecting him from the intense sunlight while outside.

At first, with the sun washing across the cylindrical service module (SM), casting stark shadows of the handrails and thruster quads, *Apollo 17*'s Ron Evans thought he might not need the filtered gold visor.

"Well, it looks dark out there. Can't even see," Evans said.

Gene Cernan was adamant. "No, you need your sun visor down, too. Bring it—one is protective, and the other is the sun. Use your own judgment. If you're in the shade, you won't need it. But if you're in the sun, you ought to have it down."

The first order of business after getting the hatch opened was to jettison some odds and ends overboard. Some items went out of the hatch into infinity on purpose, while others departed inadvertently.

CHARLIE DUKE [*Apollo 16*]: We lost a pencil. Saw it float out as I was climbing out.

JOHN YOUNG: Only one? Probably mine.

RON EVANS [*Apollo 17*]: (Laughter) Lost the—oh, there goes the pen. Yes. (Laughter) Okay. It was a felt-tip pen. No scissors.

Evans had been missing his personal scissors throughout most of the flight, and if anything turned up as a result of the new environment they now found themselves in, he would prefer that it be those, since they were critical to opening his food pouches. However, one long-lost item did happen to reveal itself during the *Apollo 16* EVA:

DUKE: Guess what I caught floating out the hatch?

MATTINGLY: What's that?

DUKE: A ring.

MATTINGLY: Oh, is that right?

DUKE: Yeah. I think it's yours. (Laughter)

YOUNG: Yeah, it is.

DUKE: Here, hold it, John. . . . Just got it going over the sill. In fact, it had already gone out and hit you and was coming back when I saw it.

MATTINGLY: (Laughter) Boy, how's that for luck?

"Well, somewhere, I don't know where in the mission, I lost my wedding ring. It came off, and I couldn't find it," Mattingly later shared. "Normally I found I could find things after a long period of time, they'd collect on the air filters, but it never showed up. I had given up and said, Well, guess I lost it." Duke recalled in 2012 that as he monitored his crewmate from the open hatchway,

> As I'm watching all this, I see this glint of gold. And I look over and his wedding ring is just tumbling, rolling out the hatch. I reached for it and I missed it and it just floated out and the only relative motion was the few feet per second that it was going out of the hatch. Ken's out there with his back to me and it slowly floated out and my first thought is 'lost in space.' And I realize it's going to hit him on the back of the head, which it did. Now a round ring and round helmet—you'd expect

> it to bounce off into space but it took an exact one-hundred and eighty degree bounce and started back toward the hatch. And about four minutes, five minutes later it floated back into the hatch right in front of me and grabbed that beauty!

Had the ring not taken the lucky bounce off its extraterrestrial owner and been caught by Duke, it would have continued tumbling in its long, weightless fall to Earth alongside the command module and been melted in an imperceptibly brief flash as it plunged into the atmosphere two days later.

Al Worden went about his work just after hatch opening on *Apollo 15* by discarding a bag of garbage and unneeded gear into space. It likely went unnoticed at the time as he narrated his actions that there was more than one bag tossed away. "Jettison bag is gone. And jettison bag number *TWO* . . ." His emphasis on the latter number was not without intent. As he explained to the audience at the 2019 Spacefest X gathering of spaceflight enthusiasts:

> I was looking at some NASA reports the other day, and one of the reports that I looked at was all of the flights and the analysis of the fecal bags that came back, right? *Apollo 15* was blank, no one else's. And the reason is when I did my EVA I asked Jim [Irwin] to hand the canvas bag out that held all of our fecal material and I just let it go by so . . . interestingly, that bag was going the same direction, the same time the same speed we were so it was outside the whole way back to Earth. We were safe because we had a heatshield, but that bag didn't, so . . .

With the housecleaning chores completed, the astronauts mounted both film and television cameras so that controllers and the world could watch "over their shoulders" as the EVA progressed. Little time was spent on anything not directly related to accomplishing the main goal of the EVA—the retrieval of the film cassettes from the SIM bay cameras.

"Nice day for an EVA, Ron," Jack Schmitt said. "Go out and have a good time." *Apollo 17*'s Ron Evans floated gently through the open CM hatchway into a surreal world of contradictive awareness. He knew his spacecraft was roughly 190,000 miles from Earth, plunging irreversibly toward it at a blistering 10,000 miles per hour, yet he held a handrail gently with just a few fingers, in stark stillness and silence. There was no reference to a horizon to be seen, no clouds or continents or oceans whizzing by below, just the sunbaked cylindrical hull of the SM that ended abruptly twenty-four feet away; deep empty blackness beyond.

The spacewalker's white-suited figure was reflected in the mirror-like Kapton tape covering of the CM as he worked toward the first camera bay on the SM. The LMP's helmet occasionally poked up out of the hatch into the field of view of the cameras as he tended to the stiffened, snake-like umbilical, fighting its memorized coil, a result of being stowed away since before launch. The blackness of the motionless world was pierced by the one distant blinding spotlight of the sun. "Thirty degrees or so off in that black infinity of space was a disk—and emphasize *disk*—of the sun," Evans recalled. "You can't tell the sun is shining, unless it reflects off or hits a body up there."

The film retrieval portion of the EVA was rehearsed by the spacewalkers endlessly in the neutral buoyancy pool. Ken Mattingly recalled that initially, it all seemed quite routine to him as he got to work: "Go work your way down the handrail. Go back to where the film is. You're just sitting there looking at this silvery side of the service module, and you're holding on so that the handrail is maybe eight to ten inches from your nose and you're kind of looking at the surface of the thing."

Mattingly inspected the exterior of the SM as he translated along the handrails, noting bubbled paint, which was seen on all three spacecraft during the trans-Earth EVAs. He had kept Young's borrowed sun visor down to block the sun's harsh rays, but it somewhat limited his view into the shadowed recesses of the camera bays and space around him. Mattingly was hindered by a stiff left wrist connector. "In order to make my left wrist move very much, I had to use my right hand to push the glove over to where I wanted it. And then I could keep it there," he reported post-flight. As a result, he couldn't carry the film cassettes back to the CM hatch as he had practiced. "I was not going to let go of the cassette. I was just going to slip along the rail with it, holding it in my hand, but it was all I could do to hold on with my left hand."

Ron Evans had no such problems as he began his first and only EVA. Unlike the business-like manner of his two predecessors, Evans was overtly jovial from the moment he left the confines of his home in space. "Hot-diggety-dog! Am I on the tube?"

He also immediately took a moment to revel in the ethereal space surrounding him. He immediately noticed Earth "right up ahead" and a nearly full moon off in the distance behind the spacecraft. "I can see the moon back behind me! Beautiful! The moon is down there to the right—full moon—and off to the left, just outside the hatch down here, is a crescent Earth." He went on to describe how the shiny blue sliver of Earth contrasted with the appearance of the moon. The glowing atmosphere that surrounds the planet refracted sunlight into almost three quarters of a circle, giving Evans the impression that it had horns, "and the horns go all the way around," he described.

Over the course of the two-week mission, Evans and his crew could "watch [Earth] rotate in that infinity of space up there with respect to the stars. It moves with respect to the stars." Gene Cernan once attempted to describe the three-dimensional aspect of it by explaining he could see "behind" Earth, as if it were a giant Christmas ornament, only no strings held it up against its starry background. But the view of the planet from such an incredible vantage point—not *altitude*, but *distance*—and without the obstruction of a thick spacecraft window was truly breathtaking.

Evans waved at the television camera, said hellos to his wife and children, and hummed songs as he went about his tasks. His brief tunes—"*Dum-de-dum-dum-dum*"—were reminiscent of Pete Conrad's musical interludes from the surface of the moon on *Apollo 12*. He was so verbally excited throughout the EVA that the press later questioned if had experienced some sort of "space euphoria" while being separated from his spaceship so far from Earth.

He made his way back to the SIM bay and firmly anchored his booted feet into the foot restraints that allowed him to work freely with his hands. "Hey, pretty stable right here," he reported. "Let go of both hands? See?" Evans spread his arms, leaned back, and briefly took in the immense universe around him. With his back to the sun, the sky spread out before him was sprinkled with an endless blanket of stars. It was a moment that would live with him for the rest of his life. "Hey, this is great! Talk about being a spaceman, this is it! Okay, back to work!"

Mattingly, fighting his balky left glove, had been working too hard to appreciate his surroundings. Then, on one of his short treks back to the CM hatch with a cassette, he turned his body so his bubble helmet faced toward Charlie Duke in the open hatchway. "That's really a neat picture," he thought. "He was standing up in the hatch looking out at me, and he was tethered to the inside. I've got the umbilical. I looked up here, and where I've told you about all the thousands and thousands of stars, looked out there and there wasn't a single star anywhere. There was this deep black picture with this silver thing [the spacecraft] that only went from there to here. If you turned and looked, you could find the moon over there. I couldn't see the earth anywhere. And that's all there was in this whole picture."

But unfortunately, there are no photos taken from the perspective of the CMP from any of the three trans-Earth EVAs. Only a handful of blurry color Hasselblad images were taken of Evans by Jack Schmitt on *Apollo 17*, but aside from those futile efforts, only still frames from low resolution film or video exist. Even if the astronauts had had more time to photograph the spacewalks, they would have been unable to capture the immense, infinite sphere of stars that surrounded them in every direction. But at the moment, Mattingly couldn't see them.

While he was taking a mental snapshot of Duke against the blackness of space, a feeling of dread swept over the interplanetary spacewalker. "For some reason, the absence of stars was really startling. I don't know why, but it really hit," he recalled. "This is unreal. I know they're there." Suddenly, he noticed that he still had the gold-plated sun visor covering his faceplate. He quickly flipped the reflective shade up into its protective shell and was profoundly relieved to find what had been there all along. "That's the only time that I had a sensation of being away. I don't know why, but around the Moon, inside the spacecraft . . . I didn't ever have a sense ever being that far away until I looked out there when there were no stars and the entire world was within fifteen feet. And there was nothing else. That was really a powerful sensation. [I've] never seen anything like it. I probably left fingerprints all over those rails."

Ron Evans and Al Worden never suffered such trepidations. Evans took great joy in narrating his every move to the ground and even playfully tossed the thin protective covers and pieces of insulation surrounding the cameras into the airless, black sky. "Whooooee!" he exclaimed with laughter as he slung the sheets in an un-Earthly flat spin off into the cosmos, making sure it was all in view of the television audience.

Toward the end of his spacewalk, Evans was making his way back toward the hatch, but with all of his film recovery tasks completed, he was in no rush to get back inside. He started an impromptu survey of the exterior of the spacecraft, describing the bubbled paint and burned areas around the thruster quads. As he clung gently to the side of the ship he had named *America*, his eye caught sight of the two large decals on the hull: "Hey, here's something. You know, the one thing that really shows up—and it makes you kind of proud—it says United States, and it's got a United States of America flag right below it. That didn't get scorched or a darn thing. That's great." He would express this immense pride for years to come.

As Evans wished aloud that he could get around the other side of the spacecraft but for the lack of handholds, his commander chimed in that all three film cassettes were safely inside. It was a subtle hint that Evans's time outside had come to an end. CAPCOM radioed up shortly thereafter: "Okay, Ron, we don't need any more spacecraft commentary. We'd like you to go ahead and terminate the EVA."

Ken Mattingly had one more job to complete before returning to the safety of his spacecraft. An experiment was devised in which a small box containing microbes would be mounted to a pole and exposed to the space environment for a short time. And this was exactly the kind of nonsensical risk exposure that commander John Young found intolerable. As Mattingly hung motionless in the hatchway, with Duke holding his legs when he needed to be moved around, there was nothing for the crew to do in the deadly vacuum of deep space except wait, as the required exposure time ticked away on their Omega Speedmaster wristwatches.

DUKE: You passed 3 minutes.

YOUNG: Seems like an eternity.

DUKE: It sure does.

YOUNG: It's gonna be the longest ten minutes in history.

Duke noted large flakes of ice "the size of pennies" floating past him into open space, the result of frozen water condensation on the glycol cooling lines throughout the spacecraft. "I mean to tell you, there's a lot of cotton-picking water in this machine," Young said. "Didn't you notice all of them bubbles leaving? That was water." He went on to express his displeasure with the EVA being dragged out for the experiment as only John Young could.

YOUNG: Listen, if anything happens during this period, the only thing we can say is that we died so that the germs may live, and that ain't no good at all. . . . First time anybody ever laid it on the line for a microbe.

MATTINGLY: (Laughter) I wish you wouldn't put it that way!

Those ten minutes of waiting did, however, give Charlie Duke the chance to poke his head out a bit more and take in the view between the two heavenly bodies he had now set foot upon. Mattingly had finally caught a glimpse of the distant crescent Earth and wanted to make sure his crewmate saw it as well. Duke assured him that he had, but then, in his distinctive youthful southern twang, radioed, "The thing that impresses me, though, is how *black* it is, Ken. Yeah, is it *black*!" It was an impression Duke had expressed several times during his three-day stay on the lunar surface with Young, and it became a source of amusement to all who heard it:

YOUNG: Charlie's only said twenty-five times it's black out there.

DUKE: What? (Laughter)

YOUNG: It really must be black out there! (Laughter)

DUKE: It's really black! (Laughter)

When the required exposure time for the experiment had finally elapsed, Mattingly removed it from the pole and handed it inside to Young, who was more than happy for the hatch to be closed and the reassuring sound of air to begin flowing back into the cabin and not back out through a jammed valve. The cockpit was a stuffed mess of umbilical hoses, film canisters, and inflated spacemen. It would take hours to get everything stowed away.

*Apollo 15*'s Dave Scott complimented his CMP as he was getting back inside after his relatively brief space walk:

SCOTT: That's very nicely done. That's about twenty-two minutes, which is just about what we figured. It's fun out there, isn't it?

WORDEN: Mmm-mmm. Should have stayed longer.

SCOTT: (Laughter) You done good. You made a lot of people back there very happy.

In all, three hours and ten minutes of time was spent outside by the CMPs of the three J-missions, and all of their objectives were met. Al Worden's EVA was the shortest, at just under forty minutes; Ken Mattingly's, with the time spent exposing the microbe experiment to space, was the longest, at one hour and twenty-four minutes. Ron Evans clocked one hour and six minutes floating alongside his spacecraft in the vast cosmos between the moon and Earth.

With the advancement of digital cameras and instruments that relay data in real time, the Apollo trans-Earth EVAs would not be required today. There may come a day when an astronaut is required to exit their spacecraft while on the way to or from another celestial body to effect a repair or to perform some routine maintenance task. But in the never-ending quest to reduce risk, these opportunities will be necessarily rare.

When Ron Evans closed *America*'s hatch on 17 December 1972, he and his fellow Apollo spacewalkers couldn't have guessed that after more than fifty years, they would still be the only human beings to have seen the universe from the perspective that a trans-Earth EVA afforded. Mankind has not left the confines of low Earth orbit since, although now, in the second decade of the twenty-first century, the expansion of the human race into the solar system is once again about to get underway.

And more are likely to follow.

# 5

# Lost in a Sea of Stars

Don't tell me that man doesn't belong out there. Man belongs wherever he wants to go—and he'll do plenty well when he gets there.

—Wernher von Braun

25 May 1973 had been one hell of a day for spaceflight rookies Paul Weitz and Dr. Joe Kerwin. It began with a thunderous ride on a Saturn IB rocket that hurled them into orbit to a rendezvous with America's first space station, *Skylab*. Now, just fourteen hours later, Weitz found himself floating out of the open command module hatch 435 kilometers above Earth, alongside the massive rocket stage that had been converted into this orbiting outpost. With Kerwin hanging onto his legs as he drifted out into the vacuum of space for the first time, he let out an exuberant greeting to the planet he had just left—"Hi, world!"

The laboratory, with its complex array of modules, telescopes, cameras, and power-generating solar panels, had been launched eleven days earlier atop the last Saturn V to fly, but as it climbed, ripping through the atmosphere at supersonic speeds, a large, curved micrometeoroid shield that was to protect half the station was torn away. As it peeled back alongside the rocket, it took with it one of the two large solar panels, and debris left from the violent event had wrapped around the remaining panel, preventing it from deploying once in orbit.

Now, for the first time in history, the crew of Skylab 2, commanded by moonwalking veteran Pete Conrad, was given the audacious task of attempting an in-space repair of a spacecraft—saving *Skylab*. After a week and a half of intense evaluation and planning, Conrad and his crew had caught up with the station, their small spacecraft crammed with improvised tools, equipment, and three different types of sunshades with which to cover the damaged area and cool the unbearable temperatures inside the vehicle.

This EVA, NASA's first Earth-orbital space walk since *Apollo 9* in 1969, might be considered near insanity by today's standards. With Weitz dangling from the open hatch of the thirty-one-thousand-pound spacecraft, Conrad had to maneuver alongside the eighty-four-ton space station, allowing the spacewalkers to get within reach of the sharp, tangled bird's nest of metallic wreckage that fouled the remaining solar wing. Using a long "shepherd's crook" device, Weitz would try to free the jammed panel: "There was a piece of bolted L-sections from the thermal shield that had been wrapped up around [the] top of the solar wing, and apparently the bolt heads were driven into the aluminum skin. We thought maybe we'd just break it loose."

But as he hooked the tool around the solar array and started exerting force to free it, he was surprised to find that in zero-g, he was pulling his massive spacecraft closer and closer to contact with the station. "I'm watching the solar panels," radioed Conrad, referring to four that had properly deployed atop *Skylab*'s telescope mount. With his view through the window partially blocked by the open hatch, he relied on his crewmate to be an extra set of eyes. "Translate down a little," Kerwin said as he guided him from the hatchway, "You're still closing on it. Go down and just one blip back."

As Conrad fought back against the gyrations of his crewmate outside, the station's systems also sensed the motion and began automatically firing small stabilizing jets as well. "So it made for some dicey times. We hosed out a fair amount of gas doing that," recalled Weitz. After a few frustrating attempts, he and Kerwin decided to try to cut the metal strap that wrapped around the panel's spine. Using a set of modified cable shears, they tried in vain to get enough leverage to make the cut. "I just didn't have enough muscle with that thing, because it was about six or eight feet out ahead of me and I was pulling on a line to try to do it, and we just couldn't get it through," explained Weitz. With his strength sapped and Conrad's fuel depleting, the crew backed away and closed the hatch, planning to dock with the station and spend the night in the CM regrouping.

But even this seemingly simple event didn't go as planned. The CM's "soft dock" capture latches failed to automatically engage and dock the two spacecraft together. The last resort was going to be what Conrad referred to as "another super-duper EVA," although they wouldn't actually exit the ship. Kerwin would float up into the open docking tunnel and cut several wires to bypass the "soft capture" latches, allowing the "hard dock" latches to close without the electrical signal that a soft dock had occurred. Weitz lamented,

"Man, if anybody had told me we would make two EVAs today . . . I'd have told them to drop dead."

With the hot-wiring job complete, Conrad was able to smoothly guide his ship in for a hard dock. He and his crew still had much to do in bringing the ailing space station back to life, but at least for now, they were home.

EVA had been a consideration for the Skylab program since its inception, with much of the hardware being carried over from the lunar suits developed for the Apollo program. At the outset, engineers presumed that the S-IVB upper stage of the massive Saturn moon rocket which would serve as the station's main structure would be launched full of fuel, to be later vented and outfitted for operations by astronauts in orbit. This concept was known as the "wet workshop," and even as the Apollo missions were headed to the moon, several of the newer astronauts were assigned to determine if this concept was even feasible.

As early as July 1966, engineers began looking into the various engineering challenges associated with a future space station from the standpoint of the tasks astronauts would be expected to perform in zero-g. Wernher von Braun and his team at NASA's Marshall Spaceflight Center in Huntsville, Alabama, had scraped together a makeshift neutral buoyancy pool out of an old in-ground tank. They surrounded the outdoor pool in a corrugated section of sheet metal from a Saturn V rocket, placed a conical roof atop it, and tapped a nearby steam line to heat the water.

"We were going to go inside this [*Skylab*] hydrogen tank and somehow get this hydrogen ice out of it and cover up all the holes in it so that we could then live in it," recalled astronaut Jack Lousma. Von Braun had devised a series of tests in the small pool with test panels, simulating the bolts of the cavernous fuel tank that would have to be removed. But Lousma, who spent countless hours with the German engineers working on the problems, also noted that von Braun insisted on leading the experiments himself. "He told his engineers, 'I asked these astronauts here to train, and to help us do better design on some of our stuff, but before that I want all of you to go in the water tank and do it yourself. But I'm going to be the first one. I'm going to lead you into that little water tank.'" And so, many of the Huntsville team gained valuable insight into the futility of such a large task with the limited EVA experience NASA had at the time.

Astronaut Alan Bean, before his assignment to the crew of *Apollo 12*, was assigned to the Apollo Applications Program from which *Skylab* was born. He was among the first of the astronauts to utilize this "small pool," diving in for the first time in September 1966 in a U.S. Navy Mark IV pressure suit similar to those used by the Mercury astronauts. Although his test run was cut short by a leaky glove and eventually, a tear under the arm of his suit which required him to be assisted by safety divers, he was nonetheless enthusiastic about the possibilities the facility provided. Bean's positive experience led to MSC management providing more astronauts to Marshall on an as-needed basis, and one more thing became abundantly clear—to accurately replicate a space station underwater, von Braun was going to need a *much* bigger pool.

While the suits for the Skylab program were the same basic A7-LB configuration used since *Apollo 15*, there was no desire to incorporate the backpack-like portable life support system, or PLSS, for operations outside of the station. The PLSS had been found to be difficult to maneuver around inside *Skylab*, and among other limitations, it required complex recharging procedures. Joe McMann, a NASA space suit engineer, had just the solution for the program's management—the portable environmental control system, or PECS. McMann and his team had begun work on the device during the Gemini program. It used a sodium chlorate "candle" that when ignited would produce breathable oxygen. "We had four candles, one in each corner of the backpack. And as they essentially, quote, burned down—they weren't really burning, but it was decomposing—you got oxygen from that."

Although the Apollo program had gone in a different direction with life support in the form of gaseous oxygen and water-cooled undergarments, Skylab presented a new opportunity for McMann to bring the PECS to fruition, but it would cost several million dollars to flight-qualify it. "So we went to the powers-that-be over in the Skylab program and said, 'We got this great system coming along. We've got it practically developed. . . . It will do your job and so much more,'" he recalled. But the management didn't *need* it to do more, no matter how elegant of a solution it was. "We've got this huge spacecraft environmental control system. All we've got to do is go out to this Apollo telescope mount and take these film canisters off. Why couldn't we use the capabilities of this spacecraft system and maybe just have, oh, some sort of umbilical system? That would be fairly cheap," they told him.

The system eventually agreed on was a stout umbilical containing oxygen supply, water cooling, and electrical lines, along with an internal tether. These fed into an astronaut life support assembly (ALSA), mounted across the abdomen of the suited crewmember and backed up by an emergency oxygen pack strapped to the leg. But McMann still took pride in the original PECS design and put together a demonstration for a group of NASA managers and astronauts. The results were quite memorable: "We made the cardinal mistake of not having a dry run first to do this. Anyway, we fired this thing up, and I mean the room emptied within about fifteen seconds. The smoke was so thick you could hardly breathe, and the thought was, 'This is your breathing oxygen?' So if we hadn't killed the system with it being overweight and over cost and too complicated for the Skylab job, that one demo would have killed it for us. So that's, again, some of my dumb tests that I've run in my career."

By late 1970 NASA had assembled a group of about fifteen astronauts who would serve as prime, backup, and support crewmembers for the Skylab program. Along with the moonwalking veterans Pete Conrad and Alan Bean were several first-time potential flyers Weitz, Kerwin, Lousma, Jerry Carr, Bill Pogue, and scientist-astronauts Owen Garriott and Ed Gibson. Like so many astronauts before them, the reality of a looming spaceflight hit them when the call came to be fitted for a space suit.

"I wore a suit before that but it just happened to be a hand me down that kind of fit me," recalled Gibson, "and you didn't realize how good it is when they finally tailor one to you because in that thing—when it gets inflated, you're really restricted. [You] can't move, and the joints have got to all be in the right place." Pool time was also a prime commodity reserved for the astronauts assigned to upcoming missions, and the Skylab crews got right down to business training and developing procedures in the newly built forty-foot deep, seventy-five-foot-wide Huntsville tank—the neutral buoyancy simulator (NBS). Not only did they have to familiarize themselves with operating in the new bespoke suits, but there was much to learn about how to move large packages of film and other hardware in the zero-g environment.

Compared to earlier Gemini and Apollo EVA simulations, the *Skylab* mockup held in the 1.3 million gallon tank was *massive* by every measure. Astronauts would stage their space walks from an airlock module barely visible from the outside, tucked down partially inside the orbital workshop of the converted S-IVB. Surrounded by a jumble of scaffolding—an Erector set of struts and beams that allowed the Apollo telescope mount (ATM) to pivot ninety degrees to the station's centerline after launch—the spacewalkers would exit the airlock via an old Gemini spacecraft hatch. "It was perfectly adequate to get in and out of," Gibson points out, and like everything else in the post-Apollo NASA, using a proven design saved dwindling government funding. Prior to being formally assigned to Skylab, Weitz and Kerwin had been the first suited astronauts to use the new NBS on 4 March 1969, while Gibson observed them conducting film transfer tests wearing scuba gear. Just the day prior, *Apollo 9* had reached Earth orbit with Rusty Schweickart, who would be conducting the first EVA of the Apollo program.

Like most astronauts, Gibson found the greatest value of the underwater simulations to be not in the imperfect recreation of weightlessness but in learning the complex choreography of conducting a space walk. "It's really good because it trains you in all of the procedures and you understand what it's like to be trying to work against the suit, and of course finally what you can actually do in a suit once you learn how to do it," he shared. The station only had certain pathways that were expected to be utilized by the EVA crewmen to do their work, which would be concentrated on the dinner plate-like topside of the ATM. In the water, they learned to translate from the airlock up the lattice of the telescope mount, replace film canisters, and transport them back to the airlock by either extendable booms or a clothesline contraption. All the while, they had to struggle with the thick, snaking umbilical that fed them air and cooling water, clamping it off at points along the way as it fought their every move.

The simulations began after the suited astronauts were weighted down with nearly three hundred pounds of weights. Without it, their inflated suits would bob to the surface like a cork. As Jack Lousma recalled, "You might have some wrapped around your legs and some around your arms and some around your chest but it wasn't a uniform distribution of weightlessness like you would have in space. So you had that to contend with." Once in the water, final weights were added to hold the crewman neutral in the viscous water. But this was just a temporary condition. "The buoyancy would change after you're in the water for a while . . . because the suit would get wet and it seemed to change its volume or something and so it wasn't a perfect simulation."

The *Skylab* crews spent countless hours commuting between Houston and Huntsville to take advantage of the NBS, and the experience was indispensable. "I wouldn't want to go into space without having that," Gibson reflected. It wasn't long after the nearly catastrophic failure of the *Skylab* space station during launch on 14 May 1973 that the nine men assigned to crew it realized just how indispensable that training would soon become.

Once the flight control team realized the extent of *Skylab*'s problems after launch, Mission Control, engineers, and astronauts across the country sprang into action to assess what could be done. The possibility of losing the station loomed large, but NASA was once again presented with the opportunity to do what it did best—solve problems in real time to turn a potential disaster into success.

The planned launch of the first crew—Conrad, Kerwin, and Weitz—was put on hold until fixes could be designed. The telemetry from the wounded bird indicated that the critical problems were twofold—a severe shortage of power due to the loss of one solar panel and the failure of the other to deploy properly, and the extremely high temperatures within the orbital workshop due to the loss of the micrometeoroid thermal shield.

Among the many astronauts who found themselves at the center of the rescue effort was Rusty Schweickart. Since his return from *Apollo 9*, he had spent a great deal of time in the Apollo Applications Program designing procedures and equipment for the future space station. At the time of the *Skylab* launch, he was also one of only a handful of astronauts to have conducted a space walk in an Apollo space suit. As he sped to Huntsville in a T-38 jet, his mind raced with the possibilities and potential pitfalls that lay ahead:

> For me it was immediately obvious that EVA was an obvious alternative. Now what to do during EVA was a wide open question. The EVA was the most likely way we were going to fix the issue if it could be fixed at all. Because no one had ever considered the possibility of something like that being required, there were a lot of questions in my mind about what options were even possible for consideration or evaluation. My mind was just filled with what do I do first, what do I do second? It was . . . get the water tank ready man; I gotta get in and get some answers here. Can I go down the side of this whale; the main laboratory itself? What kind

of use can we make of various kinds of equipment? What's the best way to tackle this whole problem?

There were more questions than answers, and time was short. Tools had to be devised and approved for flight to allow the astronauts to free the stuck solar array, without knowing what exactly was preventing it from deploying. There were no external cameras on *Skylab* that could be focused on the damaged areas; an accurate evaluation of the situation would have to wait until the first crew reached the crippled station. To address the thermal issues, some sort of both short and longer term solution would have to be engineered to shade the now unshielded side of the hull from the blazing hot sun.

As the round-the-clock effort to come up with solutions began, Mission Control struggled to orient the station in space to minimize the damage due to the intense heat within the laboratory. Schweickart recalled that there was a launch opportunity for Conrad's crew about every four days, and the thinking early on was that they would be ready to go with basic EVA equipment and procedures at the first opportunity. Fortunately for the team, "Mission Control had figured out that they could probably go eight days because they had learned a lot about how to control it and make tradeoffs to minimize the temperature gain but at the same time still save a little bit of fuel."

Schweickart was back in his element, suited up in an A7L-B space suit, counterweighted for neutral buoyancy, his booted feet just submerged in the water as he prepared to be lowered into the NBS at Marshall. He was eager to get down to the business that had consumed the space agency for the past several days. Just before he locked his helmet on, a telephone beside the pool rang abruptly and was answered by a technician. He listened for a moment, turned to the astronaut, and passed him the receiver. The gruff voice of chief astronaut Deke Slayton gave a simple order: "Get out of the water tank!" Schweickart was beside himself. "What the hell? Why should I get out of the water tank? You gotta be kidding me. Why in the world would you make that request? It was because something about the press was heading to Marshall. They had heard that I was over there and they were headed there. I don't know—obviously Deke had gotten some kind of call or orders from headquarters. I just told him 'no way.' I gotta make some decisions here—we gotta get this thing going if we were going to [fix it]."

Conrad's crew would also be dispatched to Huntsville to familiarize themselves with the repair procedures and the three different proposed thermal shield designs they could potentially deploy. "It was kind of a weird ten days," Joe Kerwin recalled. "The press was trying to get all over us and Deke was trying to keep us from the press." To Schweickart, the press was not his concern. "I don't know why anyone would imagine . . . having the press not knowing what we were doing was more important than us getting the job done," he surmised.

Satisfied with his decision, Schweickart handed the phone back and continued his preparations to get into the tank. But just before he disappeared beneath the water's surface, the technician at the phone began waving frantically to him. Slayton had called back, presumably to deliver a firmer command. Schweickart radioed over the intercom, "Hang up man. Tell him I'm underwater." The boiling cauldron of bubbles produced by his suit as he submerged soon smoothed to a ripple on the surface as his world was transformed to space.

The first order of business in the NBS was to determine how an EVA astronaut could install some sort of covering or shield over the damaged area of the workshop, but with no handholds or foot restraints, the smooth cylindrical surface presented a major challenge. Within the first ten minutes underwater, as Schweickart recalled, he realized that working in this area would be impossible. "That no longer got any consideration at all and we immediately went to operating from somewhere on the ATM to get a shield out over the thing."

Two sunshade designs promised the greatest chance for success—a parasol design, conceived by NASA's Mr. Fix It Jack Kinzler, would be extended and sprung open from a small scientific airlock on *Skylab*'s hull, then retracted to partially cover the unshielded area. This concept could be implemented the day of the crew's arrival at the station without conducting an EVA. The second, more permanent solution came to be known as the "twin pole" sunshade, which would consist of two long, assembled poles attached at the more accessible ATM, with a curtain of beta cloth and reflective Mylar spread between them. The ATM was already equipped with adequate restraint devices to anchor both the sunshade and the "orbital repairman" astronaut.

A small group of engineers tasked with designing some sort of cutter to free the solar array enlisted the help of the AB Chance company of Centralia, Missouri. The company specialized in tools used by electrical linemen working high up on utility poles and immediately offered a powerful set of cable-cutting shears. The engineers modified the stock tool with the same quick-disconnect fasteners used on the twin-pole sunshade, so the existing poles could be used as extensions to reach the debris by an EVA crewmember. Then a long length of cord was attached through a pulley to actuate the shears once in position.

Schweickart was confident that even without extensive training, the crew on-site would be able to accomplish both the sunshade installations and free the jammed solar wing. "The real goal was to get two people outside with a bunch of tools," he shared, "but the main thing in my mind that I impressed on the engineers—they already knew because they knew me very well—we'd worked together for many years on the EVA work, but convincing the management—the decision makers at headquarters—that we could do this job."

Over several intense days, Schweickart spent dozens of hours in the Huntsville pool evaluating tools and procedures for Conrad's crew to utilize. It was all hands on deck at NASA centers all over the country. "That period—I don't think there has ever been in the history of humanity a more intense engineering effort from something totally unexpected to being completely ready—the design, manufactured, tested, trained—I mean the whole thing in a matter of four days," Schweickart reflected proudly. The rescue mission would wait an additional four days while NASA management further considered all their options.

Ed Gibson remembered one final admonition from Rocco Petrone, then director of MSFC, who had oversight of the center's role on *Skylab*. "He said, 'We don't want you guys flying out at the end of the long umbilicals. Make sure you've always got yourself well secured.'" All the astronauts present acknowledged his demand with a sharp, "Yes, sir!" But space, as it was known to do, would present the unexpected in ways even Petrone couldn't simply order away.

Twelve days after arriving at America's first space station, on 7 June, Conrad and Kerwin prepared to exit the airlock and attempt to free the jammed solar array. The parasol sunshade, earlier deployed through the smaller scientific airlock, had done its job to reduce the soaring temperatures in the workshop, so now the severe power deficit needed to be addressed. As they opened the repurposed Gemini spacecraft hatch and dazzling sunlight flooded into the compartment, Kerwin warned his commander to put his gold visor down.

Still trying to unwrap the umbilical from around Conrad's feet as he exited, he noted the stark contrast between space and the NBS back at Marshall. "Man, the buoyancy is better here, Pete, than it is in the water, isn't it?"

The plan, which had been proposed by Schweickart based on his work in the pool, had the team extend the cable shears out about twenty-five feet utilizing the assembled pieces of sunshade poles. Using that, they would set the jaws on the metal strap that restrained the solar array and pull the lanyard to set them into the debris. "Now we had a handrail," explained Kerwin. "Pete could go along the handrail while I stabilized the near end of it, with another rope attached to his sleeve. When he got as far out as he could . . . he would hook that rope into the solar panel cover as far down as possible, so as to give it some leverage from the hinge."

Simply cutting the strap would not allow the wing to deploy. A mechanical damper, which was expected to be used only once, immediately following launch, was correctly anticipated to be frozen after weeks in space. An amount of force would have to be exerted by the astronauts to break the damper free and extend the spine of the array. After the strap was cut, they would use the rope, known in NASA parlance as the beam erection tether (BET), which Conrad had tied off at the far end to muscle past the locked-up unit. As an additional backup measure in case the shears failed to sever the strip of metal, the good Dr. Kerwin had even taped a length of flexible bone saw to the chest of his space suit.

The pair set out assembling the extension poles with the shears affixed to the end. They fought and struggled with the long pole assembly, tethers, umbilical hoses, and lack of foot holds to stabilize themselves. Schweickart served as CAPCOM throughout the EVA, trying to radio up his experiences from the NBS, but the mockup couldn't exactly replicate the debris that needed to be cut. And Sir Isaac Newton's laws played havoc with the spacewalkers—the air-to-ground chatter sounded like the two were trying to do a complex weightless surgery while surrounded by the swirling dance of an octopus's tentacles:

CONRAD: All right. Now turn around, if you can.

KERWIN: Which way, this way?

CONRAD: You've got to turn towards me. I've got your umbilical; I don't know where mine is. And I'm tethered, and what I want to do is get my umbilical around my back.

KERWIN: Wait a minute, let me get over here where I can hang on for a minute.

CONRAD: All right. Now, where's your umbilical?

KERWIN: My umbilical is—goes straight down, then takes one loop through yours, if you see what I mean. So you move under there a little bit.

CONRAD: I have to go under here?

KERWIN: Yes.

The duo worked through one orbital night by the glow of the EVA lights installed on the outside of the station; the airlock hatch, EVA pathways, and the top of the ATM were peppered with them. As they approached daylight again at orbital speed, Conrad and Kerwin were treated to their first sunrise while outside of a spacecraft.

KERWIN: Here comes sunrise. Oh, that's the blackest black in the universe up there!

MISSION CONTROL: Okay. We'd also like to check to make sure you got your visors down for that.

CONRAD: Got the visors down, man. We're ready. . . . I'm going to see the sunrise upside down.

KERWIN: (Laughter) I can see it right side up where I am. . . . I can see the limb brightening in a great big crest at about 180 degrees.

CONRAD: Fantastic, isn't it?

KERWIN: And stars blinking through. . . . I've got a solar panel between me and the sunrise. But I can see the rainbow behind it. Oh, brother!

The spacewalkers took a moment to marvel at the stark contrast of two blacknesses before them, one concave and one convex, divided sharply by a curved, layered kaleidoscope of colors as the sunlight prismed through the astonishingly thin skin of atmosphere. When the blinding white sun pierced the divide, the stars were washed out, and the long shadows of clouds began to stretch out on the planet far below. With the cutter's jaws finally set in place across the thinnest section of the metal strap that fouled the solar wing, Kerwin unbagged the BET for Conrad to secure to the swing arm of the solar array. Conrad floated easily hand over hand down the featureless side of the workshop trailing the rope and set about tying it off to the still folded wing. This was a task that Schweickart had assumed would come naturally to a navy man. "The idea was that he would tie off that tether in very much the same way that when you pull up to a dock in a sailboat—you wrap it around the cleat," he explained. "To me that was so straight forward that was what Pete was supposed to do to tie off the tether—he would do it that way."

Conrad and Schweickart were great friends—they had sailed together, raced cars together, and trained endlessly together throughout the Apollo and Skylab programs. "I can't think of Pete without laughing," Schweickart says today. But even to this day he cannot understand what he recalls happening next:

> For whatever reason Pete took the end of the tether and hooked that to the structure and then did the figure eight—he didn't pull it tight and then do the figure eight. He took the loose end of it and started doing the figure eight there which means when you get the last [one] done there's a lot of slack left in the line. It's not a taut line. That never entered my mind that Pete wouldn't do it the right way. But having done it the wrong way it was nevertheless secure on the end but it was a loose tether.

Kerwin struggled to gain enough leverage on the pole and then pull the rope which would activate the shears. "Conrad was trying to grab my legs with one arm and a strut with another arm, but that's not a stable enough platform," he recalled. "We went nuts for one whole day-side pass and failed, just didn't do it." During the second night pass, as they reevaluated their plan, he noticed a loop-like eyebolt on the station's hull. Kerwin had no idea what its intended use was, but he immediately saw how it could be useful to him. "We got the spare tether, and there's a hook on the front of the suit. Hooked it through there, ran it through the eye bolt, back up through the suit, tightened it up, and now I have a three-point suspension. Now I can stand. I'll place my feet on the surface of the workshop and almost straighten my knees all the way out, and suddenly I'm as stable as a rock. It was like standing in your garden at home. It was wonderful."

With all their tools now in position, the time had come to see if the weeks of effort on the ground and in space would pay off. Conrad remained on the far end of the pole assembly to report on the cutting while Kerwin used his new stable perch to activate the pull rope on the shears. "I'm in the duty cutting position," he reported to his commander. "I'm just going to stay here, contemplating rivets, until you're ready for me to cut." The rest of the task unfolded while the controllers on the ground held their collective breath:

CONRAD: All right. Go ahead and try.

KERWIN: Okay, here we go. . . . That'll take it several tries here. But I'm taking in rope every time.

CONRAD: It looked to me like it cut.

KERWIN: It may have cut.

CONRAD: Keep going. Keep pulling.

KERWIN: Man, I am really pulling! That's all the pull I can pull on that beauty.

Then, suddenly, the long, white backbone of the solar array lurched away from the curved, burnt-gold surface of the workshop. "Whoops! There she goes!" Conrad exclaimed. Kerwin laughed heartily as he added, "There she goes, SAS panel and all!" The wing only made it out to about twenty degrees before the frozen damper hung it up yet again, as anticipated by the engineers before flight. The next step would entail Conrad and Kerwin using the long lever arm of the BET to break the tension of the device and allow *Skylab* to fully spread its single wing.

Conrad and Kerwin were now faced with the quandary of the BET either being too tight or having too much slack to effectively pull the wing free from the tension of the damper. "Joe standing on the fulcrum point and lifting the tether over his head—it didn't apply the force on the beam that was needed," explained Schweickart. "And of course the two of them were sort of scratching their heads—probably thinking Schweickart's a dummy—what the hell? Didn't he realize that this wasn't going to be tight enough?" Once again, the extensive work put in by the team at MSFC provided them with one final alternative:

CONRAD: I could go stand up and try it.

KERWIN: The standup is the weapon we've got left.

CONRAD: And I've got to get standing up first.

KERWIN: I know it. And it ain't easy. I ain't rushing you. Just conversing.

Conrad laughed at himself as he struggled to get into position underneath the BET with his feet planted squarely on the workshop's hull. "That was probably the most dangerous part of the space walk," recalled Kerwin, "Pete going down there amid all that debris, but he got away with it." Kerwin then picked up his end of the BET, and from his improvised restraint added his stature, which far exceeded Conrad's, to the jerry-rigged system of tethers:

KERWIN: I tell you what. If the old doc can get his shoulder under it like this . . . Whoops! My feet . . .

CONRAD: I got it.

KERWIN: There you go. Gave it my all, except that old son-of-a-gun ain't coming.

CONRAD: Easy. Yes, here comes, Houston. Easy. If this line lets go, I'm going off in the toolies.

KERWIN: You and me both, babe.

Suddenly, the frozen damper broke free, and as the tension on the BET released, the two astronauts were unwittily catapulted off their spacecraft like two arrows fired from a bow simultaneously. "We stood up, and suddenly it released on us. We both went ass over teakettle into outer space," Kerwin shared with an often-used phrase more typical of Conrad:

KERWIN: It's coming! It's coming!

CONRAD: Yes. Let me get back down to—Okay. Pull.

KERWIN: That's it . . . Yeah . . . There she goes. Let go! Let go!

CONRAD: That's got it. (Laughter) Would you look at that?

KERWIN: I expected it to come, Pete. But I expected to lose you in the process. And, by gosh, we got you and the panel!

The steel cable tether wrapped within their umbilicals did their job and snapped the two crewmen to a quick stop before they tumbled very far. Then it was a simple process of going hand over hand back along the thick white hoses until they were safely back on the hull. "It sounded very dramatic and it was dramatic in a way," recalled Schweickart.

> There was certainly no danger in it because Pete had a hold of the rope and wasn't going to go anywhere. But he was just flipping around in the air, and I can immediately picture this thing—and it was just a great laugh—they started laughing their asses off as I did on the ground. It

> was like, what in hell Pete? He lost being a sailor at the wrong moment. The whole thing was just funny and it couldn't have happened to a better guy. Pete is the kind of guy who can just immediately roll with the punch. I'm almost glad it happened that way.

With the long white beam that housed the accordion-like solar sails now locked out at its proper position perpendicular to the cylindrical laboratory, the bluish solar cells began unfolding and producing power. As Weitz monitored the welcome indications of electricity on the gauges inside, Conrad and Kerwin began gathering their gear to head back to the airlock, leaving the BET behind. "The prettiest sight I've ever seen in my life was that solar panel cover fully deployed," remembered Kerwin.

The scale of what the spacewalking duo had accomplished would take a while to sink in. The equipment and procedures that had been put together in such short order were the critical base of a pyramid of procedures that had to go right to save *Skylab*, but the two at the top of that pyramid had just pulled off one of the most difficult and dangerous EVAs in history. After nearly three and one half hours, the tired crewmates closed the airlock hatch and repressurized the compartment:

> KERWIN: Captain?
>
> CONRAD: Yes, sir?
>
> KERWIN: Get that glove off; let me shake you by the hand. I didn't think that job could be done.
>
> CONRAD: Shake? You did it!

Sometime later, Ed Gibson had a conversation with the crew where they regaled him with the story of how they had freed the jammed solar wing. But as proud as they were at what they had accomplished, the stern warning they were given by Rocco Petrone before flight was still at the front of their minds. "He was just concerned about our safety," remembered Gibson. "If someone proposed that why don't you just push off the side of the spacecraft, and fly out to the end of your umbilical, the world would have rained down on you." As he learned from Conrad that very thing had inadvertently occurred, he could only laugh. "He would have had a heart attack if he'd saw what was going on."

Conrad was adamant in his report to Gibson about their unexpected foray from the station: "Don't tell Rocco!"

Conrad and his crew returned to Earth on 22 June 1973, having completed three EVAs totaling five hours and forty minutes. This was far in excess of the single space walk planned prior to the damage that occurred during *Skylab*'s ascent. While their historic success got the station and the program back on even footing after what could have been a catastrophe for NASA, there was still work to be done outside of the spacecraft by the next crew to live aboard the facility, consisting of Alan Bean, Jack Lousma, and Owen Garriott.

Bean's crew launched to the station on 28 July 1973 with the goal of continuing the research began by the first crew over the course of a fifty-six-day mission. But one of their first orders of business was to install the so-called twin-pole sunshade on top of the existing parasol that was previously deployed from the scientific airlock. The parasol was a relatively thin Mylar sheet that was expected to degrade over time. The more permanent shade, twenty-four by twenty-two feet in size, was spread between two assembled poles like a huge shower curtain.

On 6 August 1973 Lousma and Garriott headed out on their first EVA from *Skylab*'s airlock module, the compartment stuffed with pole segments, tools, and a large bag containing the sunshade itself. Lousma recalled his first impressions of working in open space as being familiar, yet surreal. "At first I was dazzled by the brilliance of the sunlight. When Owen and I went outside the hatch, of course, we were surrounded by the structure of the whole *Skylab* . . . so I floated out and there were handholds—places to use for mobility—handrails and so forth. We devised all of that in Huntsville earlier specifically for spacewalking. So it seemed just like when it was in training, but the view was *much* different!"

As they set about installing the sunshade, Lousma floated hand over hand a short distance along one of the tubular struts that supported the ATM and anchored himself into foot restraints that he carried out with him. "Owen was going to be stationed back at the Gemini hatch and he had the hardware . . . for making the poles, and he also had the bag full of shade fastened down," Lousma recollected. "I was standing there sort of face down. If I wanted to see the top of the S-IVB I had to change my position and mechanically bend my knees and my head so I could look up."

Garriott began assembling the eleven five-foot pieces that would comprise each pole and passing them out to Lousma, who then fastened the long, fishing pole-like rods into a connector on his foot restraint, forming a large, flexible "V." The poles each had a pull line integrated with them to deploy the shade, once it was installed. Almost immediately, Garriott ran into trouble with the complexly engineered connections of the pole assembly: "We had the aluminum segments strapped to an aluminum plate with a little elastic band wrapped around each of the tubes. We then needed to take it off of the aluminum [plate], lock it together, roll down a ring, like a screw and a nut, and then roll down a locking ring at each section joint. So there were about six or seven connections on each of the two poles."

Unfortunately, when the crew was rehearsing these deployment tasks in the NBS, the only equipment they had to use was the actual flight article, and the engineers were reluctant to allow it to be used underwater. So Garriott and Lousma reluctantly agreed to simulate the assembly without potentially damaging the expensive gear. Assembling the complex links between the poles was the only procedure they didn't rehearse underwater in pressurized space suits.

Just a few weeks later, Garriott now found himself in orbit struggling to complete the job in his bloated, rigid space suit glove. "I had to reach down and stick a finger under that little elastic band and try to lift it and at the same time get the rod pulled out. I couldn't do it," he recalled. "There was no way when those rods were side by side and a small elastic band around each of them, to get a finger in there." As he continued to find a way to accomplish the task, the crew began running behind their timeline for the EVA. At one point during a communication pass with Mission Control, Bean even suggested they may have to come back into the airlock, repressurize, and somehow attempt the task without the gloves on.

Garriott, however, was determined to get the job done and improvised a way to make the tricky connections. "That turned out to be the most physically demanding task that I did during all of my three EVAs," he recalled. "And it's because the one thing that I did not simulate under water was the thing that I had a problem with when I got to space." Years later, he would reflect that his commander likely wouldn't have made the same decision. "If that had been Alan Bean, he probably would have insisted on taking that flight unit to the water, because he doesn't like to compromise on those things. In this case he'd have been right, because that was about the only thing that gave me a real problem."

With the twin poles finally deployed, Garriott used one of the transfer booms to send the bag containing the sunshade out to his spacewalking partner. Packed by parachute riggers in Huntsville, "it was folded up in this bag and it came out kind of like an accordion," explained Lousma. "So I hooked the corners to the clips on a rope, sort of running the flag up the pole so to speak." In his face-down position, Lousma had to lean back to check the progress as he attempted to unfurl the sunshade. Once again, space presented its challenges. "I was pulling on this rope to get the shade up the poles, I noticed that the poles were kind of pulling together," he explained. "So I looked up, and sure enough, that shade was not in a nice shape, it was in a big wad."

The shade had been assembled in such a hurry before it was launched with the first crew that the adhesive that fused the multiple panels together had not had time to cure before it was packed into the bag. Lousma only had one choice. "So I reeled the whole sucker back in, so I could separate all of those folds that were stuck together." He began the arduous task of finding each section that was stuck together and pulling them free by hand, as the entire sheet blossomed in weightlessness all around him. "I'm there pulling all of these folded pieces apart, and unhooking them," he recalled. "Then as I did, I had to pull it a little more up the poles, and finally I got the whole thing to where it was flat, almost."

With the trapezoidal sunshade now neatly deployed along the two poles, Lousma was ready to put the finishing touches on the job. The whole assembly—poles, sunshade, and ropes—was angled away from the cylindrical lab by about thirty degrees. He pivoted it down over the old parasol shade and fastened the hinge point at his feet tightly. Then he floated smoothly to each end of the near side of the sunshade, pulling it with a strap to an anchor point at each corner, permanently stretching the sunshade to its intended shape.

In the now famous photographs of the single-winged space station, the accordion-like folds of the twin-pole sunshade can be seen draped over top of the parasol, the latter's sunburnt corners just barely visible poking out at odd angles. The more durable cover still browned over time in the harsh rays of the sun, but if one looks closely, there is a signature mark left behind from the intense efforts to solve one of the stations many problems. "There was one of those creases that didn't get done," Lousma pointed out. "You look at the top on the *Skylab* . . . it had turned brown, except there's one little white groove in there, brand new white, and it's obvious that I had not gotten that disconnected. It stayed glued to itself for the whole set of missions."

The thrusters of the last departing spacecraft had blown the shade around quite a bit, giving the impression on film that it was flapping wildly in some unseen breeze. When it once again settled down in the windless vacuum of space, it revealed the previously unexposed section of the sunshade. Over time, it too would slowly deteriorate in the harshness of space.

Deploying the sunshade took an unexpectedly long three and half hours. With the job complete, Garriott and Lousma went on to as many of the other tasks as they could, including a survey of the area around a leaky thruster quad on the service module of their spacecraft. As they drifted out of radio range of one of the far-flung tracking ships, Bean called down, "[Going to] give you a little appropriate music to the effort they did out there today." Just before the static drowned them out, Mission Control could hear the verses of "The Marines' Hymn"—"From the Halls of Montezuma to the shores of Tripoli . . ."—the anthem of Lousma's beloved U.S. Marine Corps.

After the unprecedented, dramatic EVAs to repair the damaged *Skylab* and make it habitable for the duration of the program, the remaining space walks settled into more routine operations. Hastily improvised, even experimental efforts gave way to the original plans, procedures, and timelines utilizing much of the equipment as it was designed to be used. The main task that required EVA was the periodic change-out of film canisters located in cameras at the top of the ATM.

Twenty-eight days into their mission, on 24 August 1973, Lousma and Garriott again prepared to enter the vacuum of space to complete their first film exchange. But after a month in weightlessness, they suddenly realized that their custom-made space suits didn't quite fit like they used to. "[By the] the second EVA, I guess, I had a hard time getting in the suit," recalled Lousma. "We had grown and didn't know it. This is the first time they measured and found out that you grew while you were in space." In the absence of gravity's unrelenting pull, the cartilage between the vertebrae expanded, adding an inch or two to each crewmember's height.

Once crammed into their now ill-fitting suits, Lousma and Garriott set up their gear outside the airlock to begin the film exchange. There were three telescoping booms on *Skylab*, one to the sun-end of the ATM, one to a central workstation, and a backup. These booms would shuttle the large boxed film cans between the operator near the airlock and the astronaut stationed high up on the ATM. Some of these were located on top of the large white disc at the center of *Skylab*'s "windmill" of solar panels. But while this area was equipped with handholds and foot restraints, a failed control unit would require Lousma to reach an area on the side of the cylindrical space station ill-suited for EVA. "The rate gyros had a problem, and so we had to take up a package of six new rate gyros. We called them the 'rate gyro six pack.' I had to hook up a long twenty-four-foot cable to bypass our rate gyros, so I had to do some external EVA into places we'd never been before, or weren't made for that purpose in order to hook up those cables."

Lousma and Garriott took turns changing out the film packages, alternating between boom operator and camera technician. "All of that worked out. . . . It was very well organized before we went up," recalled Lousma. "All of the footholds and handholds and railings and everything were all in the right place, and when we got out to the site over the top of the solar telescopes, why it really worked out just as we had planned it back in the water tank in Huntsville."

While still focused on their work, Lousma did have the opportunity to savor the experience of being outside of a spacecraft in Earth orbit a bit more. With everything going as planned and not struggling with a sunshade deployment gone awry, he finally had a chance to look around. "When you can get away from the structure and see all around you say, 'Well, gee I was missing a whole lot,'" he observed.

> When you're looking out of a window inside it's kind of like going down a railroad track on a train car and you're looking out at all the scenery going by but you don't see the other 360 degrees. But when you get outside . . . it's a total different perspective. It's like riding on a locomotive instead of in a train car. And you can see in every direction about 1,200 miles and you can see things on the ground that you couldn't see from the confines of the window you were looking out. And so you can have a better sense of speed over the ground I think when you have this more three dimensional aspect, and it's kind of like gliding along on a magic carpet into the sunrise and the sunset.

Sailing over Houston in the darkness of orbital night, the spacewalkers could see the lights of Dallas, Texas, Little Rock, Arkansas, even Cincinnati, Ohio, clear up into Chicago, Illinois. But as many moments and images that filled Lousma's memory from his two EVAs on *Skylab*, none stood out more than a brief respite from work he had during orbital nighttime. "I was out on the end of the ATM replacing the film and filters in the solar telescopes—putting in new film and bringing out the old," he remembered. In this case, he only had one booted foot slipped through the restraint bolted to the solar observatory's topside. As the station passed into darkness somewhere over the Pacific Ocean, no one thought to turn on the EVA lights. "So I'm out there hanging on to the spacecraft by one foot. I can hardly see my hand in front of my face," Lousma pictured. "And the only thing I can see is the starlight—you're kind of lost in a sea of stars until you've gone around a few times and you find the ones you know."

As the station reached its peak latitude both to the north and south of the equator with every revolution of the planet, the EVA crewmen were treated to an unobstructed view of the aurora. Long ribbons of red and greenish-blue light glowed and swirled in the rarified air far below as the sun's stream of solar particles collided with the molecules making up Earth's atmosphere, stripping away their electrons and creating one of nature's greatest spectacles. "Every light show is different. . . . They're unique," Lousma said. "It's just like every sunset is different. It's just a different picture out there of God's creation."

Alan Bean, having carried out two EVAs on the lunar surface on *Apollo 12*, finally had his chance to perform one in Earth orbit on 22 September 1973. He and Garriott ventured outside to collect the last batch of film magazines from the ATM, install several new ones for the next crew to arrive, and retrieve samples of the sunshade covering left outside during the mission's first EVA, for evaluation back on Earth. The liquid suit-cooling system in the airlock, which had been giving both crews trouble from the start, was completely empty of water by this time, so the crewmen relied on airflow alone, pumped through their umbilicals, to control their suit temperature.

Bean seemed to relish his newfound view of the world while outside the space station, taking advantage of every opportunity to comment on his surroundings. "While you're doing that, I'll enjoy the scenery, if you don't mind," he radioed to Garriott at a point where he had to wait for him to finish a task. "Incredible!" As *Skylab* sailed silently over the globe of Earth, the two spacewalkers were awestruck by the three-dimensional atlas spread out before them:

BEAN: Passing over the Black Sea; the Caucasus Mountains. Nice view, huh, Big O?

GARRIOTT: Yes, it is

BEAN: Told you I'd take you on a nice trip; even bought you lunch for 56 days. How's that? Can ride over there across the Med at the Nile River—the Red Sea, right up the coast of Israel; Syria to Turkey.

GARRIOTT: There's Crete.

BEAN: Yes, I can see the Nile.

After the quick pass over the Middle East, it was back to work for the duo, leaving Bean wishing they had put bigger and better windows on his home in orbit. They, too, were able to observe the Aurora Australis as their orbit swung them down over the southern Pacific Ocean. "Look! We got aurora!" Bean said excitedly. As he went about his work on the roof of the ATM, he speculated, "I'm the first man in history to see a southern aurora standing on his head on top of the sun end of the *Skylab*."

Bean and Garriott kept a running commentary up as they worked, describing the geography they passed over, the dazzling sunrises and sunsets, and the wondrous views of the heavens that unfolded before them with each pass through the shadow of the Earth. Not that anyone on the ground could hear much of it—the communications passes were extremely limited, and sometimes even nearly a complete orbit would go by with no opportunity to talk to Mission Control. But the reports were recorded onboard, like a traveler keeping a tape of voice notes for some future travel log yet to be written.

After a relatively brief two and a half hours, they prepared to get back into the airlock. "The view here is even better than on the moon," Bean suggested as he took one last look around. "I would guess that it might be more interesting looking down at the Mediterranean than it would be the lunar terrain," Garriott suggested. "I think you're guessing right," Bean agreed.

But Skylab 3's pilot, who had been denied a chance to go to the moon when the final three lunar missions were canceled, took a different perspective. "I don't know. I think it would be nice to see it once—*once*!" said Lousma. "You'll get your chance, Jack. You're a young man. [You] can go for 60 days," his commander encouraged him. "That'll be the day!" Lousma shot back.

Bean, ever the optimist, glanced out at the seemingly endless crescent blue horizon surrounded by an infinity of deep, shiny blackness. "Things happen."

The third and final crew to live and work aboard America's first space station would spend a record eighty-four days in space, and with the space shuttle program still years from flying, NASA planned to get everything they could from the soon-to-be abandoned *Skylab*. The crew of Jerry Carr, Bill Pogue, and Ed Gibson were to complete four EVAs over the course of the mission, including two holiday excursions, one on Thanksgiving Day and one on Christmas, which unexpectedly ended up being just over a record seven hours in length.

Pogue and Gibson were tasked with the first EVA on 22 November 1973 to accomplish several objectives. They were to take photos of Earth's atmosphere using a camera originally intended to be deployed by the scientific airlock, which was by now permanently blocked by the parasol sunshade. They also planned to repair a balky experiment antenna, which turned out to be much more complex than envisioned, causing the EVA to run nearly three times as long as originally planned.

With the airlock module's water-cooling tank replenished, the spacewalkers would once again benefit from the liquid-cooled undergarments to maintain their body temperatures while outside. Gibson never forgot his first impressions when he floated gingerly from the safety of his home in space:

> I think there's two levels. At the functional level or the operational level, I did just that . . . you open the hatch and you go out and get to work where you have to. At the same time, when you open that hatch, and all of the sudden when you look down, you're not looking at another wall or another floor . . . you're looking down at Earth, 270 miles down. So then, pushing yourself out of that hatch—you've got a natural tendency not to do that. But you know intellectually that that's not going to make any difference to you, so you just go ahead and do it. And after you've done it a few times it feels like a very normal way to live.

Throughout his first EVA, Gibson would occasionally find himself questioning whether what he was experiencing in real time was even possible. While he had what he would later refer to as "intellectual confidence" that Sir Isaac Newton's laws were sound and there was no risk of his falling back to the planet far below, at times his learned fears born from decades on a gravity-ruled planet would creep forward. *What if this Newton wasn't 100 percent correct?* The brief thought itself brought a slight chuckle to the spacewalking physicist, as he pushed it aside and got back to work.

The photography experiment failed after just five shots were taken, but rather than spend time troubleshooting the problem, the crew was told to move on to the more critical repair of the microwave antennae. At a 2013 reunion of the Skylab astronauts, Pogue shared with the audience that the work required getting into quite an inconvenient spot on the station's exterior. "Some people were saying put some lighting in there, and they said no, you'll never have to go out there," he recalled. "So where did we go? Where there was no lighting." At the worksite, the crew struggled to keep each other stabilized among the tangles of their elephant trunk-like umbilicals. Gibson affixed a portable foot restraint to a nearby three inch diameter pipe to hold Pogue steady for the work ahead.

Pogue had taken great pride in a tool he had worked on with the Marshall team to include in their EVA toolkit. The "Yankee screwdriver" was a Phillips head screwdriver that included a spring-loaded sleeve that covered the head of the screw, and when the astronaut pushed the handle down, it rotated the screw in the desired direction without slipping off. "I said I'd really like to have that Yankee screwdriver," Pogue explained. "Now this was in Huntsville, Alabama. And every time I said 'Yankee,' they reacted," he told with a theatric jolt of surprise.

The EVA tools were stored in a rolled up fabric tool banner, containing labeled pockets for each specific piece of specialized gear. Once Pogue was ready for the tools, Gibson unfurled the carrier—and immediately started laughing. "I couldn't imagine what was so amusing to him," said Pogue. "So I looked back around and looked up at the top and one of the tool [slots] where the Yankee screwdriver was supposed to be—the label had been changed—Dixie Screwdriver." The southerners of MSFC, Alabama, had gotten the last laugh.

To access the instrument, the crew first had to tear through several layers of aluminized Mylar and then remove the many screws on a cover plate over the antenna. Gibson watched as the tiny shreds of metallic foil were launched away from the station, like the fine, dry hairs of a dandelion blown in the breeze. At sunset, the orbital debris reflected the red glow of the horizon and appeared to be following the station into the darkness. Even though their suits were painstakingly sewn and glued to be airtight, over time, tiny leaks developed, and the small jets of escaping air propelled the foil away as they worked.

Gibson and Pogue struggled for several orbits of Earth to access the instrument and complete the necessary repair work. As often happened, the flight unit differed in construction from the one they had trained on, and in this case a tight fitting metal sheet only allowed one of the screws to be loosened a half a turn each time. The pair would return to the airlock after six and a half hours, having successfully completed the repairs but with bruised and damaged fingernails to show for the arduous work.

Christmas Day 1973 saw Carr and Pogue performing a record-setting seven-hour EVA to mount another experiment outside that was intended to utilize the scientific airlock and perform film change-out and repair tasks. Once again, the ability to have a human work outside of the spacecraft proved invaluable to the mission's ultimate success. The spacewalkers took dozens of photos of the comet Kohoutek, without the filtered coatings of the laboratory's observation window.

The aperture doors that covered several of the instruments on the ATM had been failing with some regularity, so the crew was instructed to pin another one open, as had been done on previous EVAs. But the bulk of their time outside was spent repairing one of the telescope's filters, another task that was not intended to be undertaken by spacewalking astronauts. Carr and Pogue used a dental mirror and flashlight to peer into the aperture of the telescope, carefully positioning the filter with the use of a screwdriver.

Pogue, realizing that this was likely his last foray outside the space station, made the most of the experience. He ventured up the EVA handrail path to the top of the ATM, stepped into the foot restraints and leaned back to enjoy the unobstructed view of the globe below. "I felt like I was doing a slow swan dive through space," he later wrote in his memoir *But for the Grace of God*. Soon, however, this distant perch, thirty feet from the center of *Skylab*'s mass, revealed another surprise for engineers when one of the station's attitude controlling gyroscopes began to struggle.

It was soon realized that the normal airflow exhaust through a valve in the space suit produced a small amount of thrust, but that tiny bit of force exerted at the end of a long lever arm high atop the complex was enough to actually rotate the massive ship.

Carr and Gibson would be assigned the final two EVAs of the Skylab program, one on 29 December 1973 and the last on 3 February 1974, to retrieve the last batch of telescope film for return to Earth. Gibson took great delight in being able to directly observe Comet Kohoutek from the outside of the space station. "The comet had . . . two tails," he described. "When it was the brightest the solar pressure would take the smaller particles and push those away from the sun and then there's other particles which were more massive which didn't move around quite as quickly. So we had what was called a sunward spike, which was the heavier particles that just didn't get swept out in the same way that the very lighter ones did."

While there would be some fairly low-resolution photos taken of Kohoutek, nothing could compare to what the spacewalkers saw with the naked eye, unobstructed by any atmosphere. "When we came back in, we said, hey, rather than try to put this in words, let's try to draw some pictures of it." Gibson grabbed a few data cards and began trying to draw out his mental images of the brilliant, fork-tongued comet with a pencil. "When we got back down, I went out and colored some of it in trying to make it look more spectacular."

On the eightieth day in space, with their mission drawing to an end, the last EVA of the Skylab program was undertaken to essentially close out the collection of science data from the ATM. The astronauts tried out the backup "clothesline" method to transfer the film boxes back to the airlock and snapped a few photos of Earth's atmosphere for later study. Gibson, now performing his third EVA, had become comfortable with the environment yet still had a healthy respect for how quickly things might go wrong in space. While the long umbilical hoses were intended to be clamped at various attach points while the crew worked, he preferred to leave his floating free. "I always wanted to be able to get back towards the hatch if something happened," he explained. "In case you started to lose pressure in your suit, you want to be able to get back there in a hurry."

As it happened, Gibson's suit began leaking cooling fluid during the course of the space walk, but he was able to switch to minimum flow to limit the leak and complete all his tasks. With the last long-duration crew essentially mothballing the station before they departed, it was hoped that the coming space shuttles would be flying in time to revisit *Skylab*. One of the last things the two spacewalkers did before heading back in was attach a micrometeoroid collection device to the exterior structure in hopes that it would be retrieved some years later by another crew.

But as Carr and Gibson prepared to close the door on an unprecedented series of EVAs, they had no way of knowing that their orbital home would be gone long before the first shuttle ever launched, and it would be nine years before another American astronaut floated weightlessly out into this orbital wonderland of space. Gibson paused for a moment and took one last look around. "I said to myself I'm probably not likely to get up here again." The only sound was the gentle breeze of air flowing from his umbilical. He could feel the warmth of the sun through his suit as he spun gently around to absorb the view once more. Even through the many protective layers of his glove he could feel the intense heat radiating from the handrail in his grasp. This was a hostile place, no doubt, but he and his fellow *Skylab* astronauts had proven throughout their three missions that humans could work effectively—even thrive—in it.

When the Skylab program was laid out, only six EVAs were planned over the three missions, each lasting only about two and half hours. By the time Carr and Gibson closed hatch on 3 February, Skylab astronauts had completed *ten* space walks totaling an incredible eighty-two-plus man-hours. Several of these excursions were quite audacious in their nature, particularly during the first crew's attempts and eventual success in deploying the solar array. But most importantly, for the first time in U.S. spaceflight history, orbital EVA had become somewhat routine.

Pete Conrad summed up the EVA experience of his crew at a post-flight press conference, stating categorically, "If you prepare correctly to do a task EVA, and give the astronaut the proper restraint, there's virtually nothing he can't do out there that you can do anyplace else." While this was indeed true, there were several cases on *Skylab* where the fidelity of the training hardware was not sufficient to provide an accurate training simulation. Situations like this would continue to plague EVA operations well into the space shuttle era, particularly in the arena of satellite capture and repair in orbit.

*Skylab* EVAs clearly demonstrated the requirement for an abundance of handholds and foot restraints, a factor designed into every spacecraft and space station since. Even satellites like the Hubble Space Telescope, engineered for orbital service by astronauts from its inception, are covered in handrails. Ed Gibson also noted in a post-mission debrief that there were areas on the exterior of *Skylab* that bristled with sharp edges. "I noticed them after I had been working around that area, actually putting my gloves in some of those locations," he reported. "That's a very dangerous situation. All it takes is one good grab with your hand and a sharp pull, and you're out of business." These too would be accounted for in the design of future space vehicles.

When the International Space Station was being designed, many at NASA looked at the use of EVA in its construction as perhaps an impossible challenge. "Now there were a few nay-sayers in NASA ... a good number of people that were saying, 'Oh, no ... no EVAs. It's all going to be done robotically,'" Jerry Carr shared in 2013. "'We're going to find a way to build this space station without using EVA. This is just too darn dangerous.'" He and Pogue argued to the contrary for several years until more and more started coming around to accepting their perceptions.

When the Skylab astronauts look back on their accomplishments throughout their months in orbit, they relished the progress made in the realm of EVA. "We developed the procedures and the techniques for doing effective space walks on *Skylab*, because we had not been really good at it on Gemini," recalls Jack Lousma. "We developed the procedures that were then used so successfully to put together the [ISS] as it's out there today and as they're doing quite often now." All the work yet to come was to be built on the lessons of *Skylab*, which gave NASA the confidence to utilize space suited astronauts more routinely. As Carr summed it up, "What we did prove was that [our] EVAs put NASA in a comfort zone to where they felt they were going to be able to do more EVAs on the [ISS]."

# 6

# Don't Overdrive Your Headlights

Tell me and I forget, teach me and I may remember,
involve me and I learn.

—Benjamin Franklin

The movie *Batman* was playing in a unique theater for a privileged few. Winston Scott, training for his first space walk on the STS-72 mission and fully decked out in his space suit, was nestled inside a vacuum chamber at the Johnson Space Center (JSC) to test the very suit that he would wear during his planned EVA. After being measured for his suit and having one assembled just for him, the next step was a six-hour stint in the chamber to ensure a proper fit and that the suit functioned properly, while hanging on a structure that supported the three-hundred-plus pounds he and his suit weighed. Suit checks are spaced out, providing lots of free time between tests, and most astronauts bring something with them to pass the time. Jeff Hoffman brought along a book. Scott brought a movie. "I remember bringing *Batman*," he recalled, "the first *Batman* [movie] to come out," and it would be Batman fighting the Gotham City bad guys who would fill the time between tests. Although Scott took advantage of the situation in the chamber to have a little fun, the business of walking in space is serious, and he and his fellow astronauts trained long and hard to ensure they would be able to perform the tasks in space that were asked of them.

The space shuttle promised to thrust NASA's space program into the future. Flying with a fleet of large, winged, reusable spacecraft and landing on a runway instead of splashing down in the ocean beneath billowing parachutes came much closer to the audacious dreams of early space visionaries such as Wernher von Braun and Willy Ley. Although the shuttle barely scratched the surface of their futuristic plans, it was a step up from the capsules that first carried men into space and to the moon, and spacewalking astronauts were going to play a pivotal role in this new era of space exploration.

NASA's early plans for the number of shuttle missions that would fly each year proved to be highly optimistic. Still, with multiple orbiters, each with the capacity to routinely carry up to seven astronauts per mission, the number of astronauts—including spacewalkers—flying into space was going to experience an upsurge. Training underwater in large pools became a mainstay for EVA preparation, whereas zero-g flights, the Air Bearing Floor, and other attempts to mimic weightlessness took a back seat, although depending on the work required in orbit, they could still be beneficial. With time and improved computer technology, beginning in the early 1990s virtual reality simulators began to creep into the EVA training curriculum.

The public enjoys seeing astronauts in space, especially an ear-to-ear grin behind the face plate of their helmet during a space walk. But there is so much more they don't see. That fleeting photo op typically lasts only seconds, and then the astronaut is back to work, laboring for hours in the alien world outside the spacecraft. Nor is the public aware of the months, sometimes years, of focused and dedicated training the astronaut had to endure in order to perform that space walk. The training can be onerous, painful, and dangerous; injuries are not uncommon. Fortunately, NASA has perfected the preparation over the years and employs professional trainers to prepare astronauts for working while they are walking in space.

Similar to the Gemini and Apollo days, simple one-g training remained a valuable method to begin training for a space walk during the shuttle program. Many crews began with tabletop sims, as astronaut Joe Tanner called them. These simulations allowed the crew, sitting around a desk in the crew office, to march through all the EVA procedures step by step and familiarize themselves with the myriad details necessary to accomplish their mission. Additionally, errors were often discovered and modifications made to improve the plan. These simulations were beneficial but could not fully prepare the astronauts for the challenging weightless environment of space.

Time and experience have shown that neutral buoyancy training underwater is the best method to train astronauts how to walk in space, although it's not perfect. The astronauts are still subjected to the laws of gravity, hanging inside their suits that are floating beneath the surface of the water. Spacewalker Linda Godwin praised the NASA EVA training program, with the caveat that "you can't really replicate the real conditions of the vacuum of space or the free fall of orbit that gives you the weightless feeling. The way that we do it comes together to produce a really good training flow. The NBL [Neutral Buoyancy Laboratory] is the workhorse of the training; we spent more hours there than anywhere else."

Those selected in the early shuttle astronaut groups in the late 1970s and early 1980s received minimal EVA training as an astronaut candidate (ASCAN), likely because at that time, EVA was not foreseen as a significant part of the shuttle program. Once NASA selected an individual for the shuttle astronaut corps, they were required to complete a training program before they were officially recognized as an astronaut. During their ASCAN training in those early days, they spent time practicing EVA techniques wearing SCUBA gear, performing rudimentary exercises, and training with some of the tools available at that time; schooling in the Extravehicular Mobility Unit (EMU) would come later for some. According to Jerry Ross, selected in the second shuttle astronaut group in 1980, there weren't many space suits available at that time for training, and the spacewalking training team was small. As the shuttle program progressed, a greater focus was placed on EVA, and by the time assembly of the ISS was on the horizon in the mid to late 1990s, most ASCANs received up to four runs in the pool wearing the EMU.

Four-time spacewalker Jeff Hoffman, selected in the first shuttle astronaut group in 1978, recalled being given an awkward latch tool while in SCUBA gear and instructed to install a piece of equipment, which was relatively easy to accomplish with his bare hands in the water. The trainers, fully aware of the differences between working in SCUBA gear and the EMU, offered him a warning: "Alright, the next time you try this you are going to be in a space suit and when you have all the difficulties with that you can't blame it on the equipment. The equipment works fine, the difference is you'll be in a space suit and not in SCUBA with your bare hands."

Hoffman recalled that they received about ten hours of EVA training in the pool, but once assigned to a mission, they were given about ten hours of pool time in the EMU for every hour of planned EVA. Joe Tanner estimated that he has slightly less than one thousand hours of suit time in the water plus probably another couple of hundred hours in SCUBA gear. Scott Parazynski logged well over a thousand suited hours, and Jerry Ross believes his time in the pool may approach two thousand hours. Many astronauts have spent countless hours in the water, so it's nearly impossible to determine who holds the record. Regardless, learning how to walk in space requires an inordinate amount of neutral buoyancy training.

Digging deep into his memory, Rick Hieb, selected in 1985 and with three space walks to his credit, doesn't recall training in the EMU during his ASCAN days. But with time, most astronauts received some basic training focused on going outside the spacecraft to correct a handful of potential failures before reentering the atmosphere. He recalled that learning how to close and latch the payload bay doors manually was perhaps the most difficult task he trained for in the pool, primarily because there wasn't a good place to hold onto while attempting to remedy the situation. With hard work and perseverance, they all learned the best techniques in the pool to address a handful of potential emergencies. Hieb often humorously asked the trainers, "Who's the best at doing this stuff? Usually, they'd tell me a couple names; like in my timeframe the names they threw out as being the best—which usually meant being the fastest—were Shep [Bill Shepherd] and Jim Bagian. Then I'd always ask, 'Who is the worst?'" With a smile, he imparted, "And I'd never tell you."

Linda Godwin, who also entered the astronaut corps in 1985, remembered minimal EVA training as an ASCAN and only occasionally after ASCAN graduation unless she was assigned to an EVA mission. They'd often make runs in the pool to test a new tool or flight hardware that was being considered for a future mission; feedback from a variety of astronauts on its effectiveness was beneficial. Hieb remembered that there was an official cadre of EVA astronauts who could be called on for these testing sessions. He was not a part of this group, but on occasion he was asked to fill in and provide his assessment.

Selected as an astronaut in 1996, just prior to the beginning of space station assembly, John Herrington remembered that EMU training in the pool was included in his ASCAN curriculum. He also recalled that as the preparation progressed, the trainers began pushing the envelope. They'd train him to carry out a specific task, and before he completed it, the trainers would inform him they wanted him to do another job. "Aw jeez, I didn't train for that one." They wanted him to learn how to use his existing skills to tackle this new obstacle. "As hard as that was, I think it paid off in huge dividends. I thought it was very well played, as frustrating as it was sometimes."

Training could easily consume an astronaut's life, especially when preparing for a difficult and important mission such as a Hubble servicing flight. While training for STS-61 in Huntsville, Alabama, Kathy Thornton recalled long days that often unofficially carried over into the night. The EVA crew frequented the hotel hot tub in the evening hours, where they talked through their procedures instead of current events or family like most normal people. She chuckled as she reminisced: "People would come along and there are four old guys and a woman sitting in the hot tub and waving their hands and talking in tongues, they're using acronyms and stuff. We probably scared people off."

Regardless of the mission, preparing for a space walk requires endless hours of dedicated training, and NASA, beginning with the Gemini program, continued to progress its EVA techniques that culminated in astronauts carrying out a wide variety of difficult and complicated assignments during the shuttle era.

Scores of EVAs would be required to construct the ISS, and in the mid-1990s, NASA was looking up at what was referred to as "the wall of EVA." Shortly after his STS-60 mission in early 1994, Ken Reightler was assigned as chief of the Space Station Branch supporting the newly approved International Space Station. Scarcely a year earlier, in 1993, space station *Freedom* had not yet been superseded by the ISS, but many in the astronaut office were aware of the EVA wall and recognized that they needed more EVA experience, including better training, upgraded facilities, and new tools. Reightler explained: "In my new position I made a strong case that we had to make rendezvous, RMS [Remote Manipulator System] and EVA fundamental capabilities. We had treated each of these as special skills which we only trained for if they were an essential part of the mission. EVA in fact was considered high risk and expensive, so needed to be avoided if possible. I argued we needed to build a highly skilled and experienced cadre of people who had the necessary skills and equipment to do the assembly tasks needed for ISS."

Initially, there were only three astronauts along with a team of engineers and support personnel in the Space Station Branch. Most astronauts were busy supporting the shuttle program, but with the space station looming on the horizon, Reightler quickly made the case to grow his team and built a "a cadre of EVA experienced astronauts," including Rich Clifford, Sam Gemar, Mike Gernhardt, Pat Forrestor, John Grunsfeld, Carl Walz, Rick Hieb, and Jeff Wisoff. Lacy Veach was also part of the team until he succumbed to cancer in 1995. Reightler also successfully recruited foreign astronauts and cosmonauts to the team, including Sergei Krikalev, Jean-Loup Chretien, Michel Tognini, and Vladimir Titov. Krikalev, Titov, and Chretien had performed space walks in the Russian Orlan suit, and their perspective on the differences in the EMU and the Orlan might help determine if some procedures might benefit from one particular suit. They would also provide advice to astronauts experienced in the EMU who might be expected to perform a space walk in the Orlan suit. Reightler was even successful in getting Krikalev an EMU run in the pool, where he joined him wearing SCUBA gear.

Actual testing of hardware in space would prove to be essential, but realistic simulations on the ground would also be critical for space station assembly. Given the sheer size of the hardware that astronauts would have to train on for EVA, one thing became abundantly clear early on. NASA was going to need a bigger pool.

The Neutral Buoyancy Laboratory at JSC is mind-boggling in scale. Measuring 202 feet long, 102 feet wide, and 40 feet deep, it holds a staggering 6.2 million gallons of water, which took twenty-eight days to fill. The first suited dives were performed by Jerry Ross and Linda Godwin in October 1996. It hosts its own simulation control center, medical treatment center including a hypobaric chamber, and two overhead cranes to position hardware. It was soon named the Sonny Carter Training Facility in honor of the NASA astronaut who helped develop many EVA techniques before his untimely death in a commuter airline accident on 5 April 1991.

Well in advance of ISS assembly space walks that began in 1998, Reightler insisted that his branch astronauts climb into the pool and test all EVA procedures; if they didn't work as planned, then the engineers had to redesign them. Once procedures were finalized, Reightler had to officially sign off on the plan, validating that the Astronaut Office agreed the EVA could be carried out as planned. According to Reightler, the enhanced focus on EVA led to many improvements, including construction of the NBL, new simulators, and virtual reality training among many innovations. Reightler progressed the group further by successfully making the case for a deputy, Ellen Ochoa, who replaced him when he retired from NASA in 1995. He explained that these efforts led to the creation of the EVA Program Office in January 1996 under the leadership of astronaut Don McMonagle, which was crucial in establishing "how the Shuttle and Station program offices were managed. The amazing thing is what we agreed to in 1995 was almost exactly what was done to build ISS."

Years before Reightler's efforts to advance EVA, Jerry Ross was appointed chief of the Astronaut Office EVA and Robotics Group. One of his prime charges was to build a space station, which would naturally require many space walks. He was aware of the looming EVA wall, and realizing that a lot of experienced astronauts had left NASA after the loss of *Challenger*, he went to shuttle management and made his case for "building up a stronger base in all of those areas to support the wall of EVA that's coming." That led to a space walk being added to STS-37, which he was subsequently assigned to, as well as additional opportunities on future missions to gain EVA experience.

> But that's how we did a lot of those space walks on a lot of those flights. We were looking [for] some places where we had some time, some weight and volume where we could put in some things. A lot of them were very low cost; we didn't add any extra hardware. We just put the guys out there with some of the shuttle contingency tools to just get out there and get some experience in zero gravity—in space walks. We also started to build up significantly our equipment, our set of tools that we could use to do the building of the station, the training in the water tank.

The early estimates of what kind of EVA effort it would take to build and maintain the station over its expected thirty-year lifespan were indeed daunting. A first attempt to come to grips with what lay ahead was embodied in what became known as the 1990 "Fisher-Price Report," named for its coauthors, astronaut William Fisher and robotics expert Charles Price. Their External Maintenance Task Team was charged with resolving the huge discrepancy between the estimated crew EVA time—1,732 EVA crew-hours annually (an astounding 144 two-person EVAs, or three per week) and the program's goal of 132 EVA hours per year, or just a single two-person space walk per month.

The team studied maintenance-reducing design changes that could be made, explored how the use of robotics might replace astronaut EVA time requirements, and revised some of the estimates that had been previously made. The study concluded that the station could be constructed and maintained with as little as 486 EVA crew-hours per year, but then the entire estimate was thrown into disarray when Congress ordered a redesign of the station.

Being over budget and overweight and requiring too many shuttle flights to assemble, *Freedom*'s design eliminated the large "dual-keel" configuration for a simpler single keel that would be assembled in prefabricated truss segments. Previous construction training in the payload bay of the shuttle would not be needed. External equipment requiring maintenance or periodic replacement, known as orbital replacement units (ORUs), would be minimized and designed to be serviced by robotic assistance to the extent possible.

By 1992 NASA's scope of EVA requirements for *Freedom* ranged from 136 EVA crew-hours annually to 384. A July 1992 General Accounting Office report noted that "even under NASA's best-case scenario, the total amount of EVA time required over the life of the station is unprecedented—estimated at nearly ten times the amount of EVA performed so far by U.S. astronauts," while observing that this number was only about 4 percent of the Fisher-Price estimate.

NASA had incorporated many of the Fisher-Price task force's one hundred recommendations to reduce EVA time in the station redesign. The number of external parts that would require maintenance by astronauts was cut by one thousand, and ORUs were reduced to about eight hundred. The program would integrate the use of several robotic systems, including the U.S. Mobile Transporter to move equipment and astronauts along the truss, the Canadian Mobile Servicing System (Canadarm 2), and two Japanese robotic arms on their module.

While the Fisher-Price group was not specifically tasked with producing firm estimates on EVA time requirements, the GAO noted that it "serve[d] as the starting point for an ongoing process of estimating and identifying methods of reducing the amount of EVA maintenance." Jerry Ross recalls that the now infamous study "certainly got management's attention, and I think in some ways it helped me to get the resources that we needed to do all of the preparatory [work] that we needed to do."

Ross was also skeptical of the limited amount of training planned on space walk procedures for ISS assembly, all geared at saving money. With his persistence, the mindset began to change. "We started getting higher fidelity full-scale mockups of each of the station segments," Ross recalled. Funds also became available for testing the hardware they would use in orbit.

Procedures were tested multiple times on mock-ups in the pool by astronauts varying in size and EVA experience, and then the effectiveness of those tests was ranked. The data were tabulated and reviewed to determine if the hardware was acceptable or if changes were in order. "I also fought for doing thermal vacuum tests for . . . station components that were going to be EVA operated to find out if they were going to work or not," Ross shared, "and we found several things that didn't and had to go back for re-works." He is adamant that this was money well spent and likely avoided a lot of embarrassment down the road. "I've said for years that ninety-five percent of the success of a space walk is determined on the ground before you launch. And certainly, the space station was a great example of that because . . . you'll remember that there [were] basically no hiccups in any of the spacewalking processes at all, so . . . very proud to admit"

Veteran spacewalker Scott Parazynski recalled that just before the turn of the century, Charlie Precourt, then head of the astronaut office, figured that he had ten to twelve people qualified in EVA who were capable of performing the work necessary to build the ISS, far too few for the amount of work needed for station assembly. He lamented that this small group of spacewalkers was basically a "good old boys club" and that the challenge of constructing the ISS with such a small group was pretty scary. Up to this time, the most EVAs that had been achieved in a single year was about ten, and it was predicted they were going to need to do about forty-five in a single year. Neither were there enough trained flight controllers to get the job accomplished. It was time to step up the EVA training program.

Precourt enlisted Parazynski to help develop the EVA Skills Program, an initiative that began by defining a "gold standard of EVA performance" to ensure there would be sufficient EVA qualified astronauts for the large and daunting challenge of space station construction. The program gave EVA trainees a regimented routine, and if required, they were allowed extra time to master skills that they were having difficulty conquering, with extra coaching from experienced spacewalkers and EVA flight controllers. Parazynski believes the program significantly contributed to the successful assembly of the ISS.

In the days just prior to the Skills Program, ASCANs were given up to four EVA runs in the pool. The first stint was for a couple of hours, and if there were no major problems, they proceeded with the remaining three runs. If they struggled, they likely were not invited back. Early in the shuttle program, the EVA cadre selected for the majority of development testing had derived some demanding hardware and difficult tasks that could not be conquered by a broad cross-section of those in the astronaut corps. Therefore, simulations on these initial runs were particularly difficult for some aspiring spacewalkers.

The intent of the EVA Skills Program was to define what was expected of the potential spacewalkers and develop a plan to bring them up to speed. Previously the expectations for EVA had not been formally defined, so it wasn't always clear to the astronauts what was expected of them, and the Skills Program corrected that oversight. The program promoted skills-based training; astronauts were taught basic spacewalking skills, which would prepare them to perform just about any type of EVA they might be assigned. In the past, training was more task based, meaning that the astronauts were taught every step involved in performing a specific EVA. The Skills Program was structured to identify astronauts' shortcomings and provide direct feedback by experienced spacewalkers and trainers on how to bring their performance up to the gold standard. One astronaut in the Skills Program excelled at most everything asked of her, but she had repeated difficulty locking her boots into the foot restraints of the robotic arm. Following a pair of SCUBA runs and one-on-one coaching, she became one of the best spacewalkers in the program.

Mike Massimino writes in his autobiography, *Spaceman*, that he also had difficulty locking his boots into the foot restraints on the end of the robotic arm; as hard as he tried, he couldn't figure out how to do it. Frustrated, he approached his friend and fellow veteran spacewalker Steve Smith for advice. Smith met Massimino at the NBL, and prior to the two of them entering the pool, Smith reviewed the procedures for locking boots into place on the arm. Once in the water, Smith showed him how the foot restraints worked, and after several failed attempts, he noted that Massimino was not getting his feet flat, essential to locking boots into the foot restraints. His toes were secured into the loops on the end of the restraint, but he wasn't pushing his heels downward to make a solid lock. Massimino practiced the procedure until it was second nature. One of the tenets of the EVA Skills Program was to give an astronaut extra practice if they struggled with a task, and the program was delivering as planned. Massimino contributed to two Hubble servicing missions and carried out four highly successful space walks.

Immediate feedback was provided throughout the program. The feedback was honest—not negative—and the astronauts appreciated the frank advice; most everyone wanted to learn new skills. Parazynski could recall only one astronaut who was reticent on receiving guidance. "It's hard; because the EVA students were all type A, bright, accomplished people. No one wants to be told they're average, or worse yet, below average. You hate that." Astronauts come from all manner of mechanical and athletic backgrounds and are highly intelligent, but surprisingly, one astronaut had to be taught how to use a pair of common Vice-Grips while suited on a training run in the pool. The Skills

Program set the expectation that you needed to learn how to use your tools prior to simulations in the water; pool time is expensive.

NASA not only wanted astronauts who could carry their own weight while doing a space walk but expected them to help others become better spacewalkers. Parazynski called it "see one, do one, teach one," a practice that he, a physician, was familiar with from the medical world.

Students often plateaued after four to five dedicated skills runs if the runs were closely spaced. The goal was at least two to three per month per individual, but the heavy workload at JSC made that difficult to achieve. Six to ten runs remained the overall target, with some students requiring more, some less. Once an astronaut attained the gold standard of performance, Parazynski believed they were "as good as anyone else in the office."

There was a six-hour final exam in the pool—affectionally known as the "the EVA from hell"—that simulated an ultra-bad day on the ISS, with multiple failures, bad communication, and all kinds of surprises. The crew was required to develop their own timeline and make real-time tradeoffs once the test began.

Parazynski admitted that there were those who seemed to be naturally made for spacewalking, but many others could get there with the support the program offered. "The folks that took this the most seriously, who really applied themselves the hardest, got the most out of it. Once you set the bar someplace, highly motivated people are going to get to that bar."

An EVA scavenger hunt in the NBL, with participants wearing SCUBA gear, added some fun and competition to the program. Astronauts were required to locate graffiti, bolts, and other hardware and answer questions about electrical crossovers and all kinds of spacewalking trivia, about ten pages of fun. But it forced students to learn the hardware and where it was located. Parazynski shared that "one team blew away all the others." Chris Cassidy and, to the best of his memory, Suni Williams excelled in the pool.

Scott Parazynski retired from NASA in 2009 but was asked back in 2012 to share his spacewalking knowledge with a group of EVA trainers, space suit technicians, and support personnel where he outlined key tenets of the Skills Program. He shared his "motherhood list" of sage advice that every spacewalker should strive to achieve. For example, the ability to accept constructive feedback—near the top of the list—was crucial to success, although he jokingly suggested that destructive feedback was acceptable too, but it had to be funny. Maintaining a light atmosphere in a stressful environment can be helpful. The motherhood list is a long one based on years of lessons learned that will continue to aid future spacewalkers.

The EVA Skills Program was a game changer in the way NASA trained astronauts for EVA, and the ISS is a testament to the effectiveness of the program. NASA made an investment in people by allocating additional training resources, and the payoff was a much larger cadre of crew, flight controllers, and engineers versed in EVA. Although the program continues to evolve, the trainers still teach astronauts how to perform EVAs for the ISS using the same philosophy—skills-based training.

The EMU, a modular technological marvel designed for EVA, comprises various parts to accommodate just about any size astronaut. The Hard Upper Torso (HUT) is mated to the lower half of the suit, the Lower Torso Assembly (LTA), and individual arms are attached, differing in size to accommodate most everyone. Suits are custom assembled for each astronaut based on body measurements and come in medium, large, and extra-large sizes. Understanding the mechanical complexity of the suit combined with sizing is crucial to how well the astronaut can function in it, whether in space or during training. According to Parazynski, astronaut Nancy Currie championed an effort to integrate a small sized HUT, and she was a real marvel in the NBL using it. She came close to convincing NASA of the need for the smaller suit, but unfortunately, her recommendation was not approved.

Story Musgrave believes that anyone can meld with the suit and tame it, but they must be willing to "learn" the suit. In 1978 the veteran Musgrave began spending a day with each new astronaut class to share his philosophy on how to take on spacewalking.

Astronauts are not all created from the same mold, and although the arms and legs of the suit can be adjusted, some astronauts get a better suit fit than others. Just like everyone else on this planet, their bodies differ in height, weight, arm reach, and many other parameters. Physical strength, endurance, and mechanical aptitude also vary among individuals, which may affect their efficiency in the suit underwater. Some astronauts have physical attributes that make pool training much more enjoyable; others are not so lucky. An ill-fitting suit can become extremely uncomfortable quickly, especially in the water. Parazynski had an advantage over many astronauts. "You'll do well if you're a goon with big long arms," he said, "and if you're fit, you can do the tasks." He sees a need for future suit enhancements to allow astronauts of all sizes and shapes to work more effectively in their space suits.

EVA flight controller and instructor Kieth Johnson has trained astronauts to do space walks for over thirty years and has seen it all. Some astronauts openly complain about the pain and difficulty of working underwater in the suit, whereas others put up with the discomfort, and no one was aware of their problems until they were out of the water. Training in the EMU is physically demanding and challenging; the human body is not built the same way the suit is designed to operate. Johnson's job was to explain that difference to the astronauts:

> Some people look like they've been made for it. They don't complain. They love doing runs. Parazynski is an example of somebody who can jump in and out of the suit and do anything you ask of him and is there to do it again the next day if you want him to. There's that side—the crew member that you never hear complain. We have other crew members that the build up for that single run is an epic struggle. They get in and they get it done and they want nothing more than to be gone and not have to worry about it until the next time.

Nicole Stott, selected as a mission specialist in 2000, disclosed that training in the NBL was some of the most physically challenging training that she participated in during her career at NASA. She remembers doing skills-based and task-based runs in the pool in preparation for her Expedition 20/21 mission to the ISS. Despite the SCUBA diver's best efforts to make her neutrally buoyant in the pool, she often had to fight the suit and gravity as she struggled to get into the various positions to accomplish her training objectives. She stressed the importance of maintaining a positive attitude and treating those training her with respect: "There are a lot of people in the pool and the control room trying to support you, but there's only so much they can and should do for you while you're training in the suit. So, unless something is actually causing harm, you need to smile and push on. I always think of the SEAL motto: 'The pleasure is the pain.'"

Linda Godwin remembered another challenge during the long training runs in the NBL: "Sometimes training you knew they were having lunch in the control room. Now that is cold!"

Tall astronauts with long arms and large strong hands generally fare better than shorter, stockier people who often must fight the suit. Physical conditioning is essential to overcoming the suit, and EVA astronauts spend a significant amount of time in the gym. However, Mike Massimino pointed out in his autobiography, *Spaceman*, that a small individual with short arms and fingers can exercise all day long in the gym, but it will not make their arms and fingers longer:

MORE MOTHERHOOD

> If you mess up, fess up—don't waste others time, although it's better to make mistakes on the ground than on orbit. Be absolutely certain that you understand the consequences of everything you do; don't guess. If you make a mistake, don't sweat it, MCC will help you out. Be careful to not make errors due to fatigue, particularly near the end of a long EVA.

Linda Godwin was one of the lucky ones who had few problems with the suit while training underwater, although she experienced fatigue in her hands and fingers. Working upside down was not her favorite position in the pool, but that was true for most everyone. The astronauts may have been perfectly neutrally buoyant, but they were still inverted, with their blood flowing downhill to their heads and the entire weight of their bodies crushing their necks and shoulders against the suit. Moving up and down in the water—especially when slightly congested—often caused ear blockage. Because it's impossible for astronauts to equalize the pressure in their ears using their hands to squeeze their nose as they descend into the pool with mounting pressure, they depend on the spongy, nose-blocking Valsalva device, which allows them to block their nostrils and then blow, which clears their ears. Godwin always had the Valsalva device installed inside her helmet.

Godwin remembered sore fingers if her gloves did not fit properly, and if her nails were not cut short, she'd get sore nail pads. Some astronauts lost entire fingernails. Kathy Thornton also had problems with fingers and kept her nails "cut down to the quick" to keep from damaging them, but she stressed, "Still there was enough pounding your fingers against the end of the suit that about halfway down the nail would be separated from the nail bed." Overall, she and her EVA partner, Tom Akers, blended well with the EMU, as he recalled: "No injuries [in the pool], sometimes had a sore spot or two the next day after a six hour training session; ... suits are not comfortable. But not painful when you're training underwater; ... [they are] more comfortable in space."

Godwin enjoyed her time in the pool, although she would have been happy to skip the time-consuming process required to ingress and egress. Fortunately for her, the JSC trainers tended to maximize the time in the pool, so most of her runs were about six hours long, minimizing the number of pool runs. The downside was that six hours in the pool was a lot of work for her: "I felt like we had a workout by the end."

The suit was hard and unkind to John Herrington's body, lending bruises and sore spots that he had to fight through to prepare himself for three successful EVAs on STS-113. "Oh it sucks," he quipped, referring to training in the pool inside the EMU. Despite learning how the suit functioned, his body was built in such a way that he had to take special precautions to get through his runs in the pool.

### MORE MOTHERHOOD

> Trust your training, but don't be afraid to challenge accepted practices. There may be a better way to do a task.

EVA crewmembers on the STS-49 mission, Rick Hieb and Pierre Thuot had another issue with the long training spells in the pool. At that time, NASA provided spacewalkers with a gelatin fruit bar positioned inside their helmets so they could turn their heads slightly and take a bite. But due to cost, these were only provided during the actual space walk, not during training. That presented a slight problem for Hieb and Thuot: they became very hungry well before the end of the training sessions. "Pierre and I learned after a while that the best way to make it through a six or seven hour run in the water tank was to fat load. I distinctly remember it. A double cheeseburger from Wendy's and fries. If you loaded up on that you could pretty well make it through a six-hour run without getting famished."

It took time for Hieb to give in to the suit; fortunately he and Thuot made many runs in the pool in preparation for STS-49 which allowed his mind and body to adjust. Thuot was the lead EVA, and Hieb was EV-2, hence, just prior to entering the pool, Thuot went through his suit checks before Hieb. While hanging on the crane sometime around simulation number twenty, waiting for his turn with nothing else to do, Hieb suddenly had a realization:

> I was totally mentally absorbed in the tasks I was going to do in the pool that day. The suit was no longer something I was thinking about. And I said, "Oh, I'm ready to go fly now!" Up until that point the suit itself dominates your thinking. It's like, "Oh I'm working against the... shoulder, I got to do something to get this lined up right, and the gloves are pinching my fingers, whatever." A million little things about the suit are dominating your perception and it's not until that stops that you really are ready to go fly. But you don't know that you're not ready until you get over that hump.

Taming the suit was relatively easy for Winston Scott, and although it wasn't a natural process, with time he gradually learned the ins and outs of working underwater. He believes that a large part of it is psychological; some trainees become claustrophobic in the suit, which obviously eliminates them from doing a space walk. Scott recalled that the air was stale inside the suit while training underwater and believes that caused discomfort for many: "Some people actually get queasy and sick under water. There was one astronaut; he got sick a couple times in the suit in the NBL. The thing is smelly and stale and you're in one-g, not in zero-g; people get uncomfortable."

Scott made a point to talk to those with experience in EVA, such as Story Musgrave, who was happy to share his philosophy of moving with the suit. Scott described it as "sort of dancing with the suit and not fighting against it." But for Scott, learning how the suit operated came in the pool: "You do it and you try something one way and that didn't work; you try something another way and that didn't work. Trying things under water, practicing, learning what works for you versus what does not work for you." He also stressed the importance of thinking through what he was going to do in the pool prior to the run, similar to how he rehearsed for his challenging flights in high performance military jet aircraft, which he referred to as "chair flying or hangar flying." As he stated, "Close the door and close your eyes and imagine yourself going through every movement ahead of time and then you try it under water. If it doesn't work then you try something else. And then you try something else until you perfect it so that it's the best way."

Scott also remembered hours spent in academic training to learn the engineering of every single parameter of the suit: primary and backup oxygen pressure, temperature range, battery current under various scenarios, the suit computer, and a myriad of other technical components necessary to safely carry out a space walk. "All the details—you knew what to expect and what to look for when you made selections on the computer panel. What happens when you throw this switch, what happens when you do this step in the checklists, what voltage do you draw?"

Training for EVA is multifaceted. In addition to learning how to accomplish their EVA objective, spacewalkers also train to recover an incapacitated EVA partner should the need arise. Depending on the situation, the healthy astronaut might be required to translate an unresponsive spacewalker through the payload bay and into the airlock, or the robotic arm operator might be compelled to pluck the debilitated astronaut to safety. "You go through all kinds of rehearsals that most people don't think about," Scott recollected.

Selected in the first group of shuttle astronauts, Jeff Hoffman was fortunate as he adapted to the suit very well. He viewed his suit as a tool, and before a tool can be used properly, one must learn how to use it:

> When you are out in your suit, it's not you who are doing the work, it's the suit that is doing the work and you have to learn how to operate the suit, essentially almost like an anthropomorphic robot. The suit doesn't move, particularly the arm motion, like the natural human body. If you're trying to do something and it's not working right, don't try to fight your suit because the suit—it's tougher than you are. I kind of appreciated all of that very quickly. I was a real hard-ass, I went through six generations of gloves during my time at NASA. I wouldn't accept gloves that didn't feel right and I have to say I never had any EVA injuries. I never had a problem.

Hoffman stressed that he paid close attention to every detail; often it's the "little things" that are most important. As Hoffman changed positions in the pool, he often called the divers over to adjust his weight to maintain his neutral buoyancy. John Herrington also relied on the divers to make the training in the pool easier. The tools he used for training were often lower fidelity than the real ones, so when he was ready to use a specific tool, it was replaced with a heavier tool, which tended to upset his neutrally buoyant attitude. Often, the astronauts requested a "zero-g assist" from a diver to help manage the heavier tool. After all, in space, the tool would weigh nothing, so why not?

Joe Tanner admitted that training in the NBL was a physical challenge, but with more runs in the pool he became stronger. He ate a hearty breakfast the morning of the training and drank plenty of water. Additionally, he made sure he drank the entire thirty-two-ounce drink bag in the suit during the training period to prevent dehydration. Just like Hoffman and Herrington, Tanner recognized the importance of paying close attention to every detail.

"The suit is relatively comfortable if it is sized correctly," Tanner remembered. "It can be very painful if it isn't! Again, the more your body is in the suit the better you're able to tolerate it." He was part of a group that highlighted the problems associated with inverted ops which led to a rule being implemented to limit them to five minutes. "We were also encouraged to not translate up or down in the water while inverted; I was happy to comply. It was even harder to clear your ears while inverted." Rick Hieb agrees with Tanner on performing inverted ops: "Yeah, pretty damned unpleasant!"

There may be an unintended benefit to working at different attitudes and orientations in the pool; Jerry Ross believes it may help prevent, or minimize, space adaptation syndrome once in orbit: "That was an idea that was tested and proved to have some value by doctors looking at that."

Regardless of how the suit accommodated each astronaut, training in the pool was crucial to learning how to perform space walks. Some sailed through the underwater training unscathed whereas other suffered. Either way, Hoffman was adamant about the value of the pool in preparing him for EVA. "But the training is incredibly important. That's the closest you can get to being in space as you can. It makes you appreciate the differences; that's where you really learn to do things."

### MORE MOTHERHOOD

> Take initiative; after all, you will soon be the expert. Ask questions of experienced spacewalkers, perhaps help out when astronauts are being suited up for the NBL; don't loiter. Find a way to get more involved and don't be afraid to take ownership.

Virtually reality (VR) has made a significant contribution to EVA training, and as computer capability continues to grow, VR will surely play a larger role in training for space walks. The EVA crew of STS-61, the initial Hubble Space Telescope servicing mission, may have been the first to use VR in their training program, according to Jeff Hoffman. Hoffman had always been interested in VR, and during a visit to the NASA lab, manager Dave Holman shared some pictures of the HST. Hoffman speculated that perhaps they could be used to improve their training regimen. With their interest sparked, the crew began using VR to help plan their runs in the pool. For instance, they'd determine the best position for the robotic arm instead of taking time in the pool to experiment with different scenarios, and then they would test it in the water tank. Hoffman recalled that while the technology was rudimentary, this simplistic approach clearly saved them valuable training time. At that time, computers were not as robust as today, and it took almost a complete second for the computer scenes to update, requiring that they turn their heads slowly—not a big hindrance, because work on EVA is best if it proceeds very slowly.

"That was pretty cool, except you don't get any torque-force feedback" explained Kathy Thornton, recalling her VR training for the STS-61 mission. "So, grabbing a tool, grabbing a handrail, you can put your hand right through the side of the Hubble—doesn't hurt at all (laughs). You lose that and maybe that has changed in the intervening years, that's a long time ago technology wise when we flew STS-61 in 1993." The technology has, in fact, progressed in leaps and bounds over the years and has evolved into a much more sophisticated virtual reality training tool.

John Herrington was impressed with his VR training for the SAFER (Simplified Aid For EVA Rescue) device, a small jet pack that is worn on the back of the EMU. The safety device uses small nitrogen-jet thrusters that astronauts can fire to maneuver themselves back to the ISS if they become detached and are floating away from the station. Inside the VR lab, Herrington donned a headset, special goggles, and gloves with small sensors in them, which allowed him to simulate being in space. For example, he could follow his motion as he reached out to grasp a handrail, and if he made a good grab, the handrail changed colors, informing him that it was safe to let go with the other hand. For the SAFER exercise, the trainers took him at night to the end of the P1 truss on the ISS and then pushed him away in a tumble at a velocity of approximately one foot per second. Herrington had been instructed to wait thirty seconds before beginning his recovery and return to the ISS via SAFER using a hand controller, which he had to first deploy prior to firing the jets. The sensation was so realistic that Herington confessed with a chuckle, "I thought thirty seconds was a really long time; I would have gotten that sucker [hand controller] out in like five seconds!" SAFER automatically stopped his rotation and then Herrington's job was to locate the station and fly back using SAFER.

Following a VR training session on how to operate SAFER, Joe Tanner had to pass a two-foot-per-second exit rate at night and return before expending his fuel. It took several runs to get the training flow, and additional runs were allowed if needed. Once tumbled, it took SAFER several seconds to stabilize him and then place him in a steady attitude relative to the station. "You may not be in an optimal attitude," Tanner explained. "You may not even be facing the station. The first thing you have to do is find the station and then begin working your way back to it."

Linda Godwin remembered that most of her VR training for STS-108 was to learn how to use SAFER. She stressed that flying back to the ISS with SAFER was not a simple matter: "If you let yourself get too far away, it was a good reminder that orbital dynamics come into play which made it very difficult, if not impossible, to translate back to the station." Tomas Gonzalez-Torres explained that the trainers typically simulate an astronaut coming off structure with a reasonable separation rate. "We do that several times to give them a feel . . . the longer you wait to activate, now you're having to deal with orbital mechanics . . . you can't just point at your target and start going because you're going to end up in a different spot." On occasion, a confident trainee might try to challenge himself with a speedier drift away from the simulated space station, but Gonzalez-Torres clarifies this is mainly to demonstrate the limitations of the system. "Train as you're going to fly," he says, emphasizing the mantra of the EVA branch. Astronauts must understand all aspects of spaceflight, otherwise even with the safety devices provided for them there are scenarios that could result in their death.

Substantial improvements in VR EVA training have been made over the years. Charlotte is a computer-based device that allows astronauts to realistically duplicate mass properties. It's a complicated tool, with metal cables strung around the corners of the VR room leading to a small box in the middle (it's reminiscent of a spider, hence the name Charlotte) and is extremely helpful in teaching astronauts how to move large mass objects in orbit. Astronaut Terry Virts, in his book *How to Astronaut*, recalls some savvy advice from experienced spacewalker Rick Mastracchio prior to his first EVA: "If you're moving slow, you're moving too fast." This is especially true when moving objects with large mass; although they are weightless, once you get them moving, they are not easy to stop. The mass of an object dictates the amount of force required to get it moving, and an equal and opposite force is necessary to bring it to rest. Therefore, astronauts are trained to move objects slowly and deliberately and to avoid unwanted collisions that might cause damage. As EVA trainer Ed Rezac stressed, "Don't overdrive your headlights!"

Rezac once had the opportunity to experience Charlotte. Spacewalker Michael Foale placed a hood over Rezac's head and before initiating the simulation warned him not to look down. Rezac was virtually situated in the foot restraint at the end of the robotic arm, tens of feet above the bottom of the shuttle payload bay, and promptly ignored Foale's wise advice. "I looked down and see what was supposed to be my feet in the stirrup and [instead] I look down two or three stories and there is the space shuttle and I immediately lost my balance and started falling over. I felt Mike grab me from behind, laughing and chiding me, 'I told you not to look down!'"

The Dynamic Onboard Ubiquitous Graphics Simulator, known as DOUG, provides situational awareness training for the spacewalkers, and can be used on the ground or on the ISS. DOUG provides a three-dimensional perspective of the entire ISS allowing astronauts to experience a bird's eye view of any location they wish to investigate. Translation paths can he highlighted, as can worksites and specific hardware that needs to be changed out. Tomas Gonzalez-Torres lauds its capabilities: "The astronauts have it on board the space station so then they can practice and go through it and get a visual before they're even outside." Having the computer capability to complement their skills-based training for EVA on the ISS is a godsend; Expedition spacewalkers are often assigned work that they have not trained for on the ground.

Joe Tanner and Carlos Noriega were charged on the STS-97 mission with mating the thirty-five-thousand-pound P6 truss and first set of solar panels to the station's Z1 truss. Charlotte provided experience in translating such a massive object. According to Tanner, the Air Bearing Floor was simply too cumbersome and time consuming to be helpful; VR was a far more effective training tool. Two-time spacewalker Bob Stewart was succinct in his assessment of the slippery table: "Yeah, we used an Air Bearing Floor. I think I was on it once. It was totally useless as far as I was concerned."

While the role of VR in EVA training will likely continue to increase in the future, Jeff Hoffman does not believe that it will ever replace training in the pool. VR certainly plays an important role in training for space walks, but it doesn't impart the sense of weightlessness as in the water or offer experience operating the space suit. Hoffman recalled pumping a lot of iron in the gym to develop upper body strength, but stressed, "Far and away the best training is to get in the suit and make a long run [in the pool]. There is just nothing like being in the water and having to move your suit around."

MORE MOTHERHOOD

> Don't move too quickly, especially on your first space walk. Mission Control and the IVA may be able to help you if you verbalize what you are doing. Communication should be "crisp and professional" and not superfluous, although it's acceptable to use the word "wow" occasionally.

The vacuum chamber is used primarily today for testing rather than training. According to John Herrington, all astronauts get at least one run in the chamber, primarily to experience a vacuum and pressurization of the EMU space suit. But NASA does add some fun to the exercise. For his run in the chamber, "they put a pan of water, a feather, and a rock in there. So you can watch the pan of water boil away. Then you grab the rock and the feather and they fall at the same rate [in a complete vacuum]; kind of neat." Linda Godwin recalled that during her training, she completed a couple of runs in the chamber to get her acclimated to a vacuum and to practice basic requirements such as moving tethers or how to open the pressure release valve on her helmet if backup air flow was needed. It gave her confidence: "You're doing this in a real vacuum and does it actually work?" She considered it excellent training prior to climbing outside a spacecraft traveling 17,500 miles per hour in the deadly vacuum of outer space.

The vacuum chamber can also be intimidating as well as dangerous. Jerry Linenger, who performed a joint Russian EVA in 1997 on space station *Mir*, recalled sitting on a chair inside the massive chamber, and when the door was closed, he felt like he was trapped inside a safe—one with the potential to implode and destroy the entire building. He vividly remembers a balloon bursting as the pressure approached a vacuum, signaling the danger lurking just outside his protective suit: "And . . . at the end of that you kind of have this feeling that yes, the suit will protect me from the vacuum of space and gives you that bit of confidence."

Spacewalker Nicole Stott enjoyed getting inside her flight suit for training in the chamber; the entire run was very comfortable for her while practicing "reach and visibility exercises." She explained that the real value of the chamber was not to run through a full simulation but rather to become familiar with how the suit operated in a vacuum.

The depressurization process in the vacuum chamber impressed Kathy Thornton: "And what that changes is the sound of your voice—it gets lower. That's sort of the first clue that when you're in the vacuum chamber or when you're on orbit and you depress that this is the real deal. Your voice changes a little bit [speaking lower and lower in a deep voice]."

MORE MOTHERHOOD

> Test everything that is installed with a "pull test" to ensure it is properly seated. Be absolutely sure of tether operations, and if you are unable to correct a tether snag within thirty seconds, ask your EVA buddy for assistance. Take advantage of body restraints at the worksite—they're your friend—and be cognizant that there may be body positions in orbit that are more efficient than what you learned in the pool.

By the time Joe Tanner was selected as an astronaut in 1992, zero-g flights no longer played much of a role in training for EVA: "You got a whoopee ride in your ASCAN run, so you see what microgravity looks like." However, it can still be beneficial in testing hardware or payloads. For example, there were many zero-g tests carried out when NASA was developing methods to repair damaged tiles in orbit following the loss of *Columbia* in 2003 due to damage sustained during launch.

Hoffman recalled a few zero-g flights in the KC-135 aircraft. He reminisced about his training in airlock procedures, a five- to ten-minute procedure done in increments of twenty-five to thirty seconds. Hoffman was on bottom (in the airlock) while Dave Griggs was on top, both in three-hundred-pound suits—no problem during a weightless parabola, but every time they did a two-g pull out, Griggs, now weighing over six hundred pounds, was slammed down on top of Hoffman. Technicians were furiously trying to keep Griggs from crushing Hoffman, but there wasn't much they could do. Fortunately, the rigid HUT protected Hoffman, and he was able to keep his appendages from being pinched between the two suits, but on the other hand, he became very concerned about the integrity of the HUT with all that weight on it.

The KC-135 aircraft was known as the Vomit Comet due to the propensity of many to become sick once the fun began. Rick Hieb remembered flying a total of about fifteen zero-g flights (some were for testing, not training) and was reasonably resistant from getting sick during the multiple parabolas, but if he indulged in too many flips, he soon began feeling ill. Fortunately, the plane carried a healthy supply of barf bags.

Training for the historic EVAs to test the Manned Maneuvering Unit was multifaceted, and there was no one perfect simulation that could cover the entire breadth of required skills. Bob Stewart remembered one particular EVA training session in the zero-g aircraft in which the old bird earned its nickname:

> My friends in the EVA department back there decided that I'm going to get really hot in there so what we'll do is we'll fill the water chest [which contained a small water pump] completely full of ice and that way Bob will stay cool and we won't have a problem on the flight. And what they left out of their thinking was that ice does not flow through the [Liquid Cooling and Ventilation Garment]. You need water. So I was as hot as I've ever been in my life! I just wanted to get out of that suit so I could puke.

Thornton agrees that diversity in training is beneficial, because each exercise lies to you. The Air Bearing Floor is great for getting the sense of being in zero gravity, but it is limited to three degrees of freedom during a single training exercise and gives you a false sense of control. Astronauts must fight the drag of water in the pool, and although they are neutrally buoyant, they are still under the influence of gravity. "You get really good orientation and choreography," Thornton says, "and what you can reach and what you can't reach. But you do get to used to viscosity and gravity to help you do things." Rick Hieb is certain that difficult exercises in the pool encourage astronauts to develop bad habits; the water helps hold them in place and although they are trained to never rely on the water viscosity for an assist, it's hard not to succumb to the water and cheat a bit. Jeff Hoffman noted that there have been astronauts who have been fooled from their time training in the pool that led to surprises in space: "Hey I did that in the water and it was easy and I got in space and I was all over the place."

Bob Stewart agrees with Hieb and Hoffman that the water can lead to bad habits:

> The thing you can convince yourself of in the water is that if you're having trouble with something and you make a quick movement, you know

like jamming something to get it seated, you can convince yourself that okay, I'm really using my body's mass as an inertial hammer, so to speak, when in fact, you're using the water. And if you try that same thing in space, it just doesn't work. So, there are some things that are accurate, a lot of things that are not. You kind of gotta piece these things together in your mind. You really can't do the whole job or feel exactly what EVA is like until you go EVA.

### MORE MOTHERHOOD

Take plenty of pictures and check your partner often. Be careful of tunnel vision; first get the big picture of the worksite, then focus on the work at hand. Be physically fit and remain hydrated at all times.

It's safe to say the pool was a challenge for many if not all astronauts, but for some, it led to more than discomfort; they sustained injuries from their time in the pool. The limited dexterity and complexity of the EMU often results in injuries to the shoulder—especially during training in the pool—including rotator cuff tendonitis, bursitis, inflamed tendons, shoulder lesions, and shoulder joint pain. Injuries to elbows, forearms, wrists, fingers, and spines are also common. Similar injuries can also be incurred during the actual EVA, and as of 2017, there was an incidence rate of 0.26 injuries per space walk, mostly in the hands and feet, some of which may have been initiated in the pool. Therefore, an NBL EMU Work Hardening Program was implemented that focused on physical fitness training to reduce the injury rate during training. The exercises are a bit more sophisticated than those learned in junior high school physical education class, with exotic names such as the dumbbell crawl, handstand pushups, push press, axle-wheel row, and farmer's walk. Fortunately, astronauts have access to the Gilruth Fitness Center, a world-class facility located on the JSC campus to keep them in tip-top shape. A musculoskeletal ultrasound injury prevention program was also incorporated in 2010, which resulted in more shoulder injuries being reported and thus treated.

NASA modified the original HUT, known as a pivoted HUT, by reengineering the shoulder joint in response to safety concerns. Introduced in 1997 it was renamed the planar HUT. According to Hoffman, "Eliminating the pivoted [HUT] got rid of a potential leak source," which could have been catastrophic in the pool or in space. An additional joint was added to the pivoted shoulder that improved mobility but at the cost of increased complexity. Hoffman was confident that the change would be conducive to increased injuries, and sure enough, a number of astronauts soon incurred rotator cuff injuries requiring surgery due to the complexity of the new planar HUT, battling the suit while trying to move their arms. "Can't fight the suit," Hoffman reported. "It's stronger than you are." The matter was investigated, and it was determined that twenty-three astronauts had incurred twenty-five shoulder surgeries; almost half of them cited the planar HUT as the cause of the injury. The authors of the investigation recommended that astronauts with injured shoulders or with the propensity to shoulder injury should use the older pivoted HUT while training in the NBL.

Kathy Thornton managed to injure her shoulder in the vacuum chamber training for her STS-49 space walk. Due to the limited movement of the EMU shoulder joint, she had difficulty closing the latch on the carrier that held a strut she had been training with. She reached the limit of the shoulder joint, and instead of investigating a more efficient way to close the latch, she gave a final "one major push," which did the trick but also caused some harm to her shoulder. Although surgery was not required, she sustained some rotator cuff damage.

"Oh yeah, heck yeah. I had to wear two moleskin patches on my back right below my shoulder blades," exclaimed John Herrington. Every time he bent his back the plenums (tubes) in the liquid cooling garment rubbed against his back, which resulted in painful bruises, so the flight doctor recommended the moleskin, which solved the problem. Herrington chuckled as he explained that every time he came out of the pool, "I'd zip the suit off and the flight doc would rip those things off and check my back. I have stripped on the pool deck!" Being an astronaut is not all glamour. Herington recalled tour groups visiting the NBL just when he was stripping down out of his suit. Kieth Johnson chuckled at Herrington's predicament and pain: "If somebody said, 'Hey John, you're gonna get to do a space walk unless you don't want to get in the water,' he'd be in the water and he wouldn't complain. But after the fact he can look back on it and say, 'Oh man! I've done my share of suited runs and I feel sore spots that I didn't know existed on my body. You get done and you're like, okay I don't want to do that again. But the next time they ask me to get in the suit, I'm in.'" Most if not all astronauts are gladly willing to put up with the discomfort and pain of training just for the opportunity to walk in space.

Herrington also suffered from what is known as EVA knuckle. The glove has a metal bar that keeps the palm of the glove from pushing away from the hand when the suit is pressurized, making it easier for astronauts to pick up equipment and carry out work tasks. The palm bar comes up around the finger and impinges on the knuckle. Every time Herrington opened and closed his gloved hand, his knuckle would hit the palm bar, resulting in a massive bump on the knuckle. Being left-handed, that hand received the brunt of the damage, and he still has a misshaped knuckle from the constant movement of the hand during training in the pool. He eventually worked out a solution—a foam pad placed on the back of his inner liner glove pushed his hand away from the palm bar. "And I tell you it worked like a champ!"

The palm bar also haunted Joe Tanner, although he believes the anatomy of his hand played a role in the problem; he developed small deposits on a bone in both hands. Additional padding in the glove helped. The growths are still present today and look a little out of place but don't really cause him any problems. He eventually developed osteoarthritis in both hands from so much work in the pressurized glove. Some astronauts incurred injuries during their EVA days that will likely remain with them for the rest of their lives.

### MORE MOTHERHOOD

> Make your teammates look good, including astronauts, trainers, and other support personnel. When the space walk is over, thank everyone that helped prepare you for EVA. Write their names on your EVA checklist just in case you suffer from a bad case of "space brain" and forget who they are.

Training in the pool, vacuum chamber, and zero-g aircraft can be dangerous. The trainers take every precaution they can imagine, but it's nearly impossible to predict every single failure that can crop up. While their mission is to train the astronauts to do space walks, their primary objective is to keep them safe; hence, they monitor the astronauts very closely during all facets of the training sessions.

Jeff Hoffman recalled that during a simulation in the pool, his suit began to slowly close in on him, and he started to gradually sink to the bottom of the pool. Once Hoffman informed everyone that he was having a problem, the divers immediately pulled him out of the water. A computer glitch had caused his air supply to be cut off, which led to the water pressure collapsing his suit. Hoffman didn't panic; he trusted his trainers! "Yeah, you're working in a potentially hazardous environment. But hey; it's the space program."

During a zero-g flight, Rick Hieb's handler noted that his breathing had become labored. The culprit was a closed oxygen valve triggered when Hieb hit the floor of the aircraft hard coming out of a parabola, a benign upset in the zero-g aircraft with a trainer closely watching over him but potentially life threatening had it occurred in orbit. Hieb added this event into his "lessons learned" toolbox just in case he might need it during a real EVA.

Astronaut Leland Melvin discloses in his memoir, *Chasing Space: An Astronaut's Story of Grit, Grace, and Second Chances*, that he had a close call during a simulation in the NBL that cost him a chance at conducting a space walk. As he descended to the bottom of the pool, he experienced an ear blockage, and without the Valsalva device installed, he had no way to clear his ears. He suffered severe pain in his ears and suddenly had difficulty hearing his trainers. He was immediately extracted from the pool, and when his helmet was removed, the trainers noticed blood trickling down the side of his face. Fortunately, he eventually regained his hearing with medical treatment and flew two shuttle missions, but he never did a space walk.

Astronauts continue to use a variety of training techniques to prepare themselves for spacewalking. Clearly, the neutral buoyancy training in the pool has proven to be the most efficient and effective method to prepare for EVA. Virtual reality has evolved into an extremely valuable method to train for EVA and continues to progress, but VR still plays second fiddle to the pool. Other methods have taken a back seat but can still add value in certain circumstances. Kathy Thornton summarized. "So all of those help you in some ways but they lie to you so you have to put them together to understand what it's really going to be like."

### MORE MOTHERHOOD

Finally, like a parent watching their child head off to college: Have a most outstanding adventure—we'll be watching with envy.

# 7

# A Deeper Pool

EVA would be fun if the suits weren't so hard to work in.

—Astronaut Don Peterson

Jerry Ross vividly remembers when, not long after his being selected as an astronaut in 1980, famed flight director Eugene Kranz came into a meeting of the future shuttle flyers and flatly declared, "We aren't going to do any EVAs. You're only one failure away from certain death." The statement, coming as they did from this larger-than-life personality, seemed absolute, and it sent a wave of shock throughout the room. Of this new generation of astronauts, most were not the pilots who would be flying the space shuttle but rather "mission specialists," tasked with doing all the in-orbit work aboard the ships—scientific research, satellite deployments, and, as they understood it, space walks.

Jeff Hoffman also recalls how EVA was perceived early in the program. "I was actually kind of surprised . . . a lot of managers were not keen on EVA." He expanded on the memorable meeting, adding that Kranz told them, "We know how to do EVA—we don't have to—there is so much we have to do now just to learn how to operate and fly the shuttle, let's not divert our attention." Years of work on the proposed shuttle space suit, the Extravehicular Mobility Unit, had been accomplished well before shuttle astronauts arrived at JSC. Kranz's decree left them wondering how the futuristic vision of spaceplanes servicing satellites and building space stations in orbit would be accomplished without them ever getting to use one.

James McBarron II, one of NASA's foremost experts on space suits going all the way back to the Mercury and X-15 programs, confirms that "very definitely at the onset of the Shuttle Program there was no EVA capability in the Shuttle system." McBarron and his boss, Harley Stutesman, creatively received funding allocated for conceptual studies by various contractors involving EVA on a vehicle like the shuttle. "We couldn't say it was for Shuttle, but *like* the Shuttle," he explained. "These contractor studies went out to all the satellite manufacturers and asked them, 'How could EVA benefit the in-flight maintenance and servicing of satellites?' That was just one thing. Then looking backwards at the success of programs like Gemini and Apollo with the use of EVA combined with the satellite repair and servicing, we were able to convince Aaron Cohen at the time, he was the program manager for Shuttle, that we should have an EVA capability in the Shuttle vehicle."

Despite being the bearer of the bad news to NASA's new astronauts, Gene Kranz himself was an early advocate for shuttle EVA. Years later, he wondered, "How in the hell did we ever get to a point in a program where there was no requirement to have an EVA capability for the Shuttle?" He too recalled pitched battles with Cohen over the various rationales for having astronauts work outside the orbiter. In addition to in-orbit repairs and the potential for transferring the crew of a stricken spacecraft to a rescue shuttle, the most obvious requirement for mission safety was the ability to latch the payload bay doors for reentry into Earth's atmosphere. If these doors failed for any reason to close and latch, it seemed logical to have a backup capability involving getting hands on the faulty gear in the unpressurized payload bay.

"EVA capability is very expensive, because you need an airlock," explained Kranz. "You need a way to get into the bay . . . but none of these things existed in the shuttle design. It was a real battle." The airlock would allow two suited astronauts to exit the orbiter without depressurizing the entire cabin. It was a massive piece of hardware—a five foot-diameter cylinder with two large, heavy hatches—which presumably would be mounted in the payload bay. However, many of the satellites that the shuttle was expected to launch already took up the entire length of the bay, so the proposed airlock was relocated to the middeck of the cabin. This dramatically reduced the amount of usable space in the living quarters of the ship. It was promoted as a mission-dependent "kit," with the capability to relocate it into the bay, yet this wasn't done in practice until many years into the program.

Despite the costs of the airlock and suit development, a May 1976 NASA document titled *Shuttle EVA Description and Design Criteria* presented EVA as not only vital for mission safety but as a potential money saver for the designers of the commercial, scientific, and military satellites that were to be flown aboard the launch vehicle: "In a period where program cost considerations are of paramount importance, EVA capability can provide the payload community a mode of task accomplishment that offers numerous returns. Cost savings can be realized by utilizing EVA in conjunction with or in lieu of automated mechanisms which could simplify design and development programs."

"So it was the evolution of the EVA capabilities [that] was probably the most difficult," Kranz recalled. "Yet to me it was the most logical." McBarron explained that in the end, a compromise was reached with Cohen and the shuttle program management. "Finally they agreed to put an airlock in . . . with the understanding that we would be self-sustaining, except for servicing things like oxygen resupply and water cooling [for the suits] while we were in the airlock, from the vehicle." These efforts from the mission operations and EVA communities would pay off many times over in the years to come. The remarkable success of the space shuttle program would be due in large part to the capabilities that EVA provided.

As Jerry Ross summed up the early debates over shuttle EVA, "Things change quickly in the space business sometimes, and that was certainly one of them."

While the benefits of EVA from the shuttle were still being debated, one of the most significant pieces of the program—the space suit itself—was in the early stages of design. As early as 1971, the Hamilton Standard Company (HS) had conducted studies on the next generation of suits. HS and its principal rival in the space suit business, the International Latex Corporation (ILC), had been forced into an arranged marriage of sorts in 1962 in support of the Apollo program. ILC was awarded the prime contract for the Apollo suit, with HS providing various parts as a subcontractor. The two companies worked together throughout Apollo, and when the program ended in 1975, six of ILC's engineers were retained by HS under a six-month contract to develop the specifications for the future shuttle space suit.

The two companies decided that joining forces would be mutually beneficial in pursuing the shuttle suit contract, so with much of their homework already accomplished, they teamed up under the banner of "Best of Apollo, United for Shuttle." But when the first shuttle-related contract was revealed to the team, it wasn't for a space suit at all—NASA was looking for a way to move marooned crewmembers from a stricken shuttle in orbit to another shuttle to effect a rescue. Unlike previous programs, astronauts were to fly on the orbiter in shirtsleeves, with the envisioned space suit being used only for EVA. With the baseline design being for only two shuttle space suits flying aboard each mission, NASA needed some sort of enclosure that could allow the rest of the crew to be transferred through the vacuum without the enormous cost and weight of additional space suits.

The HS/ILC team proposed and was awarded the contract for a "personal rescue enclosure" (PRE)—essentially an inflated spherical space suit bladder—which became known as the "rescue ball." A crewmember would enter the PRE through a zippered opening and don a portable oxygen system and mask. Once zipped up and disconnected from the orbiter's onboard oxygen supply, the astronaut could then survive long enough for the brief translation between the two shuttles, being towed along by the suited spacewalker. This entire scenario, however, depended on the ability to have a rescue shuttle ready for launch anytime a mission was underway in orbit.

At the beginning of the program, only one orbiter—*Columbia*—existed. This fact and the logistical nightmare of eventually preparing two shuttles for flight for each mission ultimately nixed the entire "rescue shuttle" concept, although it would be resurrected decades later after NASA's second orbiter loss of the program. No rescue balls were ever flown in space, but they were retained by NASA for years, used mundanely to test astronaut candidates for claustrophobia. Yet the joint work done on the PRE by the HS/ILC team was impressive to NASA and was a major factor in their ultimate selection to build the shuttle space suit.

When NASA did request proposals from the industry for a new shuttle space suit, its requirements were daunting. The suit had to be far more compact than the Apollo design and had to be sized to fit a wide range of both male and female astronauts. The days of custom-tailored suits for each spacewalker were over, but this demand would lead to many debates over the years of development and even into the operational usefulness of the EMU decades later. In another departure from NASA's lunar exploration suit, these units would be fully reusable over their expected six year lifespan.

As NASA was finalizing the requirements for the new suit design, some of NASA's veteran astronauts expressed dismay over the lack of true advancements in technology the agency was seeking. On the cover of a December 1976 report detailing the technical requirements for the EMU, John Young scribbled his frank thoughts:

1. We should not pay X million clams for a new Skylab suit.
2. We must have better gloves.
3. This is a lousy requirements document.

The language within this document, which would guide the design of the suit for the contractor, included, "As a minimum, the pressure garment shall provide an equivalent capability for tasks performance as experienced with the Skylab suit." Following this statement, Young wrote simply, "Crap!" The requirements laid out for the glove design—which would prove to be incredibly difficult—were not specific enough for Young either. He underlined the section and wrote, "This allows us the same old (twenty-year-old) glove."

The gloves for the new suits would become one of the greatest engineering challenges of the entire project. The materials used in the Apollo gloves were not intended for extensive reuse and long lifespans, so new fabrics needed to be evaluated to provide thermal, micrometeoroid, and abrasion protection under extreme conditions and for longer periods of time. The first design was further complicated in 1979 when NASA changed the operating pressure of the suit from 4.0 PSI (pounds per square inch) to 4.3. This seemingly small change made it exceedingly difficult to engineer a glove in which the human hand could do effective work without fatiguing.

Even though NASA neither required nor funded a prototype example of a suit from the competitors, HS/ILC produced one, dubbed the Shuttle Experimental #1 (SX-1), at their own expense. In essence, the suit had two major requirements placed on it—it had to fit a *lot* of different people, and it had to last a *long* time. The SX-1 was a modular design, with dozens of interchangeable arm sections and leg extensions to fit as many body sizes as possible. These "soft goods" would be custom assembled for each astronaut prior to flight and taken apart afterward to be reused.

In contrast to its off-the-shelf Apollo-style bubble helmet, the most radical design departure from Apollo was a large waist ring between the hard upper torso assembly and the soft lower torso, or legs, of the suit. While the suit was optimized for upper-body mobility using gimbaled shoulder, wrist, and waist joints, the lower torso assembly was greatly simplified. With no requirement to be able to walk in the suit—it was a true *weightless* EVA design—the mass, complexity, and cost to allow this was deemed inappropriate. As a result, lower torso mobility of the EMU was far less than that of the Apollo suit.

A much more compact PLSS backpack was fully integrated into the HUT, and a chest-mounted Display and Control Module (DCM) provided airflow control and a caution and warning system for the astronaut. The requirement for multi-mission capability drove the six-year lifespan for the soft portions and a fifteen-year useful life for the metallic hardware. With the expected number of pressure cycles in testing, training, and in space, this lifespan would encompass over 460 hours of pressurized time, as opposed to Apollo's 105-hour limit. The engineering challenges associated with these requirements would be formidable.

HS/ILC won the NASA contract handily in January 1977, and thus began several years of evaluation and design modifications with the close collaboration of NASA's Astronaut Office. While spaceflight veterans Joe Kerwin and Ken Mattingly were closely involved with these evaluations, given their extensive EVA experience, the bulk of the responsibility fell to one astronaut in particular—Dr. Story Musgrave.

Story Musgrave squatted down under the HUT, which was hanging from a metal ground support rack, wearing an oxygen mask that was used to flush nitrogen from his bloodstream. Pre-breathing came about as a result of the shuttle's cabin atmosphere being maintained at sea-level pressure, 14.7 PSI, and the same mixed gas ratio of Earth's 21 percent oxygen and 78 percent nitrogen. It was the nitrogen that was the problem; Apollo had operated at just 5 PSI of pure oxygen once in space, so the crew's bloodstreams were continuously flushed of nitrogen throughout the mission. With his arms raised above his head, Musgrave contorted his upper body into the hard aluminum upper half of the space suit with twists and grunts until his bald head poked through the neck ring like a turtle peeking out of his shell.

Although the waist ring was quite large, Musgrave noted that "the critical point in donning/doffing a properly sized HUT is not, as previously supposed, getting the upper body into and out of the shell, but rather the passing of the elbows over the lower portions of the scye bearing races." It was a real contortionist act to pinch his elbows together enough as he "dived" into the HUT and forced his arms past the metal shoulder bearings that allowed a full range of movement for the astronaut. The suit functioned very well once one was in it, but *getting* into it was proving to be a real challenge.

Musgrave had become the lead Astronaut Office representative on EVA in 1972, assuming the responsibility from Rusty Schweickart. Despite the difficulties of squirming one's body into the unforgiving hard shell of the prototype shuttle space suit, he found it a vast improvement in safety over the old Apollo suit, which he had spent considerable time working with in testing and training. "You had a four-foot zipper between you and eternity; it scared the hell out of us even though that zipper never failed," he shared. "Maybe people didn't say it but I understood." The "pressure-sealing slide fasteners," as NASA referred to them, were now gone; replaced by the metallic airtight connections on the waist ring, gloves, and helmet. "Those are hard aluminum connections with O-rings; my goodness they are not going to fail," Musgrave correctly surmised. "So we got in that system with a lot more assurance that we would not have any pressure loss."

The suit design represented a major departure from all previous space suits in NASA's history in that it would not serve the dual purpose of EVA and decompression protection during critical phases of flight. The "shirtsleeve" environment of the shuttle freed the suit engineers to develop a product tailored to just the EVA function. In addition to the simplified lower torso design, the integration of the PLSS backpack into the HUT eliminated the need for external hoses with their individual connections. The Apollo-era hoses exposed the user to potential pressure loss in the event of rupture or other damage but would also prevent the suit from meeting an early requirement to fit through the inter-deck access hatches between the flight deck and mid-deck of the orbiter.

In a 7 December 1976 memorandum to the Astronaut Office, Musgrave detailed several of his observations and concerns with the SX-1 design, prior to the awarding of the contract to HS/ILC. "Donning/doffing stations or aids must provide the crewman the capability of exerting head-ward and foot-ward forces between the body upper torso/arms and the HUT of up to 200 pounds." Mattingly underlined this requirement in his signature green ink and scribbled, "Gals too huh?" next to it. It was understood by then that the first recruitment for space shuttle astronauts would be opened to women for the first time. Musgrave noted that this was justified because several larger-sized astronauts trying out the "95% male-sized" SX-1 HUT had typically pushed up with at least 160 pounds during their tests. Mattingly, somewhat dumbfounded by this new requirement from Musgrave, circled "justified" in the memo and noted, "Demonstrated. I don't think it can be *justified*."

One of the most dismal notes from Musgrave on the SX-1 design was that "The HUT produces pain on the top of the shoulders which is no less than totally unbearable." The edges of the inner scye bearings "cut right into the clavicle area on virtually every crewmember." Musgrave conceded that "our present experience with donning/doffing the EMU has been highly dependent on gravity." It could not be assumed that a lack of gravity in orbit might make the task easier—it could even introduce its own unforeseen problems, so he advocated for zero-g testing using both neutral buoyancy and the zero-g KC-135 aircraft. Mattingly scribbled wryly, "Don't spend much time on this and do it in flight on [Orbital Flight Test] #1."

The strenuous effort it took to don and doff the suit caused another concern to Musgrave—the portable oxygen system mask just added to the misery of getting the head up through the neck ring. He wanted to remove it just before ducking into the HUT, but any breath taken during the suit donning would result in more pre-breathing once in the suit. Musgrave concluded that the current procedure was "totally unacceptable." The sharp-minded Mattingly noted that as with his Apollo experience, "The real fix is lower the cabin pressure and delete pre-breathing!" This method was eventually adopted in various forms throughout the shuttle program, but while it didn't eliminate the pre-breathe protocol entirely, it did reduce the time required.

Musgrave also agreed with Mattingly (and Young, based on his comments) that the gloves should be given special attention: "If we are to accomplish the intricate EVA operations that are already being sold to the payload world, we must have wrists/gloves which are significantly better than [Apollo's] in tactility, dexterity, mobility, and stability. Toward this end, and to give the wrists/gloves additional emphasis and attention, I'm pulling this crew effort out from the overall EMU effort and making it an independent challenge."

Mattingly updated John Young in a 24 January 1977 memo and expressed his major concerns with the EMU program at that stage. His two top line items were the EMU's sizing concepts and the development and production schedules. Early on it was proposed that only two different size suits would be produced. Musgrave disagreed and suggested that given his experience, at least four sizes would be needed to accommodate the requirement to fit the "5–95 percent male crewmember population and the large-to-medium female crewmembers and one additional size to get down to the 5 percent female."

While Mattingly thought the plan to have a minimum number of suit sizes was laudable, it might not happen in the time they had before STS-1, which at that point was thought to be only two years away. "My personal experiences with attempts to perform in ill-fitting pressure suits (I have more suited hours under these conditions than in the properly fitted configuration) indicate that they are operationally unacceptable," he wrote. He was concerned that a cost-saving plan of trying to go with just two sizes "until experience demands we add more" was going to lead to operational compromises down the road.

In Mattingly's mind, unless they fully realized the range of sizing options, NASA would have to either plan the initial flight tests of the shuttle without any EVA capability or make crew selection for the missions contingent on being able to fit in the available suits. "Neither option is attractive," he offered. In a separate, undated memorandum, Mattingly bluntly shared his thoughts on where things stood:

> The Astronaut Office is not supposed to get into the design business; however, we are the ones who have a potential EVA task in less than 2½ years. I believe we should take an entirely different approach to this and do it now while we still have time. The proposed concept will take at least two years to even define because it requires hardware for evaluation. We should go back to the custom suit concept right now and make sure we have an EVA capability for OFT #1 or acknowledge that we probably will not have it and continue this major R&D effort.

By June Mattingly noted to John Young that the working relationship with HS was good and that outstanding issues were coming to agreement. "CB [the Astronaut Office] (through Story), [Crew Systems Division], HSD, and ILC have established an exceptionally effective working relationship," he wrote. The Preliminary Design Review (PDR) was underway, a critical step in agreeing to the final EMU construction. The major issue at the PDR was still the number of sizes to produce. Musgrave was still pushing for four, while HSD was still in favor of two. "Nevertheless, everyone was in agreement that achieving good mobility with only two sizes was a very risky approach," Mattingly forwarded, "and that if we anticipate adding more sizes, now should be the time to do it."

While the team deleted all the male/female designations of the HUT (since some of the smaller male population fit the female size suit), it was also noted that the "large" size that was used by HSD for performance verification wouldn't fit most of the astronauts that tried it. "If we accept the four [size] HUT program, the selection of an appropriate additional size is still open. I believe verification of our sizing concepts remains as the biggest hurdle in the EMU program," Mattingly concluded to Young. The biggest takeaway at this point was that the team agreed "in intent" with Musgrave's position that four sizes would be the most prudent option for the program.

On an optimistic note, Mattingly observed that "everyone is working very effectively to incorporate the lessons learned during previous programs." One of those lessons would come during a suited run in the WETF involving Bob Crippen, the pilot selected for the first space shuttle mission. James McBarron recalled that "the thing that caused us the most difficulty [in Apollo] was that we did not recognize the importance of man-loads in the design of the Apollo suit." Underneath its thin Beta cloth skin, the lunar suit was laced with cables and pulleys which formed a restraint system against the pressurized bladder and allowed freedom of movement in the arms and legs.

This system experienced several problems during Apollo due to the unexpected loads that the suited astronaut was able to exert against it, so additional restraints in the legs were incorporated in the A7LB suit used on the final three lunar missions. Yet the earliest version of the shuttle EMU lacked a redundant restraint system, and the loads that were being applied by astronauts in tests were still not well understood. On one memorable day, Crippen was suited in the water tank in Building 29, with his boots secured in a foot restraint while performing a simulated EVA. As McBarron recalled, "He actually had the boot blow out [due to] a failure of a restraint connection. It was debated whether a technician failed to make the connection or it actually had failed on its own. It never was clearly in my mind defined well. So we implemented a requirement to incorporate axial restraint in all the joints of the suit at the time, which we did. We took a lesson there from what we did in Apollo."

Story Musgrave had many trials of his own during his involvement in the development of the EMU. In addition to the neutral buoyancy work in the water, he was used extensively as a test subject in the vacuum chamber at JSC. Here, in a controlled environment, the suit could be tested at its operating pressure, while the ambient pressure was as close to space as was possible. It was not uncommon for new astronauts to be introduced to the harsh reality of where they would one day be working by placing a large beaker of water on a table nearby.

As the atmosphere was sucked out of the sealed chamber and the altimeter climbed ever higher—20,000 feet . . . 50,000 . . . 100,000 . . . the water suddenly started boiling violently and evaporating until none was left. It was a stark demonstration of what would happen to their own blood should their life-sustaining suit become compromised. And for Musgrave, this became a near-common occurrence. "Early on in the development I had loads of suit failures. And it's a disaster," he recalled. "You snap your head back because when a suit bursts that neck ring is going to take your face off. When it's tearing at least it doesn't blow up—you hear the tear and then you get ready for the explosion." Fortunately for Musgrave, emergency procedures for such events were well rehearsed, and as the chamber was rapidly re-pressurized, a rescue crew in a nearby antechamber was able to reach him quickly.

Over the two year period from 1977 to 1979, NASA continued its evaluations of SX-1-based EMU prototypes with the close cooperation of the Astronaut Office. Ultimately, improvements were made to the waist ring bearing, which allowed rotation of the hips, more adjustable sizing options were incorporated, and work continued on the gloves. With the increased suit pressure of 4.3 PSI, all the restraint systems in the space suit assembly—the "soft goods" part of the EMU—were made redundant. In a report authored after the first successful shuttle EVA, suit engineers McBarron and Harold McMann detailed the challenges and fixes to the issue of getting into the HUT. The excruciatingly painful scye bearing was redesigned to allow greater mobility during the squirm up into the suit's shell.

And that shell itself was modified to better suit the new shoulders—both mechanical and human. "The complex geometry of the shoulders and neck area resulting from this design change made aluminum forming of the HUT undesirable," they wrote. "Fiberglass material was then selected. This material permitted molding of the shell and the integral layup and bonding of the gas ventilation and cooling water ducting into the upper torso structure and, thus, provided a smooth and more comfortable internal profile."

While the sizing issues could be managed to a degree by the interchangeable soft parts of the suit, ultimately, cost became a driving factor in just how many of NASA's astronaut corps would be able to do EVAs. Although some early documents noted HUTs—the only fixed-size part of the suit itself—ranging from "extra small" to "XXL," the complex, expensive component would ultimately be limited to just three sizes—medium, large, and extra-large.

By 1980 the SX-1 prototype had fully transformed into the space shuttle EMU that would be used on the first space shuttle missions.

For anyone in the EVA community, 18 April 1980 is simply known as the day of "the EMU fire." Coming just a year before the launch of STS-1, a critical test of an empty shuttle EMU was to be conducted in the Crew Systems Laboratory of JSC's Building 7, prior to a manned test that was to follow. The full EMU, serial number 3002, was lying on a table in front of a test stand, which resembled a submarine-like control panel of oversized gauges and valves and fed the suit power and oxygen through an umbilical. A small group of seven technicians huddled around the prone suit, checking connections and monitoring their subject as it slowly inflated into the shape of a human.

Electrical technician Joe Noweter watched from a short distance away from the table as suit technician Robert Mayfield switched the suit from "EVA" mode to "Emergency," feeding 6,000 PSI oxygen from two small, six-inch-diameter tanks to a regulator in the suit's PLSS backpack. Suddenly, a plume of fire exploded out of the suit in a loud *hissss*, sending the roomful of engineers scrambling for safety. Mayfield, severely burned by the eruption of flame, stumbled toward the back of the room with several others, but there was no escape from the fire and smoke. Just two technicians, including Noweter, were on the side of the table closest to the lab's only exit and quickly made it away from the inferno that had engulfed the space suit.

Joe Noweter, having made it out of the hot, smoke-filled lab, turned around and dove back into the choking, blinding cinder box to help save his coworkers.

Overhead sprinklers activated, and alarms blared, ordering the building to be evacuated. Harold McMann, a backup test subject for the suit tests, recalled, "I heard the alarm going off out in Building 7. Well, they had . . . fire drills. So we go out there. No fire drill, there's some white smoke and some people milling around." By the time McMann made it into the building to see what had happened, the fire had been snuffed with carbon dioxide fire extinguishers, and the two technicians who had been burned had been whisked away to the hospital. But the $2 million EMU was a charred mess. It lay with its arms splayed back and legs limp, a gaping, blackened hole in its chest. It wasn't hard for McMann to imagine what would have happened to him or the prime test subject, John "Dusty" Samouce, had this happened on the following test: "The next time somebody would have been in the suit, the next time that valve was open, either Dusty or myself would have been in the suit, and you would have been cut in half. There was a jet of flame that came through . . . there was no way you could have survived."

The effect this nearly tragic event had on the EVA community was immediate and profound. NASA had to find the root cause of the fire and fix it, and the clock was steadily ticking toward *Columbia*'s maiden voyage, a test flight the agency would prefer not to fly without the option of conducting a contingency space walk if needed.

Pinky Nelson was playing hooky from work the afternoon of 18 April, doing some long put-off chores. When the phone call came from George Abbey, he knew he had to immediately fess up. "Where the hell are you?" the mysterious overlord of the astronauts barked. "Well, I'm home working in the garden," Nelson sheepishly replied. "Get in here. We just had an accident with the space suit," Abbey ordered as he slammed the phone down.

Nelson had come to NASA in the first shuttle astronaut class of 1978. An astronomer by training, upon arriving at NASA, he sought out the EVA development effort because that, to him, was what being an astronaut was all about. He quickly came under the watchful eye of Dr. Musgrave:

> Story Musgrave at the time was the EMU person. So I started working with Story, and he helped check me out in the suit. [He] was a fabulous mentor in terms of just . . . physically learning how to use the suit. His depth of knowledge of the suit and the way he operated in terms of really digging in and getting to the bottom of every system, really knowing everything inside out, was a great example of how to work, so I learned a lot from just being around Story and watching him work.

Nelson was assigned to the investigative board for the next several months and would learn more than most about the intricacies of the suit's design and manufacturing. The investigation itself had little physical evidence to work with. The suspect regulator in PLSS backpack number 1002 was constructed of aluminum, and in the high pressure oxygen environment, the lightweight metal burned with the ferocity of a roman candle. It, along with much of the surrounding components and structure of the backpack, was vaporized in the first seconds of the fire. The regulator had been through nineteen previous high-pressure cycles without incident, and the investigative team ran two thousand more cycles on similar regulators from the same production batch in an effort to reproduce the failure, but to no avail.

NASA had a tragic history of working with high-pressure oxygen. No one at the agency could forget the history-altering launch pad fire that killed the three astronauts of *Apollo 1*. But there were other events—other lessons—that *had* been lost to time. In an American Institute of Aeronautics and Astronautics report titled *US Spacesuit Knowledge Capture*, several instances of oxygen-related fires were noted from the Gemini program: "During pre-flight checkout for Gemini X at McDonnell Aircraft in St. Louis, in the spring of 1966, a fire was experienced in the high-pressure check valve of an ELSS chest pack. The only result was leakage, caused by combustion of the valve seat. A few weeks later, during late-stage flight readiness checkouts for *Gemini 9-A* at KSC, a similar failure occurred. Again, no personnel were injured, and hardware damage was limited to the check valve itself."

The teams back then took steps to reduce the fill rate of the oxygen, thereby reducing temperature, and emphasized cleanliness around the oxygen fill ports. Some minor hardware changes were made to the *Gemini X* unit, but the extreme danger posed by these fires in systems utilizing 7,500 PSI of pure oxygen was never fully realized. "The fact that two such fires occurred in the matter of a few weeks without serious incident bred a false sense of safety," the report noted. "The immediate problems were 'fixed'; there was no need specified to begin searching for a systemic root cause. In short, things hadn't gotten bad enough yet."

Another event occurred during testing of an Apollo oxygen purge system (OPS), the emergency oxygen unit in the backpack of the suit. Under 5,800 pounds of pressure, there was an explosion that resulted in the destruction of the unit and injury to a technician. Yet, the report read, "the severity of the problem had not crossed the threshold of significance warranting a penetrating look at all aspects of designing for and using high-pressure oxygen. Things *still* hadn't gotten bad enough."

So a decade and a half later, NASA had to learn a hard lesson once again. As with the forgotten failures of the 1960s space suit tests, no definitive ignition source was ever identified within the suit, PLSS, or test hardware. Three potential sources were postulated by the team—heating of the thin aluminum wall within the regulator due to rapid compression, ignition of a rubber O-ring under compression, or a contaminant particle striking the interior of the regulator as it shot through the lines with the pressurized oxygen.

"We ended up totally redesigning the oxygen system," recalled McMann. "From that fire came a whole new way of designing and testing and building oxygen systems. We're still using that stuff to this day." Among several changes to the EMU and PLSS design, the aluminum components within the suit that are exposed to oxygen were reconstructed using Monel, which is far less reactive with the highly flammable gas. But the safety threats exposed by the EMU fire were not limited to the astronauts themselves. The investigative team made several recommendations to improve ground test safety, including protective clothing for technicians, improved procedures for conducting hazardous tests, and better access to egress routes should something go wrong.

Among the many procedural changes implemented, a Test Readiness Review (TRR) is conducted prior to hazardous operations such as those involving high pressure oxygen. This brings together all the key personnel to thoroughly understand the desired test objectives, and to review normal and emergency procedure checklists to be used during the test. "The [NASA] system for handling that kind of an incident really is very good," Nelson learned throughout his months on the investigation. "We've seen it with the big accidents we've had. They really can get to the bottom of a problem very well."

To date, there has not been another fire-related event involving the shuttle EMU, either in ground testing or in space. Unfortunately, like so many other complex engineering endeavors, it took a near tragedy to expose the flaws of this human-designed equipment. As McMann surmised, "Hardware always does what you tell it to do. Sometimes you don't realize what you told it to do. I think people forget what we're dealing with. People say I've had fifty years' experience, twenty years' experience. You really have only a few years of experience, and you repeat that over and over. Hopefully you get better at it as you go along, but not always."

Today the EVA group meets regularly before critical tests and astronaut training events, often in the more formal TRR setting. The meetings can be long and arduous and may even be tinged with the occasional disagreement. The walls of NASA facilities across the country are adorned with the trophies of past triumphs, but also of past tragedy—large crew patch plaques and photos of team members lost . . . *Apollo 1*, *Challenger*, *Columbia*. In the small conference room where these dedicated space suit engineers and technicians discuss the day's work ahead hangs a photo of the burned-out hulk of EMU 3002 as a reminder to all who see it to, as Gus Grissom once notably stated, "*Do good work*." The EVA team would most certainly strive to do that, and they realized that they had luckily dodged a bullet with the EMU fire. However, the intricately designed shuttle space suit still had a few lurking surprises yet to be revealed.

STS-1 lifted off the pad on the historic maiden space shuttle mission on 12 April 1981. The EMU, once feared to become the "long pole" in getting the first flight accomplished safely, was cleared for flight after the redesign without causing any delay. The sole purpose of having an EVA capability on the four shuttle orbital flight test (OFT) missions was for any contingencies that might arise during the course of the flights. No EVAs were planned during the OFT series, with crews consisting of only a commander and pilot. Any emergency space walk would be conducted by a single crewmember—the pilot.

In the case that Robert Crippen would have had to go outside to manually winch the cargo doors shut, the crew would lower the cabin pressure from 14.7 PSI to 9 PSI to prevent the bends. With two suits aboard, NASA claimed that Young, having also pre-breathed for fourteen hours with his pilot, would be prepared "should he have to suit up and give Crippen a hand with the payload bay doors." This seems unlikely, given that the decision for both astronauts to conduct the EVA would have to be made before the airlock was depressurized and would present a host of other problems should there be any issue with the airlock hatch to get back inside.

Crippen would have been under extreme time pressures, assuming that he was operating alone. Beneath the large clamshell doors were shiny, curved radiator panels that carried the tremendous heat produced by all the orbiter's systems and transferred it to the vacuum of space. Once he had the doors closed, this heat-radiating capability was lost, and the ship switched to a less effective evaporator system to manage the loads just until landing. According to the contingency timelines for the mission, Crippen would have had to ingress the airlock, re-pressurize it, get out of the EMU, and immediately get into his cumbersome pressure suit, all while an already suited commander John Young was busy preparing the orbiter for re-entry on the flight deck above.

*Columbia* returned triumphantly to Edwards Air Force Base in California after a brief two-and-a-half-day shakedown flight. By all accounts the orbiter performed better than expected, including the worrisome cargo doors, which were exercised without issue. But in a post-flight evaluation of one of the EMUs that Crippen potentially would have used on any contingency EVA, it was found that the suit's internal battery was defective. Pinky Nelson was the test subject for the vacuum chamber test, and when the umbilical power was disconnected, the battery was unable to produce enough electricity to maintain suit function.

Ken Mattingly commanded STS-4, the final OFT mission in June 1982. Given his extensive EVA experience from the Apollo days and his work with the EMU's development, he was able to convince mission planners to have him conduct a suit-donning test in orbit, just as he had suggested, scribbling on a 1976 memo. While he wouldn't depressurize the airlock, the test provided a first look at the timeline and procedures that the next crew would utilize to conduct the first shuttle-based EVA.

STS-5 was a mission of many firsts. It was the inaugural "operational" space shuttle flight, hauling two commercial communication satellites into orbit. Never before had a spacecraft launched with four crewmembers aboard, and two of them, Joe Allen and Bill Lenoir, were the first of the cadre of mission specialists who would be conducting the first EVA of America's new era in space. Allen and Lenoir had both come to NASA in 1967, but it wasn't long after arriving that Allen recalled chief astronaut Deke Slayton candidly informing his group of selectees, "You will not be making spaceflights. We do have a lot of work to be done, and if you want to work in the space program . . . we will give you assignments." He then reiterated, "But don't fool yourself into thinking you're going to be space flyers." Allen and Lenoir would both stick it out at NASA and prove Slayton wrong, becoming the first of the aptly dubbed "XS-11" group to fly after a fifteen-year wait.

The inaugural three-and-a-half-hour EVA from the space shuttle *Columbia* was scheduled to be performed on the fourth day of the flight, but Lenoir was stricken by the little-understood phenomenon of space sickness. Flight directors put the space walk off for twenty-four hours, but things only got worse from there for the two rookies. The night before they were to head outside, Allen recalls telling his commander Vance Brand, "We really had just two important things [left] to do. One was the space walk and then the second was the reentry and a safe landing. I made the observation, 'Vance, out of these two, if we have to make a choice, let's choose the safe landing.' And we all laughed about that."

After spending a day preparing their suits in the reduced cabin pressure of 10.2 PSI, Allen and Lenoir headed into the airlock on 15 November 1982 and began going through their preparations for the excursion. As Allen got his EMU powered up, however, he immediately could tell something was not right. "When one is in a space suit . . . and you power it up [you hear a very high-pitched hum] there someplace in your ear," he recalled. "When I powered mine on, the hum started, but it didn't sound like it was healthy. It sounded indeed more like an [angry] mosquito."

Allen proceeded to make various checks of the suit, and Mission Control even had him power it completely down and back on again, but nothing worked. There was a serious electrical problem within the backpack that was not going to allow them to proceed with the space walk. In desperation, the astronauts pressed controllers to let Lenoir go out on a short solo EVA, but they were told in short order, "No buddy system, no space walk." The two astronauts, after so many years of waiting and training, were crushed. "A very bitter disappointment to me, without any question," Allen shared. "It was equally bitter to Bill. Bill was really upset." But he also realized the good fortune of timing when the problem revealed itself: "So the bad news was the space suit failed; the good news was we were not outside the ship when it failed. It would have been considerably more traumatic had it failed outside. It would not have been [fatal to me], but it would have gotten [my] attention, for sure. [I] would have to scramble to get back in, button [my]self up, and get out of [the suit] before other parts of it started to fail."

Allen would one day get another opportunity to walk in space—in a manner much more daring than the brief excursion he was to have on STS-5. But in a career as unpredictable as an astronaut's can be, Bill Lenoir had missed out on the only chance he would ever have to enter space and see the world from a perspective few experienced. He never flew in space again.

"George Abbey, I think, had some people already picked out that he wanted to have the honor of doing the first space walk," recalled STS-6 mission specialist Don Peterson. "And when that [was] canceled, he said, 'Well, we'll have to slip now. It'll take months to get another crew ready.'" But when associate administrator Jim Abrahamson called and asked if he could be ready to perform an EVA on the shortened training cycle it would require, Peterson answered in a very understated manner, "Yeah." Abrahamson, along with Peterson and Karol J. Bobko, the pilot of what would be *Challenger*'s maiden voyage, all came to NASA from the canceled Manned Orbiting Laboratory (MOL) program, where they did a fair amount of developmental work on EVA techniques. The job of leading the rescheduled space walk serendipitously fell to none other than Story Musgrave.

Musgrave had spent so much time developing and refining the EMU, with over four hundred hours in the pool by that point, that Peterson surmised, "He didn't really have to be trained. He probably knew as much as anybody on the training team about that." With the whirlwind of preparation, a successful launch and satellite deployment behind them, on 7 April 1983, Peterson found himself crammed into *Challenger*'s airlock fully sealed up in a space suit to begin the three and a half hours of pre-breathing pure oxygen. On the opposite side of the cylindrical airlock hung Musgrave, his arms curved outward as Peterson's were, relaxing in zero-g. It looked as if the two spacewalkers were attempting to bear hug each other.

Bobko, with a thick checklist under his arm and reading glasses low on his nose, slipped in and out between the two, having assisted them in the suit-up process and making final adjustments to their gloves and tools. First had come the stick-on EKG sensors, then the liquid cooling and ventilation garment, laced with hundreds of feet of water-carrying tubing to regulate the body temperature of the astronaut. After squirming into the lower torso, or "pants," of the suit, Musgrave and Peterson entered the airlock and swam up into their HUTs, going through the now-familiar contortions required to squeeze into the miniature spacecraft. With the heavy waist ring connecting the upper and lower halves of the suit secured, helmets and gloves were donned, sealing the astronauts into their life-sustaining cocoons.

Through the umbilicals connected to the ship's oxygen and water supply, their suits were inflated to slightly above the cabin pressure. No space suit is completely airtight, and this assured that any leakage was *from* the suit rather than having nitrogen-filled air entering it and negating the pre-breathe protocol they were undergoing. In their cozy, environmentally controlled EMUs, and with nothing constructive to do, both of the astronauts were soon passing the time in a much needed manner. "People don't believe this, but we'd had a very busy and a long, hard mission . . . Story and I slept. I mean, I slept about two and a half hours, probably the best sleep I had on orbit, because you've got fresh oxygen coming in over your head, and it kind of makes a nice whishing sound, and there's no other noise. We turned the radio receivers way down so we weren't bothered by people talking. Got some really good sleep before we went outside."

At long last, it was time to head out. In the days before the nearly uninterrupted television coverage from orbit, the egress into the payload bay was first seen by the ground when Musgrave was about halfway out of the airlock hatch high above the Guam tracking station in the western Pacific Ocean. Musgrave was the first American astronaut to enter open space since Jerry Carr and Ed Gibson in February 1974.

Musgrave wasted no time getting his tethers configured from the airlock to the long slidewire that stretched the length of the payload bay's sidewall, known as the longeron, on the starboard side. He started translating toward the aft end of the bay as Peterson exited the airlock, pausing to look around momentarily up over the nose of the orbiter into the infinity of space beyond. But such opportunities are fleeting, and he soon forced himself back to work. The main goal of the EVA was to test not only the suits themselves, but various restraint devices, tools, and techniques for the satellite repair missions that were on the horizon. Peterson immediately found somewhat pleasant differences between working in space and training on the ground:

> The suits weigh about 275 pounds, and so you can't just put them on and walk around. You're either in the water tank or you're supported in some kind of a sling to take the weight off of you. So it's totally different on orbit. In orbit, your body floats inside the suit, so you don't have the pressure points. In other words, you're not being pressed on by the suit in various places. Of course, the other thing that's interesting, in zero-g is, your internal organs float inside your body, and that's a little different feeling. That's why you can't tell which way is up. Your middle ear is floating; it doesn't detect anything. But your stomach floats . . . and that's a little different feeling.

Story Musgrave, among his impressive list of college degrees and flying experience, was also an accomplished parachutist. While one who hasn't jumped from an airplane might think the continuous free fall is akin to being a weightless spacewalker, Musgrave dismisses such assumptions. "You never free fall. Unless you step off a helicopter and then you only get a fraction of a second of free fall." But once a skydiver jumping from high altitude reaches what is aptly referred to as "terminal velocity," there is no feeling of "butterflies" in the stomach. "Free fall parachuting is not free falling; you are resting on a blanket of air—you're not accelerating when you reach terminal velocity," Musgrave notes.

Being the naturally curious adventurer that he was, Musgrave did have an experience before he even left the confines of the orbiter that allowed him to face any fear that might lie ahead. "I was the only one I know that sleeps floating," he explained, so he set out to create the sensation of falling in his own mind:

> The way I created it was floating in the dark. There is a very familiar cliff on a mountain in Colorado—and I would stand on this rock and look down, and it's a long way down. I really got to know that rock and cliff with several visits to it. So floating in the total dark you don't know where you are over Earth and you don't know where the vehicle is around you. After a while you don't know where your limbs are—there's nothing to tell you and you're free floating, every joint is in the neutral position. In that condition I imaginably brought myself to this very familiar cliff in Colorado and I stepped off; it was a little scary! I stepped off that cliff in my imagination and I fell. I stopped it because it was frightening! I was so good in my imagination I stopped it. Damn; don't do this! Then I got my courage up and I went back to the cliff and stepped off and I fell forever. I was able to create true free fall.

Musgrave, who exudes a persona of wonderment over technical challenges few can match, explains his working in the EMU as a melding of man and machine. "You have to forget your body; the suit is not your body—it does not have your body," he shared. "It's got its body and you take your body and you put it in there and you have to integrate the two. Then you have to work them together." With the lack of the human body's "gorgeous shoulder joint—with a ball and socket [that] will go in all directions," one has to learn how to move the EMU's single bearing in new ways. "If you're moving your arm up and down, you can't move left and right. You have to rotate the bearing and then move left and right." In teaching himself how to make the suit's arms go where he wanted them to, Musgrave had a small red stich sewn into the shoulder's fabric, as a limiting mark that he couldn't rotate the shoulder past. Ultimately, he learned, "you have to make them play together—and you cannot pretend."

> You can't just reach for something; you'll miss by a foot. Because when you're picking up your cup of coffee, you don't have to think about where your hand is going to go. You do not have to research how you're gonna

pick up a cup of coffee. You just do it and it happens, because you're putting in all the forces that you've done for centuries. In the suit you reach for something you're going to miss because all the forces are different. . . . I surrendered to the situation. When I fly an airplane I'm not commanding the airplane—I don't grab it with a fist—I pull the stick with fingertips and it's a request. "Hey, come over here." I work systems; they are requests and not commands.

During another short communication pass over the Hawaiian Islands, a television camera at the aft end of the payload bay captured a gorgeous view of Musgrave, head down and feet pointing "up" toward the blue horizon of the planet below. "We've got a good shot of Mother Earth behind you there, Story," the CAPCOM radioed up. Musgrave replied comically, "This is a little deeper pool than I've been used to working in," as he maneuvered over to the camera lens and gave a little two-fingered wave. The two spacewalkers evaluated a winch on the aft bulkhead that might someday be needed to manually close the payload bay doors, and Musgrave tested a restraint method of using two tethers, one anchored to each side of the bay. He tumbled slowly in the center of the bay, strung in between like a dangling spider; a man totally in his environment, as if he'd done it all his life.

During the planning for their EVA on STS-6, Musgrave and Peterson made an unusual request of the flight control team. The space shuttle typically orbited with its payload bay continuously facing Earth. While it lent the appearance that the planet was "above" the orbiter, it didn't necessarily offer the best views of space throughout a circuit of the planet. "Story and I said, you know, wouldn't it be neat when we get on the dark side of the Earth, if we could look out . . . at the night sky, see all the stars?" Peterson recalled. Flight Control Division chief M. Pete Frank laughingly retorted, "Oh, just for you guys' amusement, you want us to roll the damned vehicle upside down?" Peterson, not sharing the various technical concerns Frank had for thermal and communications issues, replied, "Yes, you know, wouldn't that be great?!"

After some discussion with the team prior to launch, it was decided that commander Paul Weitz could simply let his ship drift in an attitude similar to that of a Ferris wheel car, where the payload bay would remain fixed toward a point in space, with the dark side of one orbit allowing the natural planetarium view the spacewalkers dreamed of. Surprisingly, the new attitude of the orbiter allowed for some unique views of the planet as well, and when they got around to the dark side of Earth, Weitz advised his EVA crewmates to enjoy the view. Peterson recalled:

> "Okay, guys, you asked for this. Now stop whatever the hell you're doing and look." So we did, and there's lot of light in the payload bay, and the helmet's got these big [lights]. You couldn't see anything. I mean, it was just too much glare. So we got over in one corner and kind of shielded our eyes, and you could see a little patch of sky . . . but that was about the best we could do. So then Story said, "P.J., why don't you turn off the lights." And P.J. said, "Not going to happen. We're not turning off any lights, because they might not come back on and then you'd be out there in the dark." So we didn't get much of a view that way. But what we could see was pretty interesting.

Looking back, Musgrave claims to have no specific memory or image that stands out from his first space walk. "I'm experiencing the whole time; I don't take one moment. I experience the work, I experience the dance," he explained. "I've also got peripheral vision—what's going on in the heavens at night, what's going on in the Earth? I'm in continuous experience."

Their work in open space continued, and Peterson was tasked with using a ratcheting wrench, specially designed to be held in his pressurized space glove, to manually pivot the large, circular tilt ring that had held their satellite payload on its trip to orbit. In a demonstration of stowing the launch ring in a safe position for landing should its motors fail, he worked furiously with one arm, his other stabilizing him by gripping a nearby structure, and his useless legs twisting behind him at the waist in response to his rotational inputs.

Suddenly, a loud alarm barked through his Snoopy communications cap. He quickly looked down at his chest-mounted display and control module, which flashed the code "$O_2$ Flow High." This indicated that the suit's backpack was compensating for a loss of pressure by cramming more air into the suit's protective bladder in an effort to keep him alive. *His space suit had a leak.*

"I've got an alarm," Peterson called to Musgrave and his crewmates inside. The warning occurred when the orbiter was out of communication with any ground station, so Mission Control could be of no assistance. "Story stopped what he was doing and came over," Peterson recalled. "The waist ring was rotating back and forth, and the seal in the waist ring popped out, and the suit leaked bad enough to set off the alarms." Without knowing that immediately, once he stopped to assess what the cause of the leak was, the seal popped back into place in the grooved waist ring and the alarm ceased.

It was fortunate for the spacewalkers that the twenty seconds or so of concern had occurred when it did. "The ground didn't know that at the time, or they'd have told us to stop," Peterson assumed. With the problem apparently self-corrected, they decided to press on with their work. When Houston was able to reach them a short time later over the Indian Ocean, Commander Weitz relayed the news: "Let me tell you first . . . Don had an '$O_2$ use high' alarm. At the time he was back at the tilt table; he was working pretty hard. We watched him for a couple minutes at the time. I think that—and Story concurs—that Don was just working hard and that's an awful lot of $O_2$ to pump through you; but he must have been doing it."

So with a shrug of the shoulders in Mission Control, the EVA continued, and the spacewalkers were able to accomplish all their remaining tasks. They tested the metal loops of foot restraints, worked with some tools from the large cabinet on the starboard side of the payload bay, and translated along the longeron slidewires lugging a simulated massive object behind them, with a bag of latch tools filling in for some future satellite part. When the ground offered to have them extend the EVA as necessary, commander Weitz declined, replying that they only needed another thirty minutes to finish up and get back safely into the airlock. A chagrined Story Musgrave reluctantly complied.

After four hours and twenty minutes, the crewmen were reconnected to their umbilicals in the airlock, signifying the end of the shuttle program's first EVA. A brief hiccup occurred during re-pressurization of the chamber, when the crew discovered they had not closed one of the pressure relief valves, but once corrected the men were soon reunited with Bobko, who assisted them in getting out of the EMUs.

NASA and the media were ecstatic. The space walk was a visual treat for viewers at home, and a technical hurdle overcome that would allow even more daring and spectacular EVAs over the coming missions. All the developmental work in space would culminate in the rescue and repair of the Solar Maximum satellite in about a year's time. Among all the hoopla surrounding the events of the day, flight director Harold Draughon, at the evening press conference in Houston, speculated that Peterson's leak alarm was a "spurious instrumentation anomaly." It appeased the nontechnical media for that night, but he couldn't have been more wrong.

"By the time we dumped the data from the computer to the ground that showed that leak, we were already back inside the orbiter," Peterson recalled. "Then they called up, and they were all upset about what happened here." Peterson ran the crew's own assumption that his workload had caused the $O_2$ flow rate to increase by some NASA doctors, and they found that theory to be highly unlikely. While his heart rate reached almost two hundred beats per minute as he fought against the stiff, pressurized suit to crank the ratchet wrench, "a guy my size can't work hard enough to breathe enough oxygen to set off the alarm that way," Peterson was told. It would be more than two years, and after several more successful EVAs, before the true culprit for the leak was discovered.

By the time Musgrave and Peterson had completed their groundbreaking space walk, the new EMU design had accumulated more time in a vacuum than all the Apollo program's suits combined. Both the lifetime certification tests of the suit and the training requirements of the astronauts had kept the vacuum chamber at the Johnson Space Center busy for several years leading up to the initial operational use.

In that vacuum chamber, several female "test subjects"—as every suited person in the chamber was known, even astronauts—had begun going through sessions in the EMU in anticipation of participating in future space walks. As part of the evaluation and familiarization in the suit, the test subject would walk on a treadmill in the near-space conditions of the chamber in order to generate a heat load and become accustomed to the odd hisses and pops of the EMU's pumps and valves. In the early days of getting the astronaut corps prepared for EVA some of these sessions could run up to three hours, and as a result, engineers began to notice a peculiar trend.

Shannon Lucid, one of the first female astronaut candidates selected in 1978, was doing the treadmill walk one afternoon with her heavy EMU suspended by a frame and cables to support the suit's weight. Even after following the pre-breathe protocol, after about forty-five minutes, she got the "$O_2$ Flow High" message. After initially suspecting that the pre-breathe time requirement might be different for the female test subjects, the engineers dismissed this theory as unlikely.

"So they knew they had a leak, in her case, and they could also see the oxygen coming into the vacuum chamber, because they were getting pressure inside the chamber," remembered Peterson. She stopped the exercise, and after a short time at rest, the alarm stopped. This intrigued the suit technicians who were monitoring Lucid from beyond the thick-walled chamber. They ended the test, and after a thorough inspection of the suit, particularly the waist ring, found no defects or debris that could have compromised the seal.

Perplexed, the engineers repeated the test again and again with different female test subjects, all with the same result—after roughly forty-five minutes, they all got the same alarm. Peterson related the story as it was told to him:

> There was a technician sitting there. These guys amaze me, but he looked at that, and he said, "You know, I've seen this same thing before. I don't remember the details, but I've seen this same phenomenon before." They went back and got the video of my flight and looked at it. He said—and this is kind of interesting—he said, "When Shannon Lucid was walking, since she's a woman, her hips swivel, and her suit was actually rotating, and we'd never seen that with a guy because guys don't walk that way." But he said, "That's the same thing that happened to Peterson's suit two years ago."

On his EVA, Peterson imparted the same rotational force to the waist ring when he was torquing his body around on the ratchet wrench as the females did walking in one-g. Without his legs secured in a foot restraint and flailing around behind him, he produced the same motion in the suit as a female's gait. This motion allowed the ball bearings of the waist ring to "bunch up," wrinkling or pinching the seal, allowing air to escape. It would require a complete redesign of the waist ring and seal, but an unknown had now become known, and no further events ever occurred.

No one could have believed in the spring of 1983 that the very same design that Musgrave and Peterson wore on STS-6 would still be in use over forty years later. While the PLSS systems and electronics have been improved over the years, and various "soft goods"—the adjustable arms and legs of the suit—have metallic interfaces in place of the old laces, the EMU, designed and first constructed in the late 1970s, has been in use on the space shuttle and the ISS for hundreds of space walks with a near flawless record of success. This is a tribute to the extraordinary efforts of the Hamilton-Sundstrand, ILC, and NASA team—the engineers, technicians, test subjects, and astronauts who worked tirelessly to produce such a robust, modular design that has stood the test of time.

EVA, once written off by NASA managers as a bridge too far for the space shuttle program, had debuted in spectacular fashion and would clearly become a mainstay of space operations. In just a few short months, NASA and its astronauts would stun the world again with their next space walk, one that realized a dream as old as human flight itself.

# 8

# Flying Free

> While spacewalking, I realized something: I used to think I was scared of heights but now I know I was just scared of gravity.
>
> —Astronaut Reid Wiseman

As the world watched the tiny white figure of U.S. Army pilot Bob Stewart seemingly stand still against the blackness of space, no one could realize just how *freezing* cold he was. He had the good fortune to be just the second human to venture untethered away from his spacecraft using a personal "jet pack," dubbed the Manned Maneuvering Unit (MMU), just an hour or so after an initial test flight by fellow mission specialist Bruce McCandless.

Stewart's sublimator, a unit within the suit's life support system that converted liquid water into vaporous gas, had failed several times during the course of the EVA, freezing up with ice and preventing it from operating properly. The procedure to fix the problem involved turning the temperature control to full cold. "That puts the maximum amount of cold water around your body; the theory being that your body heat would then warm the water up which would then melt the ice in the sublimator," Stewart related. "The part they forgot about was that you're going to freeze to death before it gets that way." When the tubing of his liquid cooling ventilation garment flushed with the chilled water, he couldn't get the knob back to full warm fast enough. "It was just unbelievably cold."

But this was just a minor inconvenience given the importance of the test flight he had been assigned. As he drifted out to nearly three hundred feet away from the safety of the orbiter *Challenger*, he was surprised to find that when he pulsed the nitrogen-powered gas thrusters, he could hear the *woosh* of the inert gas as it moved through the valves and out to the jet nozzles, imparting motion in any direction he desired.

Stewart had the distinct impression that he was flying along in his own personal spacecraft. "I've got my own propulsion system, I've got my own environmental control system, I have water on board, [and] I have a candy bar to eat. We are a spacecraft in every sense of the word." The brilliant white delta-winged space plane he had left behind hung rock-steady before him, its form slicing through the curved blue horizon of Earth, his crewmates' faces framed in the two square overhead windows of its flight deck, watching him intently.

Suddenly, Stewart was overcome with the desire to experience something nearly unimaginable to the earthbound: "I decided that I wanted to feel what it was like to be the only person in the universe," so he gave a gentle nudge to his right-hand rotational controller and slowly pirouetted around to face away from the orbiter. "I could not see any of the Earth or the moon or the orbiter, and of course in the daylight you can't see stars," he remembered. All he could see as he shivered in the cold was the deep, velvety blackness of infinity before him.

Seconds later Stewart decided, "Maybe I'd better turn around and just make sure everything is still there." But those precious seconds have remained a vivid memory for Stewart over four decades later. "I did not expect that feeling," he says today with a chuckle, "but it sort of took me." There was still much to accomplish on his EVA; a series of tests to prove out the techniques for the next mission, which would attempt a daring rescue of the failed Solar Max satellite. It was time to get back to work.

The fanciful images of daring space heroes such as Buck Rogers and Flash Gordon flying through air and space with rocket packs strapped to their backs date back to the earliest comic books of the 1930s. As EVA became a reality during the American Gemini space program in the 1960s, engineers immediately set about exploring the possibilities of astronauts to maneuver under their own control. The simple Hand-Held Maneuvering Unit used during the Gemini program provided limited control for short periods of time to maneuver outside the spacecraft.

The U.S. Air Force of the Cold War era was keenly interested in finding a role for military personnel in space. The Department of Defense was considering the possibility of crewmembers flying through space to either repair their own equipment and satellites or to inspect or possibly even disable Soviet ones. Aerospace conglomerate Ling-Temco-Vought had been quietly studying EVA propulsion for some years when it won an air force contract for the Astronaut Maneuvering Unit (AMU) in the early 1960s. Wearing this ungainly backpack, an astronaut would utilize hot hydrogen peroxide gas thrusters pressurized by nitrogen, enabling him to maneuver around the Gemini spacecraft or translate to another spacecraft under his own power.

The boxy AMU had two extendable arms with hand controllers and two thruster extensions at the top of the unit that directed hot exhaust gas away from the helmet and shoulders of the suited astronaut. Testing showed that additional protection was required to manage the high-temperature gases being expelled, so heat shields were added around the hand controllers, including eleven layers of "superinsulation" to the suit, a woven fabric of fiberglass, aluminized reflective material, and stainless steel Chromel-R cloth. This gave the EVA astronaut the unique appearance, with the heavy metallic-looking layers wrapped around his legs, like the "pants" of a medieval suit of armor.

A large chest pack, dubbed the Extravehicular Life Support System, served as an interface between the astronaut and the AMU. It contained electrical and oxygen connections to the maneuvering unit, as well as a panel of warning lights to display the status of various systems. Still, the astronaut would remain attached to the spacecraft by means of a 125-foot nylon tether.

The AMU was carried into orbit in the rear adapter section of the Gemini spacecraft, underneath a thermal cover that would be ejected by the EVA crewman. With the aid of two handrails and a foot restraint, the astronaut would prepare the unit for flight and strap into the backpack before the commander released it by firing a pyrotechnic guillotine from the cockpit, which severed the attachment bolt and fuel servicing lines. The astronaut would then maneuver around the spacecraft utilizing twelve hand-controlled thrusters, each producing a gentle 2.3 pounds of thrust.

However, despite the extensive preparation and training by the astronauts, the AMU project was destined to meet an untimely end. Two of these experimental units were planned to be flown during the Gemini program, but neither was ever ultimately tested. The first unit was left cradled in *Gemini 9-A*'s adapter section when Gene Cernan became overheated and exhausted, forcing Commander Tom Stafford to have him abandon the test flight, and the second unit intended for *Gemini 12* was removed following Buzz Aldrin's first underwater training session. While EVA successes would mount in the years to come, the idea of flying freely in space seemed to lose its momentum, and may well have ended with the Gemini program, were it not for a young air force captain named Charles "Ed" Whitsett, and an eager, newly minted astronaut—Bruce McCandless II.

The Soviets, too, had a great interest in free flight during EVA in the early days of their space program. In December of 1961, Zvezda received a proposal from the Experimental Design Bureau 1 (OKB-1) for a cosmonaut transference and maneuvering unit, abbreviated in Russian as "UPMK," as an integral part of the prototype SKV (EVA) semi-rigid space suit. This suit and its maneuvering unit were to be used for extravehicular operations on the planned Orbiting Heavy Satellite/Station (OTSST), and an air-powered engineering model was built and tested in a zero-g aircraft by 1964.

The UPMK, unlike the American AMU, was designed to utilize dangerous hydrazine fuel. Curiously, the Russians opted to store and service the unit and its toxic fuel inside rather than having the astronaut don it outside in vacuum. The large, horseshoe shaped unit clamped around the waistline of the cosmonaut, surrounding him with a chest pack of twelve thrusters and two spacecraft-like hand controllers.

The unit underwent several design changes over the years as the Soviet program progressed, plagued by continually having to keep up with new space suit designs. After the termination of the Voskhod program, which might have allowed an in-space demonstration of the unit, it was reassigned to the Almaz military space station program, paired with the Yastreb space suit. By 1968 the design of the UPMK had matured, its lethal fuel replaced by forty-two forward and aft-facing *solid* rocket micro-motors for making large translations and an additional fourteen compressed air thrusters for rotational and fine translation control.

Unfortunately, the UPMK never saw space. When the decision was made in 1969 to move to the Orlan space suit for Almaz, as well as the later Salyut space station program, space officials began to question if there were any real potential uses for the device in orbit. The Soviets would not pursue free flight in EVA again for another twenty years, spurred by the spectacular successes of the Americans.

Ed Whitsett had been studying concepts for maneuvering the human body in open space since 1961 when he was assigned to the Air Force Institute of Technology at Wright-Patterson Air Force Base. An exercise in human anatomy as much as it was engineering, Major Whitson was well-versed in human mass distribution, moments of inertia, and body movement by the time he was attached to NASA's programs in 1966. Bruce McCandless II came to NASA's astronaut corps that same year. A third-generation naval officer and graduate of the U.S. Naval Academy, he was the youngest, at twenty-eight years old, of NASA's fifth group of selectees.

McCandless was assigned to Whitsett's office in 1968. It would be a fortuitous meeting for the astronaut, as the navy pilot and air force engineer formed a partnership that would last decades. In a November 2017 speech, McCandless recalled that after *Gemini 9*, "maneuvering units acquired a bad name, I felt undeservedly, so [Whitsett] and I collaborated on trying to rehabilitate the concept of flying around outside of your spacecraft."

As the Apollo program was winding down and crew assignment opportunities dwindled, McCandless was moved to the Apollo Applications Program, which would eventually become the Skylab program. Whitsett saw the opportunity, with the huge interior volume that the space station offered, to further refine his engineering experiments for maneuvering astronauts through space utilizing several different methods. Together with McCandless as a co-principal investigator, they proposed and were granted approval for an experimental astronaut propulsion unit dubbed M509.

A bulky, trapezoidal-shaped backpack with a large spherical nitrogen tank jutting out from its metallic framework, the M509 provided a test bed for *Skylab* astronauts to evaluate different methods of maneuvering and maintaining orientation in the relative safety of a shirtsleeve environment. Fourteen jets placed around the unit provided thrust in the desired axis, and a package of control moment gyroscopes within the backpack provided an alternate inertia-based method of imparting motion. As with previous experimental and envisioned operational units, two hand controllers were fashioned using the same design and logic as a spacecraft, with a multi-axis rotational joystick on the right and a T-handle translational controller on the left. Additionally, a handheld unit, attached by a short hose on the right side of the backpack, could be unstowed and operated to provide some impulses to the astronaut, utilizing nitrogen from the same supply tank in the backpack.

Another method of mobility was the foot controlled maneuvering unit (FCMU) proposed by the Langley Research Center. A direct descendant of a "jet shoe" concept of 1965, the T020 experiment was intended to allow hands-free maneuvering while the astronaut accomplished other tasks, maintaining position using the feet. This contraption, consisting of a nitrogen tank backpack, saddle-like seat, and the pedal controllers and thrusters at the feet, offered some promise but would ultimately earn the scorn of the astronauts who flew it.

Alan Bean, wearing his flight suit, a hard bike helmet over his Snoopy communication cap, and safety goggles, floated across the huge volume of *Skylab*'s orbital workshop in a slow pirouette under complete control. As he bumped the right-hand controller to cancel out the slow spin, a burst of nitrogen popped from one of the M509's control jets, echoing off the cylindrical walls of the space station with a metallic *psssh.* Jack Lousma hovered below, filming the test flight and narrating Bean's progress to the ground.

Bean ran the maneuvering unit through its paces utilizing three different control modes: direct, which was simply pilot controlled using the nitrogen jets; rate gyro, a mode which incorporated an automatic attitude hold and thrust proportional to control stick deflection; and control moment gyro (CMG), using the reaction wheels in the backpack for angular motion in roll, pitch, and yaw. The handheld unit was used sparingly due to its difficulty of use, and a lack of realistic simulator training prior to flight.

Once Bean had completed his run, it was Lousma's turn to evaluate this new flying machine. "It handled very good," he recalled. "I flew it very precisely . . . we had a planned circuit that we would fly and probably two or three different trial routes that we would take in order to verify the accuracy of the control systems." One of the more challenging maneuvers was to have the pilot face the ring of lockers that surrounded the interior of the workshop and fly himself around the ring. This is where the engineering logic of mirroring the Apollo spacecraft's controls paid off. "We had two hand controllers. One was for roll, pitch and yaw, and the other one was for translations fore, aft, up, down, left, right, just like a capsule," Lousma explained. "We flew it the same way [we] would as if we were in a capsule, I guess. But it worked very well as a backpack."

Over the course of the fifty-six-day mission, Bean flew the majority of the test flights, including two runs in his pressurized space suit. He and Lousma not only verified the flying characteristics of the backpack but also simulated a few of its intended uses for the coming space shuttle. Bean flew effortlessly around the workshop carrying various objects of mass and even simulated approaching a spinning satellite, grabbing it, and stabilizing it using a large oxygen package as a stand-in for the spacecraft. To demonstrate a rescue of an incapacitated EVA crewman, Lousma spun himself into a tumble, and Bean approached him using the M509, secured him, and transported him to the dome hatch high overhead.

During his space walk with Garriott, Bean even lamented not being able to test the jet pack outside the station. As they admired the view over South America and on into the South Pacific Ocean, he radioed to Lousma, "Almost makes you want to fly 509, doesn't it, Jack?" The mission's pilot came right back with, "Yes. Tell you what, you can hurry up and get in here . . ." Bean jokingly suggested, "We'll get them to lay it on the schedule."

By the end of mission, Bean's confidence in the machine was sealed sufficiently to ask Garriott, "Owen, would you like to try this thing?" There was no hesitation on his part. "'Yeah, that sounds like fun. Why not?' I'd had no training on it, but it operates just like a spacecraft with translation control in this hand, operates the same way, attitude control with this hand." So after a brief discussion with Mission Control to gain their approval for the untrained science pilot to make the attempt, Bean got his novice flyer squared away in the jet pack.

With minimal training, Garriott easily translated around the spacecraft. "It is surprisingly intuitive. . . . So it was very pleasant to see how easy it was to fly, and perhaps useful to the designers, as well, to see that a person with essentially no training could learn to fly it so quickly."

Mere months later, the Skylab 4 crew arrived at their home in space and also had the opportunity to evaluate the futuristic rocket pack. Although they had less time available to fly the machine, Both Jerry Carr and Bill Pogue logged over six hours throughout five sessions of shirtsleeve and suited operations. By the time they left the M509 behind on the mothballed space station, five crewmen had amassed nearly fourteen hours of flight time testing its capabilities.

Of the three methods of control, the CMG was found to be the most stable and the easiest to fly, but the gyros consumed an enormous amount of battery power from the unit. In Bean's opinion, the direct mode of pilot input right to the thrusters was "all that was really required for an operational maneuvering unit," although he agreed with the other crewmembers that some mode of automatic stabilization would reduce workload. On the other hand, the foot-controlled T020 experiment was found to be far from adequate for any further consideration.

"I give a zero, which means non-controllable," Bean recorded during the flight. "I would say that this vehicle right here is unacceptable for EVA operations." Known for being one of the more introspective astronauts, he went further: "Nobody even asked the question. Here you say to yourself before you go out, 'What if number 2 thruster sticks on? What am I going to do? Am I going to die then and that's the end? I'm betting my whole life on that thruster down there?' Can't have it. That's why I don't think you want a foot controlled maneuvering unit. You want to bet your life on your feet? Your hands are the precision instruments of your body."

Whitsett and McCandless proudly assessed the test flights of the M509: "The utility of an EVA maneuvering capability has been adequately demonstrated to justify its inclusion in the baseline configuration of EVA for future programs." They went on to suggest that "an operational AMU should find use in the shuttle era in three major areas: unplanned operations, contingency operations, and payload support."

The M509 experiment—the first successful astronaut maneuvering device to be flown in space, although limited to the interior of the spacecraft—advanced the gathering of critical engineering data that would drive designs for an operational unit. Bean, in closing out his remarks about the tests he was involved in, drew on his experiences and challenged the engineers of these futuristic rocket packs on their task ahead: "Having been up here, flown this thing a number of times . . . looked out the window a lot, going EVA . . . we've been on the lunar surface, I've thought about this command module we've got out here with a couple of failed [thruster] quads and seen ourselves swirling around the earth at 18,000 miles an hour, 270 miles up, is you have got to have the best system you can build, because you're hanging your neck on that system."

Whitsett and McCandless—the engineer and the astronaut—would set out to do just that.

In the over seven years between the end of the Skylab program and the first launch of the space shuttle, the concept of astronauts flying free of a spacecraft using a maneuvering unit was not even a remote priority in the halls of NASA. In a time when the idea of regular EVA operations from the shuttle was still under debate, the project, which came to be coined the Manned Maneuvering Unit (MMU), seemed to be a solution looking for a problem. "It was supported by Martin Marietta dollars a lot more years than it was NASA dollars," Bruce McCandless told author Bill Harwood in his 1993 book, *Raise Heaven and Earth*.

In the 1978 *Manned Maneuvering Unit User's Guide*, Martin laid out a variety of normal and contingency tasks that could be accomplished using the MMU. These included supporting shuttle operations with inspection and repair, payload operations such as servicing free-flying platforms, and even crew rescue, in which an EVA crewman would fly from a stricken shuttle to the safety of another ship with pressurized "rescue balls" containing unsuited crewmates. Fortunately, for McCandless, Whitsett, and a small group of dedicated engineers at Martin that had kept the project alive, it seemed their efforts were about to be rewarded. In February 1980 NASA awarded a $26 million contract to Martin to develop and build a qualification MMU unit, two MMU flight articles, a simulator, spare parts, and tools. Given the desired use of the MMUs, NASA also wanted the flight hardware delivered in time for the first shuttle mission, which at that time was scheduled just nine months later.

While engineers scraped together various fuel lines, valves, and electrical components from the *Viking* Mars lander and other Martin programs in the charge to build the three MMUs, it was hardly a crash program. Martin's Bill Bollendonk, who took over leadership of the MMU in 1982, recalled that the team wanted to verify the hardware in ways far beyond what the contract stipulated. When the decision was made to put the completed flight units through thermal vacuum testing, several issues were identified that would have gone unknown until they were put to use in space.

Martin's MMU was a large self-contained backpack that was latched over the shuttle EMU's PLSS. Its hand controllers mirrored the M509 prototype, and two 3,000 PSI gaseous nitrogen tanks were contained within its aluminum structure, which also housed two silver-zinc batteries and all the required electronics and plumbing. Four rectangular towers jutted from each corner of the backpack to position the cold-gas thrusters in their optimum locations, allowing the forces they imparted to act through the center of mass of the MMU/astronaut combination. One major difference between the MMU and its predecessor was the elimination of the momentum wheel gyroscopes to maintain attitude—they required far too much battery power. They were replaced with smaller gyros which simply fed attitude data to the thrusters to hold a fixed position.

The decision was eventually made to eliminate the requirement to carry the MMU on the orbital test flights of the shuttle, which gave Martin Marietta some breathing room in the fabrication of flight-qualified hardware. Once again, the MMU was without a mission to perform, but the team still believed in the utilization of the rocket pack. A satellite launched in February 1980 serendipitously provided them with the opportunity to prove their years of work worthy of the time and money spent.

The Solar Maximum Mission satellite was the first to be designed as a multi-mission modular spacecraft, which could be used across a wide variety of scientific, weather, and communications satellites. As such, Solar Max, as it was known, also incorporated design features that allowed it to be serviced in orbit; replaceable electronics boxes surrounded its base, and it even had a remote manipulator system (RMS) capture mechanism attached to facilitate recovery by the space shuttle's robotic arm.

One of Solar Max's scientific instruments failed shortly after launch, and after less than a year, the satellite's primary attitude control system also failed, leaving it in a slow spin and rendering much of its data-gathering capabilities severely degraded. The project's managers decided to shut down the spacecraft to preserve what they could of its mission, in hopes that once the space shuttle began flying, it could be resuscitated. But before NASA committed to the daring repair attempt, they had to test all the hardware and specialized equipment for the mission in space, including the MMU. When the crew of STS-41B was named in February 1983, the shuttle program's first EVA had yet to be performed. But after a successful first test of the shuttle space suit on STS-6 in April, the path was cleared for Bruce McCandless to finally fly his jet pack in the vastness of space.

Bob Stewart, a helicopter test pilot from the U.S. Army, would be the second EVA crewmember to fly the MMU.

> What was influential in my selection was that I was an experienced experimental test pilot. . . . Bruce had been working on the MMU for about sixteen years so he had a lot of pride of authorship. The thought was that . . . Bruce is gonna think it's great because he helped build the thing. Let's put Bob in there because Bob is used to taking a flying machine and evaluating it with no strings attached. So we made actu-

ally a pretty good team. Bruce could say it's flying like we designed it to fly. I could say whether that design was good or bad.

A computer-controlled space operations simulator was built by Martin Marietta in Denver, Colorado, to practice maneuvering up to a satellite. The ungainly mechanical contraption held the suited astronaut in a mockup of the MMU and allowed him to use replicated hand controllers to feed commands through the computers, which were then translated into somewhat jerky motions along the tracks that the machine moved on. McCandless recalled spending over three hundred hours on the device during development, but for later crews, this was reduced to fifteen hours. After so many years of refinements and testing, Stewart quipped that much like the M509 experience on *Skylab*, "In five minutes, you know how to fly the thing."

The EVA plan for STS-41B was to test all the equipment and procedures needed for the Solar Max repair mission in February 1984. *Challenger*'s payload bay carried the Shuttle Pallet Satellite (SPAS), which would serve as a stand-in for the Solar Max satellite during the MMU tests. A new piece of gear, the trunnion pin attachment device, or TPAD (pronounced "tee-pad") was designed to grasp onto a protruding metal tube on the side of Solar Max by the free-floating astronaut, firmly attaching the astronaut to the unstable satellite and allowing him to slow its spin using the MMU's thrusters.

Following launch, *Challenger* was almost immediately beset with a string of failures. Two satellites were deployed but the solid-fuel booster rockets failed on both, leaving them stranded in low, useless orbits. Following that, an inflatable Mylar balloon that was to serve as a rendezvous target for the crew to practice approach techniques burst at its seams after deployment, scrapping that exercise for the mission. It seemed for a time that STS-41B was a cursed flight, but the events of the upcoming days would leave the world in awe.

Bob Stewart's problems on his first EVA all started with a yawn. As the air vented from the airlock with an audible hiss, he employed the age-old method to alleviate the pressure change in his ears, and suddenly the chin strap securing his communication "Snoopy cap" to his head popped loose. In his self-published memoir, *To Fly*, Stewart wrote that the tiny snap had failed several times during training sessions and was of great concern leading up to the flight. Stewart complained about it, "but we were told that this comm carrier served men on the moon! It is perfectly safe! We really had no grounds to argue."

Unable to reach into his bubble helmet with his hands, Stewart blew the strap around and nudged it with his nose and tongue in an effort to get it out of the way. But then he realized that the factory-new, unstretched material of the cap was now creeping up on his head and pulling his carefully placed microphones away from his mouth. This had the potential to jeopardize the entire EVA, so Stewart was hesitant to say anything to Mission Control that might cause them to overreact and perhaps cancel his first space walk and the first flight of the MMU. After a brief discussion with commander Vance Brand, it was agreed that he could manage the problem, but it would make the next six hours quite difficult for the astronaut.

After so many years of waiting, McCandless was keen to get outside. After the hatch was opened, he went to work preparing his MMU for flight. Unlike the battered and chipped mock-up gear used for training in Houston, McCandless noted, "Everything looks nice and new in this big water tank in the sky," as he configured his tethers. Meanwhile, Stewart tethered himself off and floated to an equipment box to begin assembling the manipulator foot restraint (MFR), one intended to be grasped by the manipulator arm to maneuver an astronaut around as if on a weightless cherry-picker. Here, Stewart found another issue that would cause him some misery during the EVA. His right boot heel wouldn't fit properly into the fixed foot restraint at his workstation due to some of the suit material from the lower leg slipping down and partially covering the heel. With just one foot secured into the horseshoe-like restraint, Stewart's hands and arms had to do a lot more work to hold him in position, which made it difficult to accomplish tasks with his hands while trying to remain stabilized.

As McCandless and Stewart went about their separate assignments, they couldn't help but take in a glimpse of Earth on occasion. "It was just an all-encompassing experience," Stewart recalled. "You can tell that the earth is going by tremendously fast. You're just fascinated by the whole thing. You're even fascinated over the ocean because there's a lot of things you can see in the ocean." It was a high-definition display of beauty compared to looking through the smudged, dirty windows of the orbiter for the three days prior. Everything was going to plan, with the exception of the few inconveniences

Stewart was dealing with—which would mount over the next several hours—making his first EVA a real test of the rookie's skills.

Every major TV network in America, and many around the world, broke into their programming to broadcast the live video of the first free flight of an astronaut in space. Banner headlines read "Human Satellite!" and "New Challenge in Space!" Gene Cernan, who himself had been denied his own Buck Rogers–like flight on *Gemini 9-A*, sat in as a commentator for ABC News's live coverage. As viewers tuned in, McCandless was virtually hidden, surrounded by the MMU, its extended control arms, and the flight support structure that cradled both man and machine. *Challenger* was still in orbital darkness, so the payload bay was bathed in a soft yellow glow from floodlights; a camera mounted on the elbow joint of the RMS provided an overhead view of the activities of both astronauts.

Over the incessant interruptions of the TV reporters, viewers at home might have wondered why one of the astronauts seemed to be yelling. Stewart was continuing to struggle with his communication cap, recalling that "the microphones were up about my eyeball level. I was having to scream to be heard at all." He went about his work with one foot securing him to the worksite without complaint, while just behind him, history was about to be made, though he barely noticed it at the time.

McCandless had pulled two release straps to free himself and the MMU from its support structure but waited until he was back in radio communication with CAPCOM Jerry Ross before beginning his test flight. It was a surreal moment for McCandless and millions of viewers around the world. For nearly two decades, he had relentlessly pursued the realization of this science fiction-like dream. He then pushed gently up with his legs to free himself from the cradle. Floating slowly up into the empty space of the cargo bay, he barely nudged the left-hand controller, sending one of 729 possible command combinations through the MMU's processor logic, which then released an invisible 7.5 Newton squirt of dry nitrogen from each of the upper thrusters to bring him to a stationary hover.

In the darkness high above the Hawaiian Islands, McCandless plus the MMU's three hundred pounds of mass floated effortlessly over the payload bay as he exercised the controls in every axis. The MMU's motions appeared crisp but slow in response to his inputs, far different from the comic book images of heroic explorers rocketing about on a tongue of flame. Two black, snaking booms wrapped around his helmet to place indicator lights from the MMU's electronics in front of his faceplate, providing a visual cue as to when the thrusters were firing.

While he waited for orbital sunrise to begin his first sojourn away from *Challenger*, McCandless took an impromptu tour of the cargo hold, translating aft over top of the SPAS satellite, which was secured amidships spanning the width of the bay. "You might pass to Ed Whitsett that we sure have a nice flying machine here!" he radioed to Mission Control. Jerry Ross, who would soon become one of NASA's premier spacewalkers, agreed. "It looks like you did a good job with all of that engineering work over all of those years, as well as Ed and the rest of the crew."

As McCandless maneuvered back to the forward end of the bay, he allowed himself to drift into a slow tumble but was under complete control. Incredibly clear video of the inverted astronaut against the blackness of space was beamed to a worldwide audience, his spectacled face clearly visible through his clear visor. With a signal acquired for a long communication pass in daylight over the United States, McCandless thrusted straight up out of the payload bay, giving a quick wave to his crewmates inside.

A TV camera revealed the silhouetted image of McCandless against the curved blue horizon of Earth as white clouds sped by below. He pitched forward to face the payload bay as he moved, exclaiming, "Hey, this is neat!" Having raced across the American continent in just over twelve minutes, the free-flying astronaut suddenly recognized their launch site as he floated high above it at over seventeen thousand miles per hour. "It looks like Florida. It is Florida! It's the cape!" he shouted out. "I gotta see if I can get a picture of this."

The sun swept across the distant, white suited figure of the astronaut as he moved out of the shadow of the orbiter, the blinding glare too much for the cameras to handle as the image blossomed on the screen. Commander Hoot Gibson pointed another camera, a Hasselblad still camera, toward his crewmate, and when he looked through the viewfinder, he was struck with a sudden sense of awe and fear—fear of failing to capture the most amazing scene any Hollywood special effects wizard could dream up. "Oh my gosh, I can't believe what an image this is," Gibson thought. "We knew the MMU flight was going to be interesting, but I don't think I had really appreciated how spectacular that was going to look and how fascinating it was going to be. I remember as I was doing this I said to myself, 'Wow, if I don't mess these pictures up, I'm going to get the cover of *Aviation Week*.' I actually got three covers of *Aviation Week*."

McCandless floated gently out to 150 feet away from the orbiter, paused, and returned with little effort. Then he once again set out on another traverse to twice that distance. The only maneuver he made during the coast was a slight roll to align himself with the earth rather than the orbiter. Vance Brand, keeping an intent eye on his crewmate who had now become a separate spacecraft of his own, radioed, "Looks like you may get the label The World's Fastest Human Being going around there at four miles a second, Bruce."

"It's a record that will only stand I guess for the next hour or so. Although I guess to break it, Robert is going to have to go 10 percent faster," McCandless replied. Bob Stewart, considering the most velocity change the MMU was capable of, shouted back through his errant microphones, "I don't think I've got enough delta-V!" Back in Mission Control, the exchange sent Jerry Ross and nearby astronauts John Blaha and Pinky Nelson into a fit of laughter.

Stewart, with just one foot in the restraint and his arms severely cramped due to the three-point stance he had to retain to do any useful work, was growing impatient. "I can't wait until my turn to fly the MMU," he later wrote, "though I think I'm going to have to shoot Bruce down to get the chance."

Stewart pulsed the nitrogen jets and spun slowly back to face his home in space after his ever-so-brief moment spent contemplating the universe. He had been saying as little as possible given all his problems. In the dry, 4.3 PSI atmosphere of the EMU, his voice had grown hoarse and his throat dry. He even considered the nearby drinking straw too much of a risk to reach for, fearing that any excessive head motion might cause his Snoopy cap to migrate farther north on his head. "The ground knows that something is wrong and I'm not telling them what's wrong so that's when they get curious," he related. "They don't get the idea that I'm a big boy and I do not have a death wish. I have a problem. I'm dealing with it. Leave me alone."

The cold continued to plague Stewart throughout the EVA. He and McCandless were hardly exerting themselves, which meant they produced little body heat. Stewart recalled:

> What they did, it appears to me . . . when we were flying it was a fairly early stage of development that—they used the same theory that they used for the lunar suit, as far as the cooling and things like that. But what they forgot about was that on the moon the guys were working very hard and the temperature of the lunar surface is about 250 degrees Fahrenheit. In the MMU . . . the background that we were radiating to was like three degrees absolute, so there was quite a bit of difference. We did not need cooling as much as we needed heating. Particularly in the hands—hands got very, very cold.

Stewart improvised a new procedure to combat the frigid temperatures and simply turned his cooling system completely off. "It's pretty quiet in the suit without those fans going," Stewart shared. After a few minutes, the suit began to get "a bit stuffy," and he turned the cooling back on. It did the trick, as the ice had cleared, but the cycle would continue for the remainder of the EVA, with both astronauts manually deicing their backpacks' heat exchangers with periods of eerie quiet in space.

Stewart felt very relaxed as he drifted free in orbital daylight. He pulled some "really high-class instrumentation" from his chest pack to determine his distance from the orbiter. "We had a stick with two notches cut in it," he shared with a laugh. "You had to hold the stick out at arm's length and the length of the payload bay—that's the length of the two notches you were at 100 yards!" Another very basic method of orbital navigation was utilized as well—one that relied simply on "eyeballing" the approach back to the orbiter, without losing the battle with the counterintuitive complexities of orbital mechanics.

Stewart explained that as he began flying back toward *Challenger*, "if the forward bulkhead [of the payload bay] looked like it was edge on to us, in other words we were looking down the vertical of that bulkhead then we were at the same altitude. And as if you're at the same altitude you're going the same speed." But even at a seemingly miniscule one foot per second thrust toward the shuttle, the astronaut would accelerate to an orbital apogee *above* the ship and begin to slow down. If he allowed this to go uncorrected, after one orbit the astronaut would end up 1,800 feet *behind* his target! Intent to remain at the same orbital altitude as the shuttle, Stewart relied on his visual cues: "You thrust toward the shuttle and you start to climb you see the bulkhead tilting, showing that you were going up so you . . . just punch the translational hand controller down, get back down onto the same altitude. You're still going the same speed—you're still going too fast and you start to climb again and you thrust down again . . . we knew that was nothing fancy, just looking at that forward bulkhead."

While Stewart completed his two traverses out and back to the orbiter, Bruce McCandless evaluated the cherry-picker-like MFR, strapping both feet into the foot restraint that was grappled by mission specialist Ron McNair controlling the orbiter's robotic arm. McCandless bounced up and down on the end of the fifty-foot boom, testing to see how far the arm might deflect given different inputs of force.

After nearly six hours of unprecedented, historic work in open space, McCandless and Stewart packed up all their gear and stowed the MMU in its cradle. With their primary task of test flying the jet backpack completed, Stewart felt that it was now safe to mention the problems they were having with the cooling systems of their suits. As he suspected, Mission Control's first reaction was to order an immediate ingress to the airlock, but Stewart informed them that he had been working the procedure to address the icing issue. What he didn't tell them was that *it wasn't the procedure that was in the checklist*. That news could wait until after the mission. They still had another EVA to do, and there was still a risk that the powers that be could nix that one.

It was a day of historic firsts, and while the astronauts returned to the safety of the orbiter's airlock, the public and the media were still catching their collective breath. The images would be replayed on the evening news for days to come, and the world would be shocked again when, after the flight returned, Gibson's quintessential photo first hit the newsstands. McCandless had been rewarded with a flawless maiden flight of the machine he had helped to create over so many years. As Ed Whitsett told the *Washington Post*'s Thomas O'Toole, "Nobody has left his stamp on any instrument in space like Bruce has left his mark on the backpack."

Sixteen years had passed since McCandless had first become involved with Whitsett and his wild ideas of astronaut maneuvering devices, and he had been met with a fair amount of resistance—and some criticism—along the way. But on this day, even Gene Cernan, the last man to walk on the moon, told ABC's television audience, "I gotta admit perhaps now more than in any other flight of the shuttle, I'm just a little bit envious."

Two days later, McCandless and Stewart headed out into the payload bay once again to conduct more tests of the MMU, and this time, Stewart took no chances with his Snoopy cap. He wrapped several loops of surgical tape under his chin and over his head to secure the headset in case the poorly designed snap gave him any more trouble. They had planned to use a slowly rotating SPAS satellite to practice capturing Solar Max, but problems with the robotic arm's wrist joint precluded this operation, so McCandless had to settle for conducting the demonstration on the SPAS while it was firmly anchored in the bay.

Each astronaut took turns using the chest-mounted attachment device to clamp onto a trunnion pin, supposedly identical to that on the Solar Max, which was installed on the SPAS structure. On the actual rescue mission, the astronaut would then use the MMU's thrusters to counter the spin of the Solar Max, stabilizing it for grapple by the shuttle's robotic arm.

No matter how McCandless approached the pin—right-side up, upside down, or face down—the operation was found to be easy and uncomplicated. It gave the astronauts and engineers on the ground a great deal of confidence going into the next historic mission to retrieve and repair Solar Max. Although this EVA lacked the spectacle of the astronauts flying out and away from the orbiter, it was nonetheless a critical test of the hardware, and still drew national attention.

As Stewart was stowing MMU #2 following his final test of the device, one of the mobile foot restraints inadvertently came loose and began floating out of the payload bay, far beyond the reach of the astronauts. Stewart was not yet out of the MMU and offered to retrieve it when Brand radioed down to Houston, "We can go get it with the orbiter." A proposed astronaut rescue demonstration for the mission, with Brand and Gibson flying in to scoop up a "stranded" crewmate into the orbiter's payload bay, was denied for safety concerns, but the lost piece of hardware provided the commander with an opportunity to test his skills.

With some guidance from McCandless, Brand pulsed the orbiter aft, drifting the restraint forward in the bay toward the astronaut. McCandless reached out and grabbed it with his right hand to the surprisingly wild applause from Mission Control. "At least it won't be reentering in a few days," Brand commented over the communication loop. McCandless, considering that for a moment, replied, "I hope it will be reentering in a few days! The rest of us are planning to!"

Watching the awesome events unfolding on STS-41B, President Ronald Reagan decided to congratulate the crew during their final EVA of the mission. With a rigid timeline for every EVA, there wasn't a single minute allotted for such an interruption, but when the call came up from Houston to the crew, "Standby for the president," Stewart could only recall thinking, "Yes sir! We'll standby for the president!" Stewart was in the bright sunlight and warmth of one of the orbiter's curved radiators, but McCandless floated in the deep shadows of the open cargo hold. "What the shade was doing for him was cutting off that three-degree background radiation," recalled Stewart. As they had learned when they flew to great distances from the orbiter's environment, they became extremely cold, which their newly designed suits couldn't properly manage. "When you go down into the bay it was like pulling a nice warm blanket over you," he explained. But to get a clear image of McCandless for the president's call, McNair asked him to maneuver slightly higher up out of the warmth of the payload bay so the television camera could see him.

Stewart remembered that they had to wait longer than anticipated for Reagan's call, and McCandless became unbearably cold. "When we were talking to President Reagan, he was shaking violently. I mean I could look out there and I could just see his whole body trembling. He was hanging out with the space background, and what I did was, I just went to the side of the bay and hung out over the radiators, so I was warm and toasty."

Through chattering teeth, McCandless managed to get in a pitch for his MMU, telling the president, "We believe that the maneuvering units . . . we're literally opening a new frontier in what man can do in space, and we'll be paving the way for many important operations on the coming space station, sir."

*Challenger* returned to Earth after a nearly eight-day mission, leaving behind in orbit two errant satellites, but bringing home a profound sense of accomplishment and a new mindset of what humans were going to be able to do in space. Ironically, the two lost satellites would soon provide yet another opportunity to demonstrate the usefulness of the MMU, as engineers almost immediately began thinking about how they might be retrieved and brought back to the planet for refurbishment and future use.

McCandless expressed great satisfaction in the capabilities of the MMU, and in a post-flight press conference, he contrasted it with the predictive simulations he had done on the ground: "We were very pleased with the way the MMU handled. It was very much like the simulations at the Martin Company in Denver. Except that we did find that when translating with the attitude hold system activated, since we had built it to conserve propellant by turning some of the thrusters off, that you started to pick up a rotational rate. This hadn't been modeled in detail in the simulator, and . . . here you could feel every little thruster coming on and off."

Of course, no simulation could compare with the reality of flying in zero-g. "Here we had complete freedom to turn summersaults and maneuver freely about all three rotational ax[e]s," he reported. Stewart could not get over just how well the MMU handled. "As a test pilot I rate flying machines on a scale of one to ten with one being the best possible, and ten being I'm going to lose control of it. I have never given an airplane a handling qualities rating of one. But that's the MMU—it was a one. Just could not be any easier. If I had anybody I could teach them to fly it in an hour."

Stewart took a few other lessons from his opportunity to work outside of a spacecraft in the unforgiving environment of space. "There are no trivialities in EVA operations," he wrote. The seemingly minor detail of a snap on a chin strap, which should have been addressed prior to flight, turned what was to be an unforgettable first EVA into "six hours of misery" for the rookie spacewalker. The other was for the flight controllers: "When a crewman is isolated and chooses not to chat, he probably is successfully dealing with a problem," Stewart lamented.

The way had been cleared for Bob Crippen and his crew to chase down, retrieve, and repair the Solar Max satellite on STS-41C in the spring of 1984. In the new reality of astronauts rocketing through space like Buck Rogers and Flash Gordon, that job would fall to George "Pinky" Nelson, and James "Ox" van Hoften.

The Solar Max satellite hung against the blackness of space like some sort of shiny gold Christmas ornament, slowly spinning around with bluish solar wings extended. Rotating at barely a degree per second, it appeared to be an easy target for the spacewalking crew of Nelson and van Hoften and RMS operator Terry (T. J.) Hart. *Challenger*'s commander, Bob Crippen, and pilot, Dick Scobee, had flown a flawless rendezvous to bring their shuttle to within less than two hundred feet of the derelict satellite.

Now within reach of their target, Nelson waited for sunrise to allow better visibility then pitched up with his back toward the payload bay, the TPAD pointed at Solar Max. "Ready when you are, Crip," he radioed to his commander. "You have a GO," Crippen replied. "Here we go . . . one potato, two potato," as he counted off the inputs on the translational hand controller.

From the elbow-mounted camera on the remote arm, Nelson seemed to ride right up the bright blue crescent of Earth's limb in sunrise as he slowly translated out of the bay. "Looking good Pinky!" Crippen transmitted in an effort to encourage NASA's latest spacewalker. As he slid across the void between *Challenger* and Solar Max, Nelson pulsed to the left once to correct his approach. His helmet-mounted camera caught close-ups of the rotating satellite as the invisible nitrogen gas exhaust from his thrusters punched at the gold Kapton insulation. He effortlessly matched the slow spin rate of the satellite and edged in between the two solar arrays.

But when he moved in to place the TPAD on the trunnion pin, he bounced off. "The jaws didn't fire," he reported flatly. He tried again, but with the same result. After a third attempt, things went from bad to worse, as his repeated thumps against the satellite sent it into a slow, multi-axis tumble. "Everything worked perfectly until I got to the satellite and flew up to dock with it, and then it didn't work," Nelson recollected. Giving up on the TPAD for the moment, he set out to correct the tumble by maneuvering out to the end of one of the solar arrays in an effort to grab it and use the MMU to slow the rates.

At one point, rather than using the MMU, he walked hand over hand along the edge of the wing. With the white underside of the wings tumbling in front of the camera he tried to both hold the wing while making inputs with the hand controllers. "The MMU again performed flawless[ly]. It was a fairly easy; just fly up, grab ahold of the solar array and walk out to the edge of it and try to stabilize the satellite. And when I let go of the satellite we all thought we had it stabilized, but there was some energy in there someplace."

Crippen, watching intently from the flight deck, suggested that Nelson consider grabbing the grapple fixture. But Nelson had been using an exorbitant amount of fuel trying to stabilize the satellite. "How much you have left?" Crip asked. Nelson had already begun to back away from Solar Max as he read off his diminishing fuel reading. "Okay, c'mon back in Pink," Crippen ordered. A dejected Nelson spun back around to face the orbiter and return without having accomplished his mission.

"He was so disgusted that, you know, that was a real low point," van Hoften recalled of his EVA crewmate.

Nelson managed to stabilize the tumbling spacecraft enough before letting go of it so that Crippen and Hart were able to make three attempts to utilize a back-up procedure known as the "rotating grapple." The commander precisely maneuvered *Challenger* closer and attempted to match the satellite's rotational rate, so Hart could capture it with the RMS. These efforts, too, proved unsuccessful.

Crippen backed away from what had turned out to an untamable beast of a spacecraft. Mission Control told the crew that engineers would try to further slow the spacecraft's spin using its onboard systems. For now, all they could do was hope the engineers would be successful and give them a second chance at the rescue and repair.

Over the course of two days, Goddard Spaceflight Center engineers had managed to slow the Solar Max's spin rate to even less than when Nelson attempted to capture it, enabling T. J. Hart to snag the satellite with the robotic arm on his first attempt. Wild cheers erupted from the control room as fists pumped in the air when Crippen announced, "Okay, we've got it, and we're in the process of putting it in the FSS [flight service structure]!" Jerry Ross responded with, "Outstanding!" over the bedlam of the normally reserved MCC.

With Solar Max in the payload bay, Nelson and van Hoften would set out the following day to complete the repair for which they had trained so long. They replaced a large white electronics box containing the failed attitude control system on the exterior of the satellite with a new unit, and Ox had to complete some orbital electrical surgery within the spacecraft, all while wearing restrictive pressurized gloves. This too went off without a hitch. "The repair itself was a kick," remembered Nelson. "It was so much easier to work in space than it is on the ground." Ox remembered the excitement of that moment: "When I went out for that second day, we were pretty bullish. We knew what we had to do, and we did it so fast. We had always planned the timeline on this thing very concisely, but we planned it with a lot of contingency. I think we had like a five-hour window to do the repairs, and I think we did the whole thing in about two hours or less."

With the extra time, van Hoften was given an unexpected opportunity:

> I didn't go out planning to fly the MMU, but when we got done with the repair—they all knew I wanted to, so they just said okay. I don't even know who on the ground authorized it, but they said, "Oh yes, why don't you go fly it?" So actually I didn't even have all the apparatus out there—we left it inside—to go fly it, but I knew I could do it without it, so we just did it anyway. That thing flew exactly like the simulator, so that part was really rewarding.

Van Hoften remained within the confines of the payload bay yet had the once-in-a-lifetime experience to fly freely in space, untethered.

NASA and the crew of STS-41C were ecstatic with the success of the capture and repair. The next day, Hart gently released Solar Max from the end of the RMS with a new lease on its orbital life to explore the cycles of our nearest star, the Sun.

It was later revealed that Nelson had done everything perfectly in his attempt to grab Solar Max with the TPAD and MMU. But a slight difference in the trunnion pin on the satellite and the training mockups prevented the jaws of the TPAD from being tripped to automatically fire as they were intended to.

The successful capture using the "rotating grapple" back-up procedure demonstrated an unproven feature of the space shuttle. Without even knowing it at the time, Crippen's delicate, nimble flying and T. J. Hart's precise handling of the arm to save Solar Max was the first nail in the coffin for the future of the MMU. That combination proved to be a powerful, flexible method to retrieve derelict spacecraft designed for such a recovery, one that ultimately would convince NASA that the MMU was an unnecessary risk to astronaut safety.

But in the heady days following history's first successful retrieval and repair of a satellite that would have otherwise been written off, there was still more work to be done for the MMU. Its next calling would be even more daring than the mission to Solar Max.

Glynn Lunney, the legendary flight director who was now head of the space shuttle program, stared at the somewhat over-engineered model on his desk that director of engineering Tom Moser had just brought into his office. A standard space shuttle pallet, stuffed with scale miniature tools and gear, was topped with the cylindrical barrel of the Hughes HS-376, a common satellite design often deployed by the space shuttle, standing upright. Moser, with an intrigued Lunney assisting, flipped the model satellite around and demonstrated the use of the various tools with what seemed a giant spacewalker's hand.

It wasn't long after the loss of the two communication satellites, Palapa and Westar, on STS-41B that NASA began contemplating how to turn the dual failure into an opportunity for the space shuttle program to flex its newly found muscle. Although left in unusable orbits by the misfiring of their boost motors, both spacecraft were still able to be reached by the shuttle, which presented NASA—and some very eager insurance underwriters—with a chance to retrieve the satellites and return them to Earth for future launch. It would be the first "salvage operation" of the space age.

The smooth, solar-panel-wrapped drums of the spacecraft, along with the high fifty-RPM spin rate that stabilized them, were major challenges that stood in the way. Unlike Solar Max, these satellites were not designed to be serviced in orbit by astronauts, given that their planned geosynchronous orbits were far above what the shuttle was capable of. But Hughes was able to slow the spin remotely, which made capturing the stray machines more plausible. Lunney's team now just had to figure out how to reach them, arrest the remaining stabilizing spin, and haul them into the payload bay. At the very outset of Moser's work on the problem, it became obvious that this would be a job for the MMU.

After dismissing the thought of grabbing the satellites by their fragile antennas, engineers came up with a procedure by which the spacewalkers of STS-51A, Joe Allen and Dale Gardner, would attach themselves to a rocket nozzle at the base of the spacecraft. To accomplish this, JSC designed and built a device known as the "stinger," which, as Allen explained, "resembled a folded umbrella which one could put up inside the rocket and then open such that the tines of the umbrella would now stick against the side, and essentially the stinger would now be locked in the rocket engine of the satellite." The device was originally conceived by Gardner, and as the engineers advanced the concept, the elegance of the solution became clear.

With the stinger attached to Palapa, the free-floating astronaut would begin rotating with it and then simply select "Attitude Hold" on the MMU and bring it to a stop. Once stable, RMS operator Anna Fisher would then grasp the satellite by a grapple fixture affixed to the stinger and bring the combined spacecraft and astronaut back into the bay together. An enormously complex sequence of steps would then be required by Gardner to clamp the satellite to a pallet in the payload bay for return to Earth, but given the design of the satellites, it was the best NASA could come up with. A few days later, Allen and Gardner would exchange roles, and the whole process would be repeated for Westar.

Joe Allen and Dale Gardner, 185 nautical miles above the earth, could look up and see the Palapa satellite spinning in the blinding sunlight high above the payload bay. Allen had waited many years for this, the first EVA of his career, after being stymied by an electrical problem in his EMU on STS-5. With his full kit of space suit, MMU, and the bulky stinger device attached to his chest, "we looked for all the world like a space-age medieval knight entering a jousting contest," as he described it.

Free-flying astronauts in an MMU had by this point become a familiar image to the public. As Allen began to bridge the gap between *Discovery* and Palapa, no one expected the capture itself to be the most complex task to be accomplished; that would come later as the satellite was flipped around and configured for its ride back to Earth. With all of his training and tips he received from Bruce McCandless and Bob Stewart, the short flight across to the satellite was effortless. "I had confidence that I could maneuver wherever I wanted to go at whatever speed and achieve the first objective, which was to affix the satellite," Allen related.

Bathed in intense sunlight against the earth, he yawed to his left and lined up perfectly on the centerline of the spinning cylinder that was his target and almost immediately realized he had a problem. The blinding glare now silhouetted the spacecraft, making it impossible for the knight-like astronaut to spear his opponent. He explained the situation to pilot Dave Walker, and almost immediately commander Rick Hauck chimed, "Joe, not a problem. I'll move the Orbiter over and shadow the satellite." The commander deftly maneuvered his ship so that the nose cast a shady blanket over the target, affording Allen a clear view of the rocket nozzle.

"It was beautiful, clever as could be," Allen recalled. "I could then see the bull's-eye . . . very easily, threaded it like I'd done it all my life, deployed the tines of the stinger, tightened down the clamp, and voilà, there it was." Palapa tumbled slightly as the stinger and six-hundred-pound mass of suited astronaut and MMU bumped into it, momentarily giving those on the ground flashbacks to Solar Max. But as the enthusiastic Allen firmly locked onto the satellite and set the MMU to bring it to a standstill, he let out, "Stop the clock! I've got it tied!" like a rodeo cowboy roping a calf.

Allen then maneuvered the satellite around into a position where Fisher could see the grapple fixture and grab it with the RMS. "From a MMU crewman's point of view, it's interesting to see the large arm coming up right over your shoulder to suddenly grasp the satellite," Allen reported post-flight. With the satellite now firmly held by the RMS, Allen disconnected from the stinger device, and slowly glided back into the payload bay to stow the MMU.

Unfortunately, the rest of the plan to bolt the satellite into the payload bay had to be abandoned; it was discovered that a piece of structure on the spacecraft that was not noted by engineers interfered with the design of the clamp. Instead, the spacewalkers relied on a backup procedure to berth Palapa.

Two days later, the spacewalking duo repeated the recovery procedure—only now with the refined procedures learned from their experience with Palapa—and safely stowed Westar in the payload bay. With two satellites recovered for return to Earth, refurbishment, and eventual re-launch, it was an unprecedented victory for the space shuttle program and held great promise for the future. Allen and Gardner took a ride on the manipulator foot restraint high up over their handiwork and posed hanging above the satellites against the jet-black of space and the sharp blue horizon of Earth below. Gardner unfolded a small, handmade sign that read simply, "For Sale."

Even for the crew, it was hard to believe what they had accomplished, given all that could have and did go wrong. As Dale Gardner recollected, "We had to, several times after flight day 7 . . . go up to the windows back there just to make sure both satellites were there. There were times when none of the five of us would really believe that we actually pulled it off, and it was certainly comforting to look back there and see both of those satellites staring at us."

When the crew of *Discovery* arrived by NASA aircraft back at Ellington Field in Houston after landing at Edwards Air Force Base, they were met by scores of NASA officials, family members, and employees. One sign held high in the crowd read "EVA is here to stay!" It was a heady time for NASA, with three spectacular missions featuring free-flying space walks utilizing the MMU, and now three stunning satellite retrievals with two returned to Earth, a whole new world of possibilities seemed to be on the horizon for this rapidly developing capability.

While EVA certainly was to remain a prominent feature of the space shuttle program until its conclusion and on into the ISS, the Manned Maneuvering Unit had seen its final mission. By the summer of 1985, the flight-used MMU's had been returned to Martin Marietta in Denver and found to be in "near-perfect condition." Joe Lenda, system engineering lead for the MMU, told *Space World* writer Gary Graf, "It has demonstrated what it is capable of doing; now we're just waiting for the next reason to use it."

At the time, the entire fleet of orbiters had not been configured to carry the MMU's and their support cradles. Unfortunately, many larger payloads, such as the Hubble Space Telescope, took up so much payload bay space that the MMUs and their respective cradles could not be flown to support deployment operations.

The entire justification for the MMU once again came under close scrutiny following the *Challenger* disaster in January 1986. In the two and a half years of reengineering and self-reflection on safety priorities, the cost of recertifying the MMU for flight was deemed too high, and the entire program was scrapped. "It was just a nice machine . . . I just couldn't think of any reason not to use it," Bob Stewart shared. "I have no clue why NASA decided to abandon it. In my opinion there should be two of them on the space station right now."

The decision, in part, came as a result of the unexpected versatility of the shuttle and RMS capability that had been demonstrated on several missions featuring satellite retrievals. Those capabilities would be showcased many times over in the years following the loss of *Challenger*.

"It [MMU] was a great concept and everything, but it was like a refrigerator on your back," opined Jerry Ross. "And once you got somewhere, you couldn't do much in the way of real work with it on your back." Early on in the conceptual phase of space station *Freedom*, it was envisioned as not only a space laboratory, but also an orbital repair shop for satellites, with free-flying astronauts hauling spacecraft into a huge hangar structure for on-site work. But over time, as the project morphed into the International Space Station in the mid-1990s, this requirement was eliminated, leaving only assembly, maintenance, and repair of the station itself as a justification for EVA.

After he left NASA for the academic world, EVA trainer Tomas Gonzales-Torres put the question of the usefulness of a MMU-type capability to a class of his engineering students. In looking at the risk versus reward of utilizing such a device on the space station, the more they studied the issue, the more it validated NASA's decision from three decades prior:

> We have enough capabilities with translation and safety tethers . . . that the risk and cost associated with designing this is probably not worth it. It may be like a neat thing to have but I don't know if it's buying you down enough capability. [I]f it's taking you an hour to translate somewhere, and this jet pack could do it in 15 minutes then yeah maybe, because you're buying back a lot of EVA time and you can get a lot of other tasks done. That's not really true based on the size of the space station and the activities that we are doing.

So after nearly two decades of design, development, and eventual utility, the MMU concept championed by Bruce McCandless, Ed Whitsett, and countless others became a minor, although historical, piece of the human spaceflight story. But the stunning, iconic image of one of its inventors, floating serenely in the space between Earth and infinity, still serves as perhaps the most inspirational photo to come from the shuttle era.

Today, the Number 3 MMU resides in the National Air and Space Museum's Udvar-Hazy facility in Dulles, Virginia. It is strapped to a fully suited figure and suspended high above the space shuttle *Discovery*, which flew it into orbit on the STS-51A mission. The Number 2 unit has been similarly displayed at several different museums around the country, including the U.S. Space and Rocket Center in Huntsville, Alabama.

For the decade or so between its final flight and the beginning of NASA's cooperation with Russia on the ISS, beginning with the Shuttle/*Mir* program, the long-discarded notion of free-flying astronauts working around a space station wasn't given a second thought. But the technology developed from the MMU would find a second chance at life. With a space shuttle docked to an orbiting space station and unable to affect a rescue or if an ISS spacewalker were conducting an EVA with no shuttle present, what if he or she became a free-flying astronaut . . . *accidentally*?

The Soviet Union, after almost twenty years of internal controversy and false starts, finally saw its own version of the MMU launched into orbit aboard the Kvant-2 module that would expand the *Mir* space station in November of

1989. The 397-pound *sredstvo peredvizheniy kosmonavtov* (SPK), or "cosmonaut maneuvering equipment," dubbed 21KS, was similar in many ways to its American counterpart but appeared slightly larger and more like a cushioned armchair, with its thick thermal blankets wrapped tightly around the backpack. Despite the many doubts Soviet space leadership had about its utility in orbit, they were eager to demonstrate their technological prowess equaled or exceeded that of the Americans.

The 21KS design abandoned the toxic fuels of earlier Soviet efforts for simpler, safer pressurized air. The air was fed to thirty-two thrusters utilizing hand controllers similar to those of the MMU. It too had an automatic attitude hold feature to reduce pilot workload and fuel consumption. Cosmonauts trained to fly 21KS on their air-bearing facility.

Cosmonaut engineer Aleksandr Serebrov first donned the unit, informally tagged with the name Icarus, the legendary son of Daedalus who flew too close to the sun, on 1 February 1990 for a test flight of about forty minutes in length. He made three short translations away from the *Mir* space station before again moving out to approximately 110 feet and returning. Without the ability to rescue Serebrov in case of a failure, he remained tethered to *Mir* throughout the excursion away from the station.

Four days later, commander Aleksandr Viktorenko flew the 21KS slightly farther out for forty-five minutes and carried a large chest-mounted radiation monitoring experiment. He was able to complete a full lap around *Mir* at a distance of about 150 feet, roughly the same distance as McCandless and Stewart's shorter excursions on STS-41B. It was reported that the cosmonauts found the unit to be quite cumbersome and had difficulty moving while tightly clamped into its rigid frame around the waistline.

Following these two brief sojourns, no further work on SPK was ever pursued by the Russian space program. Like NASA's experience, it was found that it was a machine without a mission, as much of the work it was envisioned to do could be done more safely and cheaply without it. The unit was left unused until February 1996, when it was unceremoniously stowed outside of the Kvant-2 to make room for other equipment. It remained there until *Mir* was de-orbited in March 2001.

"Whoa! That's way too fast!" Mark Lee was tumbling wildly over the cargo bay of the orbiter *Discovery*, having just been spun by fellow spacewalker Carl

Meade. Using a small joystick protruding from a box attached to his chest pack, Lee selected the automatic attitude hold mode (AAH), sending commands to a new experimental jet pack mounted below his PLSS backpack. Short bursts of nitrogen squirted from the computer-selected nozzles, and the unit brought him to a stop with minimal effort. With the full moon hanging over his shoulder, Lee noted, "I'm glad it's dark or that would have been scary."

The aptly named SAFER device—into which NASA acronymologists forced the designation "Simplified Aid for EVA Rescue"—was a high-tech life jacket of sorts. In the event that an EVA crewperson would become "detached from structure" and started floating off into space without a tether, this smaller MMU-like device could be employed to maneuver back to the safety of the space station. In this first flight test of SAFER in September 1994, STS-64 mission specialists Lee and Meade would put the device through its paces in preparation for the upcoming Shuttle/*Mir* program, where EVAs were to be conducted while the orbiter was docked to Russia's aging orbital outpost. Meade had worked on all stages of SAFER's development, from conception to flight test to production.

Post-flight, Meade reported, "EVA has been described as a series of continuous tether manipulations occasionally interrupted by useful work." When considering the robust mechanical components of the tether systems, their likelihood of failure was expected to be extremely low, yet "despite this assumption, it must be noted that NASA has experienced at least one tether hook failure in flight," which was discovered and corrected before the crew exited the airlock.

Human error was considered the most likely scenario for an EVA crewmember to become detached, and while it was not known to have happened previously in the U.S. program, "we have had a least one crewmen lose grasp and be retrieved by the automatic wind-up of the . . . safety reel." Astronaut Jerry Ross was a vocal proponent for a rescue device of some sort early on, arguing that even at a distance of just a few feet, if the astronaut could not reach a handhold, he or she was as good as dead. He warned managers, "Okay, what are you going to tell somebody's spouse, when you could have built this and you didn't, and they're now drifting over the horizon, their battery and oxygen are running out, what are you going to tell them?" Over the course of the hundreds of EVAs it would take to build and maintain the ISS in the decades to come, it was a credible threat, but how to go about mitigating it was the subject of great debate.

"Before we could convince the program to build the SAFER, we had to prove nothing else would work," recalled astronaut Rick Hieb. Five different devices were conceived—two consisting of extendable poles driven out by the Pistol Grip Tool (PGT) attached to the end, an inflatable pole, and the aptly named "astro-rope." NASA stated that the "astro-rope uses an approach similar to the concept of a bola-type lasso. The astro-rope is thrown by hand and is meant to wrap around an element of the space station structure." All of these concepts were only effective to a range of fifteen to twenty feet.

The fifth device was a handheld propulsive gun similar to the one tested on *Skylab*, which, Hieb recalled, "we got out of the Smithsonian." Several astronauts tested the contraptions out in the zero-g KC-135 aircraft. "It was pretty much a disaster. The bolos would tangle when you tried to get them out of the pouch," Hieb explained, and by the time the PGT could be attached, the astronaut was too far away from [the] structure to reach it with either of the pole concepts. NASA managers soon realized that an MMU-based design might be the most suitable solution.

The SAFER backpack is a small, self-contained unit that includes two small nitrogen tanks, a microprocessor, and associated valves and plumbing. The device is fitted beneath the PLSS unit and features two vertical towers that extended flush along either side of the backpack, positioning thruster nozzles at the top of it. Like the MMU, SAFER features twenty-four gaseous-nitrogen thrusters, but given its limited fuel supply, it can impart a total of only ten feet per second change in velocity, or "delta-V," as compared to the hefty sixty-six feet per second of its older cousin. SAFER was designed to be simple and lightweight and to not be a hindrance during normal EVA operations. It lacks the redundant systems of its predecessor and has no telemetry to the ground but records data onboard for later analysis.

It was crucial to the program that the evaluations were made on a shuttle mission where an astronaut could be rescued if SAFER should fail. Additionally, the tests would be conducted untethered, in the unlikely case that a "stuck on" thruster might wrap the tether around the astronaut—not a good outcome.

Continuing the "self-rescue" demonstration above the payload bay, Meade repeatedly spun his crewmate in each axis, and SAFER dampened out the rotations before Lee slowly jetted back to the RMS to grab on. RMS operator Susan Helms then backed Meade away from the now stabilized Lee to simulate the astronaut having drifted away from a space station. "It was easier to simulate a translation by moving the RMS rather than throwing a 300 pound-mass EVA crewmen into deep space," Meade quipped. He repeatedly spun his crewmate in each axis, and SAFER dampened out the rotations before Lee slowly jetted back to the RMS to grab on.

The astronauts gained confidence in the device as the EVA continued. He next traced a path up along the RMS, which was positioned high above the payload bay, to the elbow joint, turned, flew out to the end effector, and then pivoted around and reversed course back to the base of the arm.

A faint blue glow of reflected earthlight illuminated the right-hand side of Lee's space suit as he precisely glided along the length of the arm. The gently curved horizon of Earth sliced through the upside-down "V" position of the arm, completing the "A" image formed by space-age technology and nature. Pausing briefly at the elbow joint to wave at the camera, he rounded the corner and continued out to the RMS's "hand," while spectacular television images of the untethered astronaut beamed to the ground.

SAFER would never be used this way in practice, but the evaluation validated its ability to return a wayward spacewalker to safety in an emergency. Meade took a turn at flying the same course, and the duo even had the ability to replenish the units with more nitrogen during the EVA in order to extend their flight time. Joe Allen, who was assisting the public affairs officer in covering the EVA for NASA TV, commented, "It looks like an Olympic figure skater, it's so smooth."

Fuel consumption was a primary concern, but battery power had never been a problem during training. "During the flight it was a different story," Meade reported. "In fact, by the time the SAFER pilot reached the elbow to begin the approach to hover, battery power was down to 18 percent . . . unusually low." By the time Meade had made the final input to return to the refueling station, SAFER had shut down completely, its batteries depleted in the extreme temperatures of the vacuum of space. Despite this unexpected data point, the SAFER tests were considered to be a complete success.

The astronaut's evaluations led to several modifications to improve SAFER's utility and reliability in rescuing an adrift spacewalker. Batteries would be upgraded, the hand controller would be stowed inside the propulsion module and be deployed only for emergency use, and a Velcro attachment to the chest pack would allow one-handed operation. When deployed, the AAH mode would be autonomously activated to immediately stabilize a tumbling astronaut. And the manual thruster isolation valve would be replaced with a single-use pyro-technic design. But NASA would soon find that the SAFER design still needed more work.

On 27 March 1996 Linda Godwin and Rich Clifford conducted the first U.S. EVA from a space shuttle while it was docked to a space station, Russia's *Mir*. Both wore SAFER devices similar in design to those tested on STS-64 without yet having been extensively modified. It wasn't until October 1997 that cosmonaut Vladimir Titov and astronaut Scott Parazynski flew with an updated design while conducting the first joint U.S./Russian EVA during a space shuttle mission. Both were wearing American EMUs, and the EVA also occurred while docked to *Mir*, so the SAFER was necessary should one of them come loose.

On STS-86 Parazynski and Titov continued a series of EVA development flight test (EDFT) space walks, designed to evaluate equipment and techniques to begin construction of the ISS. Parazynski also conducted SAFER operations near the end of their five-hour EVA.

Floating tethered to the cargo bay bulkhead, Parazynski lifted a lever on the right side of SAFER that deployed the hand controller and immediately indicated that the AAH mode had activated. This was a change recommended by the STS-64 crew; there was no need for a tumbling crewman to waste precious fuel trying to cancel out a multi-axis spin when this feature could do it quickly and automatically. Parazynski then slipped into a foot restraint and verified the operation of the thrusters with LED indications on the controller. "I've got thruster firing, and the AAH light is still illuminated," he reported. "Looks like we've got a good unit," he said. "Can you hear the thrusters firing?" Commander Bob Cabana asked. "Yes, I can," he replied, as the sound of fuel valves opening and closing transmitted through his suit. "Could you feel them pushing at all?" Cabana continued. "Nope, not at all." This was no surprise, given that less than a pound of thrust was being generated by each of the twenty-four nitrogen thrusters.

It wasn't until *Atlantis* returned home and the data from SAFER were analyzed that engineers realized there was a *big* problem—Parazynski's SAFER hadn't worked at all. His nitrogen tank was found to be completely full after the mission, no doubt sending shivers down the spines of NASA EVA engineers and astronauts. Had he or Titov become detached during their EVA at *Mir*, there may have been no saving them.

A pyrotechnic initiator that was supposed to fire and open the valve of the nitrogen supply tank failed to do so. This same device had been used for many years at critical points in the shuttle's systems with virtual perfection. An extensive investigation led by astronaut Steve Smith determined that a change in the type of batteries used in SAFER required an avionics design change. The new circuit didn't allow enough power to reach the pyro to fire it reliably. According to the Failure Review Board's report, "Only two firings . . . were attempted with the avionics redesign. Although both firing attempts were successful, they are considered 'luck' based on probabilities. The 'successful' firings were therefore misleading. Both firings were conducted on the Certification Unit; no firings were ever attempted in either of the flight units."

Among dozens of recommendations on technical fixes, process changes, and program management made by Smith's board, the final one would prove fortunate for one experienced spacewalker. "A flight test should be manifested at the first opportunity," it read. "SAFER is an emergency piece of equipment that will be used for many, many years and hundreds of EVAs. A simple flight test would be relatively inexpensive and low risk, yet the return would be high; complete verification of the design."

Given the primary mission of STS-88—joining the first Russian and U.S. modules of the International Space Station together in orbit—surprisingly little was said pre-flight about the potential to conduct another SAFER test flight. It was seen as a low-priority test objective that would be accomplished only after all the other major goals of the mission's three EVAs were completed. With the potentially disastrous failure on STS-86, veteran spacewalker Jerry Ross lobbied hard to get the test added to the flight. "When we flew on STS-88, [SAFER] was a new piece of hardware," he recalled. "The other stuff had been prototype . . . we had what was the flight configuration. It had new electronics, new logic, some new hardware pieces in it, and I thought to do some flight testing on it."

Some managers in the EVA office disagreed, assuming that the previous tests done in orbit and on the ground since the modifications were sufficient. Ross ran up against a brick wall with one in particular, arguing that "you used to be a test pilot . . . whenever they made any [modifications] to your airplane, you always wanted go out and test them to make sure they hadn't changed anything." But it was to no avail. It wasn't until NASA veteran Bill Gerstenmaier came into a leadership role shortly before STS-88 that Ross found a sympathetic ear, and he was able to convince him that the test flight would be a prudent thing to do. "I made one more push with Bill to do a brief, non-interruption—no additional testing or training or anything like that—and he agreed to do it."

Following completion of their exhausting EVAs to configure the U.S. Unity module with Russia's Zarya, Ross got to fly his brief test flight. "We did that in a very simple way by my crewmate just holding on to my tether loosely and letting me float free out in the payload bay," he recalled. "Fortunately we did get a chance to do that test, and there were a couple of issues that we discovered that had to get corrected before we started using it for follow-on flights."

STS-92 launched in October 2000 on another ISS assembly mission, and on the fourth EVA of this critical flight, Mike López-Alegria and Jeff Wisoff were tasked with a final checkout of the production model SAFER. Once again remaining tethered to the orbiter, they took turns flying precise tracks toward a TV camera that was used as a target. SAFER was praised with high marks from the spacewalkers, and it was at last considered operational for the ISS program.

Despite being exposed to the same risks as their U.S. counterparts during EVA, Russian cosmonauts today have no SAFER-type system to utilize in the dire event of becoming detached; they rely solely on strict tether protocols. Like so many other space projects of the former Soviet state, budget shortfalls led to its demise after NASA ended their financial participation in 2000. Russian cosmonauts continue to operate in open space with this one distinct difference in safety philosophy, one that nearly led to tragedy in a little-known event aboard Expedition 25.

On 15 November 2010 cosmonauts Fyodor Yurchikhin and Oleg Skripochka completed a six-and-a-half-hour space walk to install various equipment and experiments to the exterior of the Russian segment of ISS. Their American crewmate, Scott Kelly, remained inside but recalled in his 2017 memoir, *Endurance: A Year in Space, a Lifetime of Discovery*, that when the two spacewalkers completed their EVA and began removing their suits, "both looked shaken." Neither mentioned that anything was amiss to Kelly or to flight controllers on the ground, and Kelly wrote it off to the elation of having worked out in the ether of orbital flight.

It wasn't until five years later, during Kelly's next long-duration mission, that one of his crewmates told him what had supposedly happened—Skripochka reportedly became untethered on his first foray into space. As he floated helplessly away from surface of the station, only a fortuitous collision with an antenna bounced him back toward a handrail and safety. The story has never been publicly acknowledged by Skripochka or Roscosmos, but it serves as a stark example of a spacewalker's worst fear. Kelly wrote that it was "not something I wanted to spend a lot of time thinking about as my own space walk was approaching."

To date, in the hundreds of space walks throughout the two decades of construction and operation of the ISS, SAFER has not been called on to rescue a marooned astronaut. NASA's training and safety protocols for EVA are so ingrained in the crew's muscle memory that no U.S. or European astronaut has ever become detached from the station and necessitated its use. But if this event ever comes to pass, today's spacewalkers have the ingenious legacy of Bruce McCandless, Ed Whitsett, Bill Bollendonk, and countless other visionary engineers literally on their backs to protect them.

SAFER is the result of decades of work aimed at accomplishing a different capability from what eventually came to be. It is the ultimate spinoff of new space technology—the MMU—that directly benefits the astronauts who risk their lives to work outside their spacecraft in the unforgiving vacuum of space. Today, if one looks closely at any astronaut conducting a space walk, one might notice an oblong emblem stitched into the thermal covering of the SAFER backpack. It features a spacewalker, under precise control, piloting the jet pack back to a distant ISS. Two large, feathered wings protrude from the astronaut's PLSS, the guardian angel coming to rescue the spacefarer and return him or her to the safety of their home in space.

1. Mission specialist Kathy Thornton performs the second space walk on the STS-61 mission along with Tom Akers in December 1993, charged with repair of the Hubble Space Telescope. Thornton believes she was selected for a space walk on her second mission—STS-49—because of her positive attitude and hard work in learning how to function affectively in the EMU despite her small stature. Courtesy NASA.

2. The crew of *Voskhod 2*, Alexei Leonov and Pavel Belyayev. Leonov performed history's first space walk, spending little more than twelve minutes outside the spacecraft on 18 March 1965. Courtesy Gagarin Cosmonaut Training Center.

3. *Gemini 4* pilot Ed White during the first American space walk. Holding the hand-held self-maneuvering unit, he maneuvered himself away from the spacecraft. White recalled: "I decided I would fire a pretty good burst . . . and I literally flew . . . right along the edge of the spacecraft, right out to the front of the nose, and out past the end of the nose. I then actually stopped myself with the gun. That was easier than I thought." Courtesy NASA.

4. Neutral buoyancy training in the McDonogh School swimming pool was crucial in training Buzz Aldrin for his successful *Gemini 12* space walk. Aldrin conducted five sessions in the pool, and Gene Cernan, his backup, made two passes. Courtesy NASA.

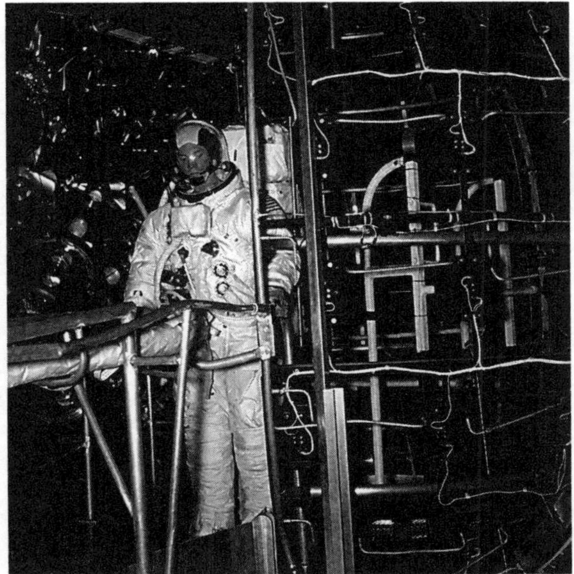

5. *Apollo 9* lunar module pilot Rusty Schweickart trains inside vacuum chamber A in December 1968. The chamber familiarized Schweickart with EVA equipment in an environment like the one he would have to contend with in orbit. Courtesy NASA.

6. Rusty Schweickart points his camera during the first Apollo-era EVA in March 1969 on the *Apollo 9* mission following a short bout of space adaptation syndrome. Had he not recovered in time to carry out the EVA, NASA would have likely inserted a new mission to test the EMU space suit and PLSS before moving on to the *Apollo 10* dress rehearsal for the first moon landing in 1969. Courtesy NASA and Arizona State University Digital Repository.

7. "Talk about being a spaceman! This is it!" *Apollo 17* CMP Ron Evans conducts the final trans-Earth EVA of the Apollo program to retrieve film cassettes from America's scientific instrument module bay. Courtesy NASA.

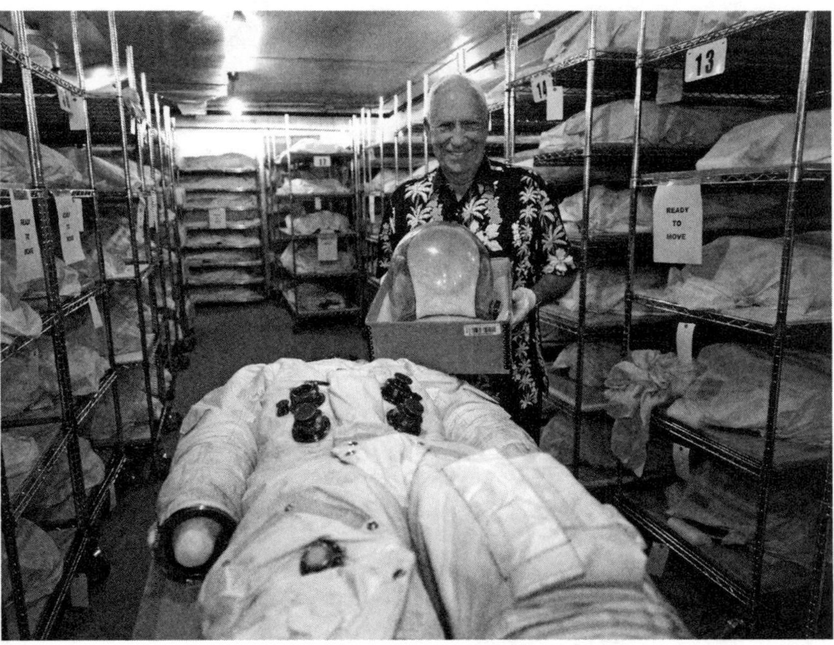


8. *Apollo 15* CMP Al Worden, pictured at the National Air and Space Museum's Garber facility in July 2011, is reunited with the space suit he wore during his trans-Earth EVA. The 5 August 1971 space walk lasted thirty-eight minutes and was one of only three ever conducted beyond Earth orbit. Courtesy National Air and Space Museum.

9. Immediately following the *Skylab* space station's trouble-plagued launch, teams of astronauts and engineers went to work around the clock to determine what could be done to save the crippled vehicle. *Apollo 9* veteran Rusty Schweickart and *Skylab* crewman Ed Gibson raced to the Marshall Spaceflight Center's neutral buoyancy simulator (NBS) to develop procedures to free the stuck solar array. With one solar panel completely gone, it was critical to the station's survival to get the remaining one deployed, allowing electrical power to feed the station's systems. Courtesy NASA.

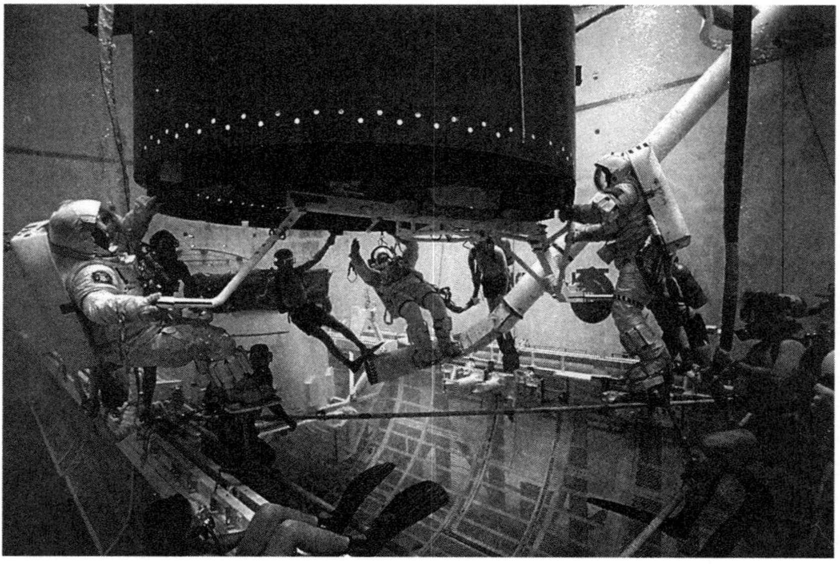

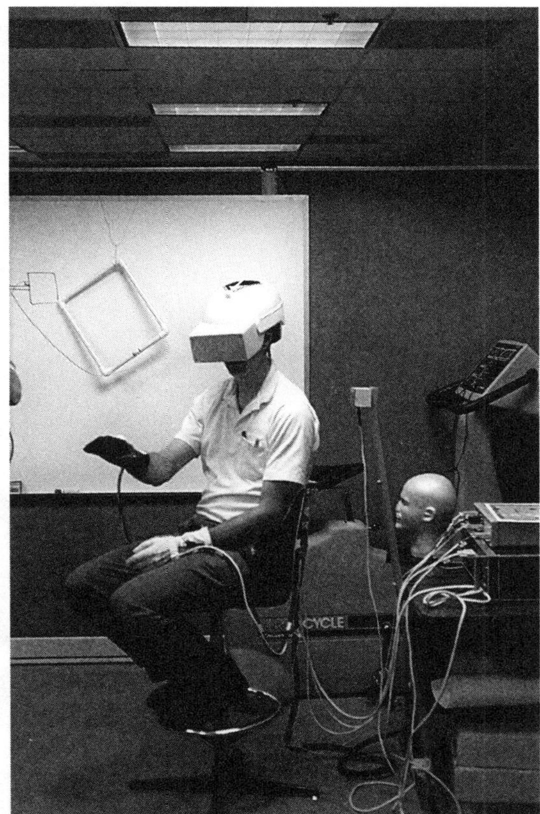

12. The STS-61 mission to repair the Hubble Space Telescope may be the first time that anyone employed virtual reality to prepare for a space walk. Jeff Hoffman trains in the VR lab wearing a special helmet and gloves to simulate EVA. Although new and rudimentary, the EVA crew was able to utilize VR to plan the positioning of the robotic arm, which could then be tested in the pool. Courtesy NASA.

10. (opposite, top) Skylab 3 pilot Jack Lousma during the 6 August 1973 EVA to deploy the more permanent twin pole sunshade over the orbital workshop. With Dr. Owen Garriott assisting him, Lousma wrestled with the hastily packed shade and its many folded sections that were stuck together with dried glue. Eventually, the pair was able to fully deploy the shield and lower it over the first "parasol" type shade deployed by the first crew. Courtesy NASA.

11. (opposite, bottom) Neutral bouncy training under water became a mainstay for training astronauts for EVA beginning with *Gemini 12* in 1966. Here, astronauts Rich Clifford, Story Musgrave, and Jim Voss—in the weightless environment training facility—are testing, in May 1992, whether three suited astronauts could capture the Intelsat satellite following multiple failed attempts by STS-49 just days earlier. Courtesy NASA.

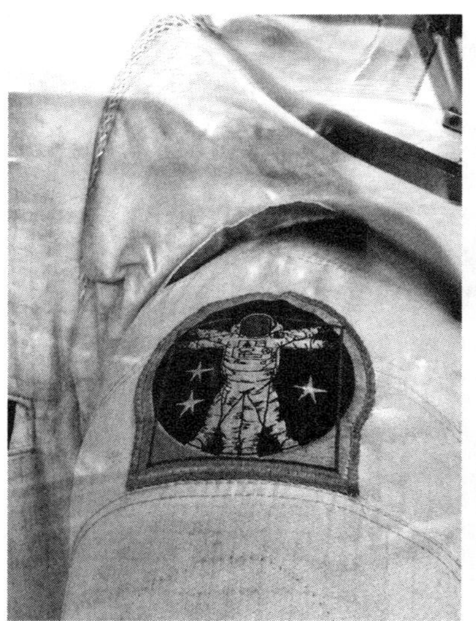

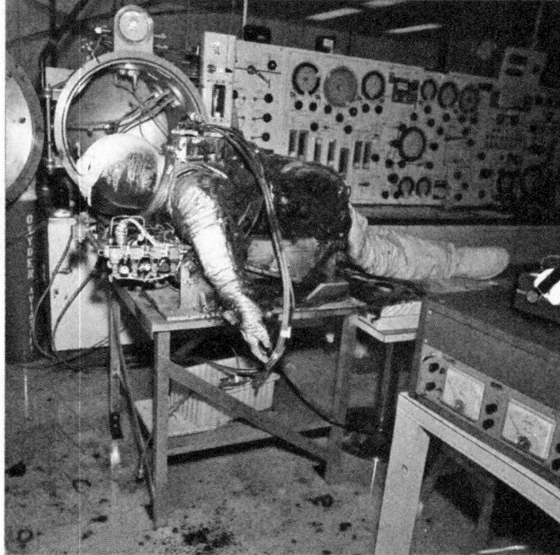

13. (left) The EVA patch, based on Leonardo da Vinci's informally titled "Vitruvian Man," has adorned NASA space suits since the first shuttle space walk. The three stars represent the pioneering EVAs of *Gemini 4*, *Apollo 11*, and Skylab 2. Later, two more stars were added to honor the first space walks of the shuttle and ISS eras. Since no company logos were permitted on NASA equipment, Hamilton Standard's Fred Keune and Walter J. Wick, MD, conceived the emblem to represent the design work by the company throughout the U.S. space program's history. Photo by John Youskauskas.

14. (right) EMU 3002, destroyed by a flash fire during testing on 18 April 1980. Even though NASA had been working with high-pressure oxygen systems for decades, several lapses in safety led to two technicians being seriously injured. No definitive ignition source was ever identified, as most of the evidence was destroyed by the inferno. Several technical and procedural changes were made, and no other such failures have occurred in the four decades since. Courtesy *Spaceflight Safety Magazine*/NASA.

15. Don Peterson, STS-6 mission specialist, being assisted with his EMU by pilot Bo Bobko as he prepares for a training session in the weightless environment training facility in December 1982. After the canceled space walk on STS-5, Peterson and Story Musgrave were assigned the first EVA of the space shuttle era with the newly developed space suits. Courtesy NASA.

16. Story Musgrave, who was instrumental in the development of the EMU, carries out its first test in space from *Challenger* on 7 April 1983. Lasting just over four hours, it was the first U.S. space walk since Skylab 4, nearly a decade before. Courtesy NASA.

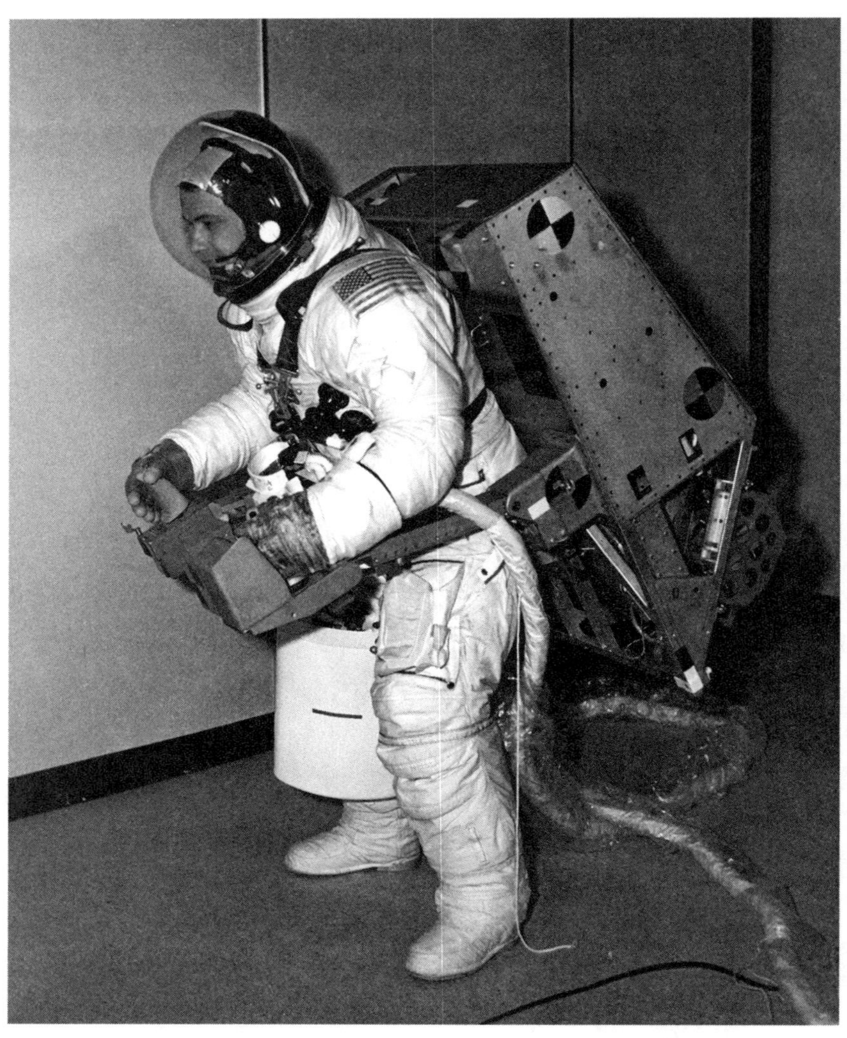

17. At Martin Marietta Corporation's Denver facility, Bruce McCandless II checks out the prototype astronaut maneuvering unit dubbed the M509 experiment prior to *Skylab*'s launch. The backpack was flown by several members of the Skylab 3 and 4 crews inside the cavernous orbital workshop, demonstrating the fine maneuvering ability of the system and contributing data for development of an operational maneuvering unit for the space shuttle era. Courtesy NASA.

18. (opposite, top) "It was a piece of cake. Actually, I felt really relaxed." Robert Stewart, bathed in sunlight on one side and reflected Earthlight on the other, performs a flight test of the manned maneuvering unit (MMU) from the orbiter *Challenger*, February 1984. Courtesy NASA.

19. (opposite, bottom) "I got it! Looks like I gave it a good knock." Dale Gardner closes in on the spinning Westar VI satellite on 14 November 1984 using the aptly named "stinger" device. He and the satellite briefly tumbled end over end before he was able to stabilize it using the MMU's thrusters. It was the second spectacular satellite retrieval of the highly successful mission but also marked the final use of the MMU in space. Courtesy NASA.

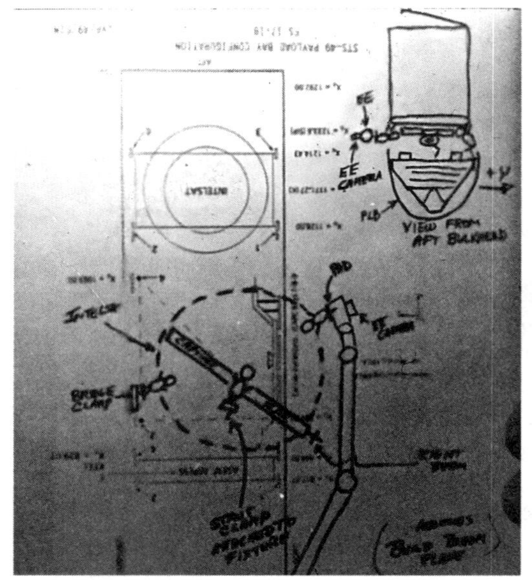


20. The checklist page that the STS-49 crew used to sketch their plan for the only three-person space walk in history necessary to capture the Intelsat VI satellite. The upside-down page illustrates the view from above the payload bay and from the side (upper right corner). Courtesy Rick Hieb.

21. Rick Hieb, Tom Akers, and Pierre Thuot secure the Intelsat VI satellite during the only three-person space walk in history. The ground wasn't sure three astronauts could fit inside the shuttle airlock; Hieb believes it could hold four astronauts. "Yeah, we could have fit four people in there. I'm convinced of that. It never seemed really crowded." Courtesy NASA.

22. Joe Tanner at work on his STS-97 mission at the Unity connecting module. The EVA began from the ISS airlock, and as Tanner recalled, "Egressing the ISS airlock was more visually challenging than the shuttle because there was nothing between the hatch opening and Earth. You forget that you can't fall . . . it sure looked like a long way down! The shuttle hatch opened into the payload bay so you felt less exposed." Courtesy NASA.

25. (left) Scott Parazynski has a bird's-eye view as the ISS P6-4B solar array is deployed following his successful repair job on the STS-120 mission. Five "cufflinks"—thick pieces of wire—are visible on the array, which provided the necessary strength to fully deploy the damaged array. Courtesy NASA.

26. (right) Jerry Ross floats outside the ISS Zarya module hanging on with one hand. One of his objectives on the STS-88 mission was to test the SAFER device developed to provide spacewalkers a means of returning to the ISS if they became detached. "We did that in a very simple way by my crewmate just holding onto my tether loosely and letting me float free out in the payload bay . . . doing just a very quick test on it." Several issues were discovered that had to be corrected. Courtesy NASA.

23. (opposite, top) Jeff Hoffman and Story Musgrave—STS-61—are poised to install the new Wide Field and Planetary Camera 2 (WFPC2) into the open cavity visible on the Hubble Space Telescope. The WFPC can be seen just to the right of Hoffman's striped legs. Five space walks on this mission corrected a flaw in the telescope's mirror, resulting in images coming from the cosmos with vastly improved resolution. Courtesy NASA.

24. (opposite, bottom) John Herrington, inside the ISS Quest airlock in November 2002 preparing for his first of three scheduled STS-113 space walks. He recalled that his entire focus was on the job at hand: "Do your mission, do it well, and when you're done, if you've done it well, then let's relax and have fun. After that, we were taking pictures, playing with our food, making jokes." Courtesy NASA.

27. Winston Scott sports a fulfilling grin while in the payload bay of *Columbia* on the STS-87 mission where he helped capture the SPARTAN solar research satellite. "That was the most stressful thing we did in orbit. But I slept like a baby the night before. I knew that we were going to be successful—we were going to get that doggone satellite. You've got the A team!" Courtesy NASA.

28. (top) EVA instructors Tomas Gonzalez-Torres and Christy Hansen train the STS-125 crew on hardware they plan to install on the Hubble Space Telescope. Left to right: Gonzalez-Torres, Mike Massimino, John Grunsfeld, Hansen, and Drew Feustel. Hubble mockups that went into the NBL for the training runs can be seen in the background. Courtesy NASA and Christy Hansen.

29. (left) Ed Rezac demonstrates the Mini Power Tool in Tucson, Arizona, in 2016, developed for the STS-125 crew to repair the Hubble Space Telescope. It has a slimmer profile than the Pistol Grip Tool, which makes it easier to access hard-to reach places, turns at a more efficient higher speed, and comes with an LED light source. Like working on your car in your garage, having the right tool is essential to getting the job done in space. Courtesy David Chudwin.

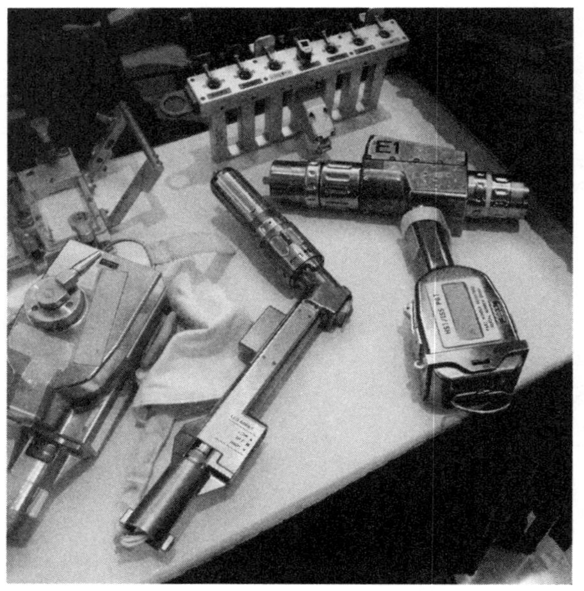

30. Tools of the EVA trade. Far right: Pistol Grip Tool. To its left lies the Mini Power Tool and its battery pack. In the top left of the photo is the fastener capture plate used for replacing the advanced camera for surveys on Hubble. The bit caddy is in the upper right-hand corner; it contained the various tool bits that would fit into a variety of fasteners necessary for the ACS and space telescope imaging spectrograph servicing tasks. According to Ed Rezac, once the crew was named, detailed procedures were developed necessary to accomplish the task, and in parallel identify the required hardware, including special tools. Courtesy Melvin Croft.

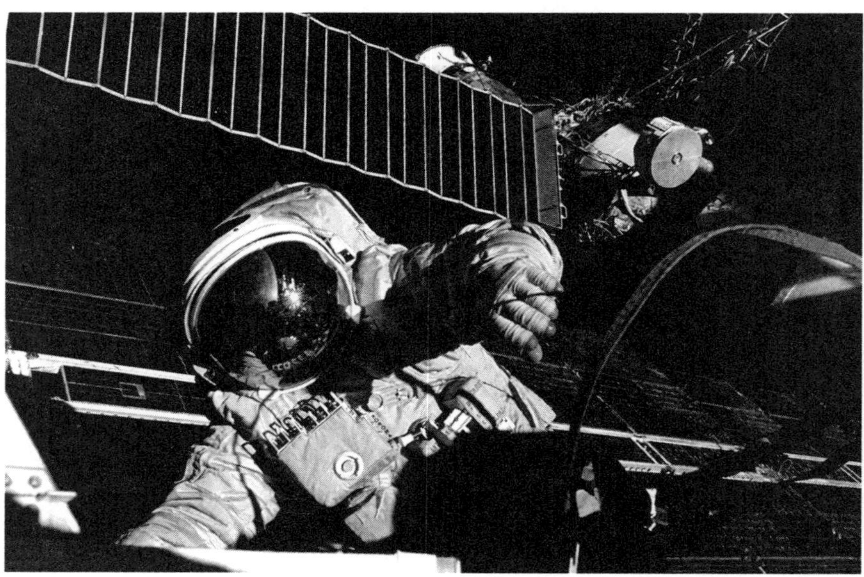

31. U.S. astronaut Jerry Linenger became the first American to conduct a space walk in a Russian Orlan-type space suit when he exited the *Mir* space station on 29 April 1997. Linenger spent 132 days aboard the station, facing not only the challenges of a long-duration spaceflight but frequent electrical power failures and even a serious onboard fire that filled the orbiting outpost with dense smoke. Courtesy NASA.

32. In the early days of the International Space Station program, U.S. and Russian crews utilized both the EMU and the Orlan interchangeably, depending on which segment of the station required external work. Expedition 12 commander William McArthur (*left*) and flight engineer Valery Tokarev pose with their respective suits in the Destiny laboratory module of the ISS. Courtesy NASA.

33. With New Zealand as a backdrop 220 miles below, Robert Curbeam (*left*) and ESA astronaut Christer Fuglesang conduct the first of four space walks to install the new P5 truss segment to the ISS. The December 2006 STS-116 construction mission delivered the truss and rewired the station's electrical system to its permanent configuration. Courtesy NASA.

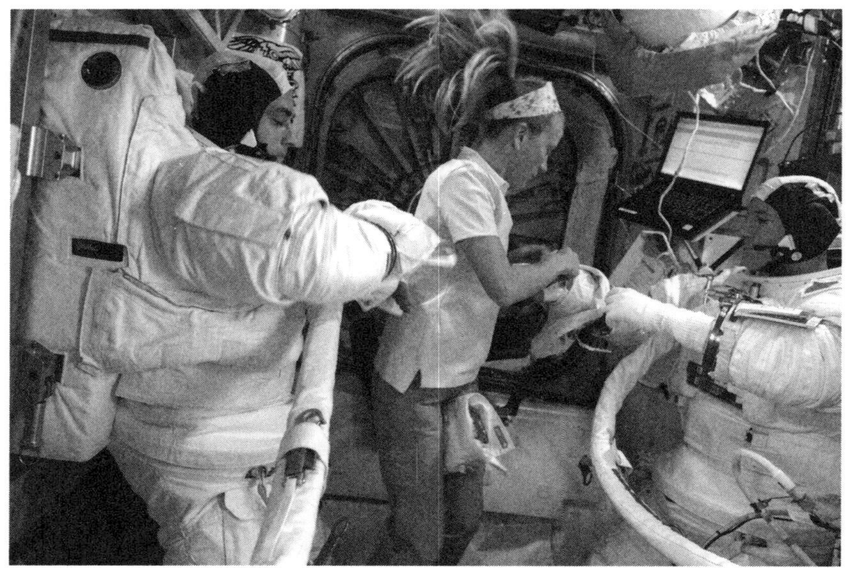

34. NASA's Chris Cassidy (*left*) and ESA's Luca Parmitano prepare for a space walk on 9 July 2013 in the equipment lock section of the Quest airlock, with Karen Nyberg assisting. A small amount of water was noted in Parmitano's helmet following the EVA but was dismissed as having come from his drink bag inside the suit. The following week, the pair's next space walk would become far more dramatic after that erroneous assumption. Courtesy NASA.

35. The immense scale of the ISS is evident as ESA's Thomas Pesquet (*left*) and Akihiko Hoshide of the Japan Aerospace Exploration Agency (JAXA) work to install a new Roll-Out Solar Array (ROSA) to increase the station's electrical generating ability. The 12 December 2021 EVA was the first U.S.-based space walk that involved no American astronauts. Courtesy NASA.

# 9

# We Deliver

Nothing that is worthwhile is ever easy.

—Indira Gandhi

"Hell yeah, let's send all four of us out" was Rick Hieb's sarcastic response to Bruce Melnick's audacious suggestion to send out three astronauts to capture the Intelsat VI satellite on the STS-49 mission following two earlier failed attempts. Never in the history of spaceflight had three astronauts, much less four, performed a joint space walk. STS-49's primary mission was to capture the Intelsat VI satellite that had been abandoned in low Earth orbit in 1990 and then boost it into a geosynchronous orbit as originally intended. The crew also planned to test the ability of spacewalking astronauts to perform basic construction techniques utilizing an engineering experiment known as Assembly of Station by EVA Methods (ASEM) in preparation for the planned space station *Freedom*, as well as devices that could be used to attempt to rescue an astronaut that somehow became detached from the station and was drifting off into the void of space to a slow and certain death.

There were four space suits but only three helmets on this maiden flight of *Endeavour*, so it would have been impossible for four astronauts to perform a space walk at the same time. Each of the three space walks planned for the mission comprised only two astronauts, as on all previous shuttle space walks, which required only two helmets. Fortunately, the crew had the foresight to bring along a third just in case, because after failed attempts to capture Intelsat on two space walks, to accomplish their primary objective—capture the satellite—the best plan anyone could come up with required three spacewalking astronauts.

A specially designed capture bar had been developed to allow Pierre Thuot to grab hold of the satellite while attached to the end of the Canadian-built robotic arm. Once Thuot had Intelsat secured, Melnick would then grapple it with the arm and re-berth it into the payload bay, where it would be attached to a launch device—a Perigee Kick Motor—which would later propel it out to a geostationary orbit of 22,300 miles.

Thuot and Hieb trained in the pool and on the Air Bearing Floor, where a full-mass simulator of the nine-thousand-pound satellite had been built to approximate the conditions they would encounter in orbit, or at least as best they knew. Hieb recalled that with the satellite resting on the Air Bearing Floor, they were supported with bungee cords "where forces were being imparted and then drive motors to make this spacecraft react the way the physics models say it should react [in orbit]." Thuot successfully captured the satellite a few times, and then Hieb snared it—as he recalled—on his first attempt, surprised that it was so easy:

> Put the capture bar up gently against the bottom of the spacecraft and let it slide around until the levers; you couldn't come in any time you wanted because there were a couple rocket nozzles sticking out the bottom of the satellite and you had to wait for those to pass. Then slide the capture bar in and then the little lever on the bar would get snapped by a couple of those tangs on the bottom of the spacecraft and that would fire and then the capture bar would do a soft dock. Then you get out a Pistol Grip Tool and put it on in the middle of the capture bar and drive the hard dock and then hold onto it while Bruce [Melnick] would reach out with the arm and grab the grapple fixture which is now attached. And it was easy, really. I was like I did that—I think—the very first time I tried it. I asked, "Gosh, are we simulating this right? I mean it seems too easy." The answer was, "Yes . . . we're simulating it correctly."

*Endeavour* caught up with the satellite on the fourth day of the flight. Thuot, on his first space walk of the mission, secured himself to the end of the robotic arm while Melnick slowly maneuvered him toward the satellite that was rotating at the relatively slow rate of less than one revolution per minute. Once he was in position, Thuot attempted to grab Intelsat using the capture bar, but to his dismay, he failed three times to secure it, and the first space walk was ended to rethink the plan. The next day Thout and Hieb ventured outside for another attempt to snag the wayward satellite, but after some minor adjustments to their plan and five more failed attempts, the dejected spacewalkers admitted defeat and returned to the safe confines of *Endeavour*.

Astronaut Ken Reightler had been assigned as the lead CAPCOM for STS-49 along with Sam Gemar and John Casper, and then assigned Gemar as the flight lead CAPCOM due to his extensive training and knowledge of EVA. It was common practice for the flight leads to be on console for most of the major events during a mission, which is why Gemar was on duty for the EVAS. Reightler, CAPCOM for ascent and entry, recalled that everyone in the Mission Control Center was pulling so hard for the crew to capture Intelsat that the entire room seemed to be exhibiting a collective body English to help Thuot snare it. Reightler recalled that after the first EVA failed, they began getting "a lot of help," and as lead CAPCOM, he began to spend more time in the control center to assess what additional support was required. A number of astronauts also began drifting in, some to offer help and others just curious. Over the next three or four days, Reightler left the control center only to catch a little sleep. Stress levels were running high. When they began having problems during the second EVA, he wanted as much astronaut perspective as he could muster and assigned CAPCOMs to the RMS and EVA consoles as well as next to the flight director. Each failure led to more solutions and corrections. Following the second EVA, they realized that they were going to have to come up with a different approach.

A "tiger team" was formed that immediately began looking at possible alternatives to grab hold of the satellite, and Bill Readdy was assigned as its official Astronaut Office representative. One piece of information they needed to know was where they were allowed to touch the satellite. The capture bar had been developed because the manufacturer would allow them to make contact with the satellite only at its base, but the team needed to know if there were other places where they might be able to grab it. According to Reightler, the plea to the manufacturer was, "How badly do you want your satellite fixed?"

> That is when we found out about the three jack screws that drove the concentric solar array up and down, that were arranged 120 degrees apart. One of the big issues that we worried about with the three person EVA was what would happen once they grabbed it. The satellite needed to spin to provide stabilization. We tried to calculate the force needed to stop the spin and what that would mean on the suits. When the spin stopped, the fuel inside would continue to spin for a while. We also tried to simulate the gyroscopic forces that would result and deter-

mine if they could be overcome and how long it would take for all to dissipate. It turned out that the estimates were accurate.

Twenty-eight years later, Hieb recalled that Thuot was likely applying less than four pounds of force on the satellite when he attempted to attach the capture bar, and based on their training on the Air Bearing Floor, that should have been acceptable. Surprisingly, that small force "was enough to cause the satellite to just torque out of the way—every time." Hieb admitted that

> the first day maybe he [Thuot] wasn't smooth. I don't know; people may have an argument about that one way or the other. The second day, I had the best perspective of anybody in the world. I was where I could really see; he was absolutely smooth—he did it absolutely perfectly. And if you listen to the tapes you'll hear me say in anticipation, "I do believe . . ." and I was about to say "got it" when it rotated [away]. It was perfect. He could not have done it any better. It just wasn't going to happen.

Hieb vividly remembered the night after the second unsuccessful space walk. He was responsible for collecting the exposed film onboard *Endeavour*, labeling each one, and then replenishing the film for the next day's activities. Inevitably he was usually up late with those chores. Pilot Kevin Chilton was also up late on the flight deck that evening, clearly frustrated at their inability to secure Intelsat, and approached Hieb to think about what else could be done to solve the capture conundrum. Hieb was tired and just as exasperated as everyone else, and he was tempted to inform Chilton: "Kevin, we have looked at this every possible way that we could pre-flight; there is nothing new under the sun, there are no new ways to do this, I need to go to bed, this is a waste of time." But he knew Chilton was a very smart guy and a team player, and ignoring his enthusiasm wasn't good teamwork: "You don't stomp on somebody's ideas—you work through them." The two of them began kibitzing about the problem while gazing out of *Endeavour*'s back windows.

Hieb rationalized the situation; they had to work with materials on board. Clearly, the retrieval of Intelsat held priority over the two planned space station construction EVAs, and these simple givens led to possibilities, so they began dreaming up wild ideas. He and Thuot had already assessed all gyrations of the two of them mastering the satellite, but there were simply too many obstacles—sharp edges, deployed antennas, and the rotation of Intelsat that complicated their ability to secure the satellite. "If you tried to grab it on the ends it would roll out of your hands. If two guys tried to grab it at the bottom it would tip one way or the other." Nevertheless, in the back of Hieb's mind, something told him to think out of the box. With their excitement growing, Chilton's and Hieb's voices became louder as they continued to assess the options, and the rest of the crew began drifting back up to the flight deck to join in the brainstorming session. That's when Melnick came up with the idea of sending out three astronauts to snag Intelsat, and it was at that moment that Hieb realized, "I knew we were going to get the satellite."

The night grew late as their creative juices flowed into idea after idea, although the ground kept telling them to go to bed, which was the last thing they were going to do. They turned off the closed-circuit television so the ground wouldn't know they were still awake and continued the session for another hour or two. They considered building a structure using the ASEM materials—one astronaut attached to one end and another at the other side of the structure, then grab both ends of the satellite. Eventually they settled on using the bottom portion of the ASEM structure but were convinced that three spacewalkers were necessary to catch the satellite.

They put their plan together, complete with an annotated checklist page that they could send to the ground with a special new camera on board, comparable to a modern-day PowerPoint presentation. Commander Brandenstein then instructed Chilton to call the ground and pitch their plan, and he insisted that Chilton not stop talking until the entire scheme had been laid out. Brandenstein finished up the presentation by encouraging the ground to strongly consider their proposal; it may have helped that he was still head of the Astronaut Office. The die had been cast, and it was time to go to bed and let the ground mull over their recommendation.

Although the plan was modified by Mission Control—Reightler shared that the ground was also contemplating a three-person EVA—two of the three spacewalkers would be strategically positioned on the base of the ASEM platform assembled in the payload bay and one on the robotic arm which would allow them to grapple the satellite by hand. Brandenstein and Chilton would have to fly the orbiter close enough to Intelsat for the three astronauts to have a chance to capture the satellite.

Kieth Johnson, flight controller and instructor at JSC, recalled that initially, everybody thought the problem was their inability to control Intelsat.

The ground had remote but limited command of the satellite and had slowed its rotation rate significantly prior to the first attempt to grasp it. On the second space walk, they planned to modify how they approached the satellite and attempt to stabilize it, to no avail. Johnson was part of the suits group and was quite concerned about what the astronauts were going to grab hold of to secure the satellite on the third EVA. Were there sharp edges or other protuberances that might damage their suits? And how long would they be required to stay outside?

Story Musgrave had already been assigned as an EVA crewperson for the upcoming STS-61 mission to repair the Hubble Space Telescope, so he often spent time in mission control as a spectator, especially during space walks; he wanted to learn as much about spacewalking as he could, and watching others perform EVAs was an effective learning process for him. He was already an experienced spacewalker, having performed the first shuttle EVA along with Don Peterson on the STS-6 mission in 1983, plus he had extensive knowledge of the suit having been heavily involved in its development. But he knew he could always learn more. When Thuot and Hieb failed to grasp Intelsat on the second space walk, Musgrave, trying to stay below the radar screen, remembered that most everyone in Mission Control slowly turned in his direction, and Musgrave, as best as he could remember, believes it was Randy Stone (mission operations director at that time) who told him, "Story. . . . Fix it."

"Okay! I will need my suit! As you know we don't have custom suits. They put pieces—legs and arms and all different things together to form your suit." Musgrave knew the first step was to arrange for a suit to be assembled that would fit him; this would take some time. When he called to begin the process, he was told, "Story, your suit has been ready for twenty-four hours." They had been expecting him.

On the way to Building 9, full of mock-up hardware where astronauts train, Musgrave's mind was focused on what had gone wrong. Upon arrival, he was greeted by a man wearing a red suit who grabbed him and pinned him to the wall—completely surprising Musgrave—insisting that he knew what the problem was. "Oh my God, this little man in the red suit he had the whole answer. Mission Control didn't have it, I didn't have it, but he had it. He had the answer."

The floor had misled everyone except this fellow: it didn't have the fidelity required to accurately duplicate the conditions of a weightless environment.

Following the mission, Hieb was informed that there was about one pound of friction generated on the floor for every thousand pounds of mass, so for the nine-thousand-pound satellite, there were about nine pounds of friction. Musgrave illuminated what he was told by the red-suited man: "You can bang it [the training Intelsat] in the face all day long and it will go nowhere." This gentleman had requested money to build a precision Air Bearing Floor, one with less friction than the one they had, but his request had been denied. Musgrave rationalized the mindset: "What the hell; satellite not going anywhere; I'll be able to get the bar on. So, they designed the tolerances of grabbing that satellite very close because Pierre . . . is going to put the capture bar on [and] will have total control."

Kieth Johnson remembered the event as if it were yesterday:

> So, we listed out all the challenges that we had to face in preparation for sending three crew members out the door. We checked in the WETF to see if we could get three suited crewmembers in the airlock. The tasks people went off and figured out where we could put crew members that would be best to grab this thing by hand. We went out and looked at the mockup to figure out what they could hold onto, and then clasp on that grapple fixture. It was a huge challenge; we knew that they were going to be out there for a long time.

CAPCOM Reightler assessed how the three spacewalkers would communicate at the same time—not an easy task. "We did simulate the comm on the ground with three EMUs and felt it would work. We also did talk about hand and voice signals the crew could use. The only issues we saw were if two people tried to talk at the same time. We emphasized that potential and to keep voice comm minimal."

Along with Musgrave, astronauts Jim Voss and Rich Clifford were sent into the WETF to test all facets of the new strategy. Although slightly different from the one the crew had devised in orbit, it was a good plan and the crew was eager to put it into action. Reightler was confident and described the atmosphere in the Mission Control Center as one of growing optimism. Despite the need for close proximity of the satellite to the three astronauts in the payload bay, he felt the plan was a cautious approach and that the pilots could quickly and easily back away from the satellite if they had to. "I had total confidence in Dan's flying, Bruce's arm work and the EVA team."

Randy Stone informed Musgrave that he trusted him and the plan, but the dynamics people were not convinced it could be accomplished safely. Musgrave remembers Stone sending him to a meeting to defend the scheme: "I went in there and they said, 'You're going to kill people. There is no damned way people is [*sic*] going to grab nine thousand pounds.'" Their primary concern was how close *Endeavour* would have to maneuver up to Intelsat to accommodate three astronauts in the proper position to grab it. But Musgrave was convinced otherwise, all due to what he referred to as "tool chases." He had seen two of them, one being the errant foot restraint on 41B but had never done one himself. "When you go in the toolbox," Musgrave explained, "it's [the tool] tethered to the box. You put your tether to the tool and then release the tether from the box. Tool discipline; everything is always tethered so it won't float off." Spacewalkers have to be aware of myriad gremlins that can raise their ugly heads during a space walk, and despite their training, sometimes the tool-tethering protocol takes a back seat, and tools escape. "The pilots inside they want to have some fun," Musgrave chuckled. "Now, you see your tool floating away and all of a sudden, the tool stops and the tool starts coming back to you. And the tool ain't flying; the shuttle is flying you to the tool! This one fella, he just put his hand down—he's looking in the window at the pilot and says, 'You guys think you're so damned good, try this.' They put the tool right in his hand."

By this time in the shuttle program, the pilots had become extremely adept at flying the orbiters in close proximity to other objects, explained Musgrave, so he had no problem with sneaking up on a nine-thousand-pound satellite within inches: "You see it's the tool chases that gave me the confidence that we could come up with three people going out and grabbing it. And if you're in trouble you [the pilots] just slam the handle down—the big jets push you away. It's a stupid maneuver but it's an escape maneuver. We came up with that and darned if it didn't work."

Thuot, Hieb, and Akers began the third EVA by building the bottom plane of the ASEM structure within the payload bay. Portable Foot Restraints (PFRs) were then attached to two poles reaching several feet above the bottom of the ASEM base, and Akers and Hieb were stationed on these PFRs, Akers in the center of the payload bay and Hieb along the starboard side. Thuot was then translated on the robotic arm opposite Akers and Hieb to form a triangle, which would allow each of them to grab a gear box or motor on Intelsat all at the same time.

A slow dance ensued between the three spacewalkers and the satellite as Brandenstein and Chilton delicately flew the orbiter into perfect position—within mere inches of Intelsat—in preparation for the three spacewalkers to grab hold of the slightly gyrating orphan. As the satellite came within reach, Hieb finally pronounced, "Let's do it." It had taken several attempts, but with precise communication, often with hand signals due to the limited voice communications, the three spacewalkers captured Intelsat! "Got it!" Someone then cautioned, "Easy, easy." Brandenstein confirmed to the ground, "Houston I think we've got a satellite." Thunderous applause erupted from the ground, prompting Brandenstein to caution them, "It ain't over yet." They still had to berth Intelsat inside the payload bay. Tom Akers was experiencing some discomfort: "I'm really in a bind here holding my knees down . . . but I can do it for a while." Fortunately, by lowering Thuot on the arm, they were eventually able to position the satellite to accommodate everyone until the capture bar could be attached. Teamwork among everyone involved in capturing Intelsat in orbit coupled with a tremendous knowledge base from those on the ground led to success that day.

Once EVA capability had been demonstrated on STS-6, an increasing number of space walks began to creep into the shuttle mission planning. The space shuttle was a highly versatile spacecraft designed specifically for work in LEO, and NASA desired to squeeze every ounce of potential as was possible out of their futuristic space planes. It took years for the EVA capability of the shuttle to be realized, but once established, it surpassed all previous space programs. But progress came slowly. Only thirteen space walks—one unplanned—were featured on the first twenty-five shuttle missions. By this time, the orbiter pilots had demonstrated their mastery over the machine with the most delicate and precise flying in orbit. They could snuggle up to within inches of satellites, chase down and rescue a spacewalker if needed, or track down a "dropped" tool that a spacewalker inadvertently lost in space.

Over two and a half years passed following the loss of *Challenger* in January 1986 before flights were resumed, and there were no EVAs carried out on the shuttle in the next two and a half years following the return to flight on STS-26 in 1988.

Jerry Ross and Jay Apt returned NASA to the EVA game in 1991 with two highly successful space walks during the STS-37 mission. Still, spacewalking from the orbiters had yet to hit its stride. During the next seven and a half years—prior to the beginning of ISS assembly—there were only thirty-one space walks undertaken, but NASA took full advantage of these spectacular forays outside the orbiters and built an impressive résumé of spacewalking experience. Tool evaluations were conducted in preparation for a plethora of Hubble Space Telescope servicing missions. Modifications to the EMU were tested, hardware and tools were evaluated in anticipation of construction of a space station, and large-object handling techniques were practiced. The dividends were huge. NASA also learned much about performing free-floating EVAs; some part of the body needs to be securely attached to some element of the spacecraft to provide stability. If not a foot restraint, a stable handrail is a must. Otherwise, executing tasks are exceedingly difficult if not impossible.

Most shuttle astronauts downplay the risks they accepted leaving the orbiters and climbing into the deep coldness of weightless outer space. Long-duration space walks—some lasting over eight hours—taxed many physically and presented challenges that required creative responses to unforeseen problems. The STS-49 mission is unique for being the only time three spacewalkers ventured outside of a spacecraft, but it is representative of the teamwork and creativity that allowed NASA to overcome the unknowns associated with most EVAs. With over 180 space walks from the shuttle during its life, stories abound about the EVA exploits by scores of spacewalkers—disorientation, sweat, bruises, difficulty, risk, thrill, and the unexpected—exciting accounts of the likes that may never be encountered again in the future of space exploration.

The minds of some astronauts play tricks on them during their space walks, including periods of perplexing disorientation, the sense of falling, or a flipped frame of reference.

Jeff Hoffman was finishing up his fourth spacewalk—and the last one of the STS-61 Hubble servicing mission—when he had a few spare moments to experiment while free-floating at the top of the telescope. His tether was not set to retract automatically, so it was floating loose and not tugging on him. Holding on with one hand, he had the impulse to feel the sensation of floating freely alongside Hubble, so he decided to let go with the hand that was providing him a sense of security:

> I'm holding on with one hand and I thought, alright I'll let go. Then I realized I hadn't let go. Well, come on; let go. It was just something that deep in my consciousness that didn't want to let go. I finally said, "Come on stupid, let go." And it was incredible because psychologically I disconnected from the shuttle; it wasn't just physically. It was an amazing feeling, particularly when I turned my back so I couldn't see the shuttle. It was like I was just alone there and it was—an incredible moment. I'll never forget that. But it was hard letting go.

Winston Scott colorfully described an unexpected turn of events on his STS-87 mission when he ventured outside along with Takao Doi to capture the SPARTAN satellite that had malfunctioned shortly after the crew had released it from *Columbia*'s payload bay. A highly experienced naval aviator, he suffered one of the worst cases of vertigo of his career. "I can remember Kevin [Kregel] needed to rotate the shuttle port [left]," Scott explained, "and the shuttle begins to rotate out of the corner of my eye and I can see the Earth tilt. When you see the Earth tilting, you get the sensation of falling. Instinctively I tried to right myself. I'm attached to the shuttle, there's no way I can right myself. I can remember thinking; my God this is terrible! This is very difficult—serious vertigo here." The sensation was so real and disturbing that he wasn't certain how to proceed.

Played out over mere seconds, the solution was to rely on his years of training as a jet pilot, which told him to focus and trust his instruments. He was confident that he could remain calm. Assessing the situation further, he realized, "I don't have any instruments up here. [But] SPARTAN is stable, SPARTAN is not going anywhere. I will focus all my attention on SPARTAN, and mentally I tuned out what the earth was doing in my peripheral vision." His remedy saved the day, and he and Doi successfully captured SPARTAN and returned it to the payload bay to be returned to Earth for repair.

Kathy Thornton never felt like she was falling on any of her three space walks, but she sensed some slight disorientation while outside in the payload bay of *Endeavour* during her STS-49 mission. She remembered that while fellow spacewalker Tom Akers was stowing the K band antenna, she was hanging out nearby waiting for him to complete his task. "I think I was looking over—up on the forward bulkhead, maybe looking over the nose or the side of the orbiter, and I instantly had this sensation; I didn't know where I was.

Nothing here looks familiar—where am I?" Like Scott, it took her only seconds to manage the situation with no impact on her completing her assigned tasks.

Rick Hieb had to contend with his frame of reference when he was working in the payload bay during the STS-49 mission. He was occupied with the ASEM support structure—situated down in the bottom of the payload bay—which formed a forty-five-degree angle from the side of the bay to the center of the bay bottom. Focused on ASEM, it became his internal frame of reference. "When I turned to do whatever I was going to do next I couldn't figure out where the hell I was. My whole world had gone forty-five degrees out of kilter on me. Yeah that's definitely a real thing."

While attached to one of the ISS trusses on her STS-108 flight, Linda Godwin experienced a slight disorientation. She felt comfortable in the shuttle payload bay, but once on the truss—with limited view—she recalled being slightly unsure of where she was on several occasions, although she doesn't remember ever going in the wrong direction. She found it very helpful to look at renderings of the entire ISS and visualize the layout before heading outside. "But it's true once you get out on the truss there is a little bit of a limited view with the helmet and you're thinking—this direction, that direction."

To Godwin's surprise, she was thrown a curve ball on her STS-108 EVA while working on the ISS solar arrays. The main solar arrays have mechanisms that allow them to rotate, ensuring maximum capture of solar energy essential to power the station. Godwin and Daniel Tani were tasked with adding insulation—to manage the extremes of heat and cold—to the two devices located on the very end of the port side of the ISS. Godwin explained the layout of the structures in the NBL:

> One had been in an upright work position and then the one below it—to duplicate what we did on the top one—we had to turn upside down. I was sure that was an issue only on the ground. We got up there and when I went to do the one in the water tank that had been upside down—of course I had to go 180 and change body position and point down. But I was in space, right? I did feel upside down the whole darn time and I could not shake it. I don't understand it and I'm kind of annoyed because I know it was in my head. I would like to have felt right side up. I was kind of relieved when it was over and I flipped back right side up. Our brain is the final frontier I guess—I don't know how to explain it.

John Herrington remembers his first foray into open space on his STS-113 mission to the ISS. Herrington explained that it's more difficult to start your movement in the water during pool training than in space, but it's much easier to stop. It's the exact opposite in space—easier to begin moving, but more difficult to stop. Therefore, the first-time spacewalker had been assigned a translation adaptation exercise—customary at this time in the program—which gave him time to adapt to the new environment prior to beginning work. Herrington planned to climb around the airlock and translate up to the Z1 truss to a toolbox where he was to grab a tool. The Z1 is the massive central truss that would anchor the station's backbone as the complex was expanded. Departing the crew lock of the ISS, he climbed over to the Z1 Truss where the toolbox was located. Using an "ice cream scoop" tool that allows astronauts to grab hold of irregularly shaped objects, while just coming out of a night pass, he grabbed the toolbox and was startled when it shook in his hands—it was loose! No one had informed him that the fitting the box is mounted to was designed to accommodate the extreme changes in temperature. "I grabbed it when it was cold and it goes 'dugga dugga dugga,'" he shared, laughing. "I thought it was going to come off in my hand, [as] I floated backwards." Faithfully tethered to the ISS there was no way he could have floated away into the deepness of space, but when the unexpected occurs during the beginning minutes of your very first EVA, it can be momentarily disconcerting.

Herrington remembered another event where his mind played tricks on him. Climbing from the airlock to the front of the station on the Crew and Equipment Translation Aid, which holds carts that spacewalkers use to transport equipment, Herrington was looking up, acutely focused on his hands. Suddenly, he was looking down at his hands. "My mind told me I was looking the other way. That was a shocker. Because I was so focused on just looking at my hands, I lost perspective of what was around me. And that's when your mind goes 'OK blip' and flips you upside down. Then you look around and no, I'm in the same spot; my mind doesn't think I am. Once you realize you're in the same spot your mind kind of flips you back."

These harmless phenomena affect some spacewalkers, whereas others seem to be immune from such incidents, emphasizing the unknowns that astronauts routinely encounter during EVAs. Regardless, astronauts are well trained professionals who are able to work through such incidents and accomplish the job.

Most space walks are complex and hence challenging, some more so than others. In general, tall bodies, long arms, and large strong hands make EVA easier. Unfortunately, not all astronauts are blessed with these physical attributes, and being height challenged with short arms seems to make EVA more difficult for some. Additionally, not all space walks play out as planned—surprises are the norm rather than the exception.

Joe Tanner clarified that it's not easy to describe the most difficult part of working outside; what one spacewalker finds challenging might be a simple task for another astronaut—everyone is different. He believes it's essential to focus and concentrate on the job at hand to squeeze in as much work as possible in as short amount of time as you can, all while expending minimal physical effort. Conserve energy and preserve your strength. Minimize the amount of oxygen you use and don't overtax the carbon dioxide scrubbing capability of the suit. Maintain a steady work pace and do not rush. "We used to say 'slower is faster' because making a mistake by rushing through a task required more time to fix it than it would have to slow down and do the task right the first time."

Tom Akers found himself on the short end of the stick during the three-person EVA on STS-49. There were no plans on how to send three people outside for a spacewalk—initially, it wasn't certain that three people could fit inside the shuttle airlock. Was there enough room? There were only two umbilicals inside the airlock, meaning one of the three astronauts was the odd man out, and that was Akers. Without an umbilical, Akers had no external suit cooling capability, and he was also on internal battery power, so his suit was being depleted of power and other consumables with no way to replace them. The ground decided that Hieb's suit should remain pristine because in addition to training for his tasks, he was Pierre Thuot's backup, and Akers was his backup. So Hieb was trained for any job. All this meant that Hieb was not to share any of his consumables with Akers, but Thuot was allowed to. Chortling, Hieb recalled that Thuot was hesitant to give up any of his umbilical time to Akers because that meant he was shortening his own EVA time. "So, Tom was the guy—he was probably sweating like a dog—but he was such a good guy he wouldn't complain about it. But at some point, he would [ask,] 'Pierre can I get a shot of cold water?'"

Winston Scott remembers that everything about capturing the wayward SPARTAN satellite by hand on STS-87 was difficult, and it was the unknowns that worried him the most, and rest assured, gremlins would raise their ugly heads once he and Takao Doi ventured outside. The satellite was slowly spinning, and he and Doi couldn't be certain of its orientation or that its motion would dampen enough for them to grab hold of it. Nor were they sure where to grasp it; sharp edges are always a concern. Working with the ground, they had to address these issues before they tackled the three-thousand-pound colossus. They had practiced a few simulations on the Air Bearing Floor but had not handled a mass that size in zero gravity. Additionally, this was Doi's first space walk, and neither he nor Scott had trained specifically for this unscheduled EVA. Outside and in position to assess SPARTAN's condition more closely, Kevin Kregel flew the orbiter into close proximity to SPARTAN while Scott gave him directions as it slowly drifted into various orientations for some unknown reason. As EV-1, it was Scott's call on when and how to secure the satellite. He and Kregel had to be on the same page so that Kregel knew that when Scott instructed him to move upward, he intuitively knew which direction Scott meant—upward relative to what? Scott illuminated: "We had all kinds of things going wrong that we had not anticipated." Ultimately, the two spacewalkers captured SPARTAN, but it took over two hours to finally move into the best position to make it all happen, during which time Scott had been mulling over how they would capture the satellite. But it was premature to announce success; SPARTAN had to be stowed inside the payload bay where Scott encountered another problem: an ingress aid had to be removed before the satellite could be docked, and it required much more force to loosen than expected; Doi had to assist him. Post-flight, it was discovered that the aid had not been properly lubricated.

Often it's the little unexpected occurrences that complicate space walks. Joe Tanner shared that prior to his missions, he was often asked what he was concerned about. "My answer was always the things I hadn't worried about because I had a plan of action in mind—or in the procedures for everything we had thought about." In spite of all the planning, including contingency plans, Tanner recalled that unforeseen problems were common.

Tanner and Carlos Noriega encountered problems on the STS-97 mission with one of the many tasks they were charged with—installation of the first section of solar arrays to the ISS. While deploying one of the solar panels, the solar array tensioning cable became dislodged from the reels. Immediately, the ground went to work to develop a solution for the two astronauts to execute. Tanner wasn't particularly distraught: "Carlos and I got an extra EVA out of that unexpected problem."

A bolt that Tanner was removing from a launch restraint cover capture feature failed on his STS-115 mission to the ISS. Unfortunately, he could not capture the bolt along with its spring, and they both absconded into space. He also recalled that Daniel Burbank and Steven MacLean battled a stubborn bolt—on the same mission—that secured a solar alpha rotary joint launch lock. If the bolt could not be removed, the solar array would not be able to track the sun and provide the needed power for the ISS to function at full capacity. Fortunately, strong muscles and tenacity prevailed as they successfully removed the bolt, but it had taken over an hour and a half to accomplish what should have taken minutes.

Space station *Freedom* had been approved by President Ronald Reagan and announced in his 1984 State of the Union address. The project was wrought with overly optimistic cost estimates, and ultimately, the U.S. Congress was unwilling to fund it. NASA still held optimism for approval of *Freedom* in 1992 and planned two space walks on STS-49 to test basic construction techniques in support of *Freedom*, including assembly and maintenance capabilities, with the ASEM experiment. The astronauts planned to build a truss structure in the payload bay, evaluate their ability to maneuver large mass objects with the Multiple Purpose Experiment Support Structure (MPESS) pallet, and then dissemble the truss structure on the final EVA of the mission. Assembly of the truss would be facilitated by a newly designed truss joint, which could be locked very easily by simply rotating a collar. Due to the two unplanned EVAs to capture Intelsat, only one ASEM space walk was carried out—by Tom Akers and Kathy Thornton.

Thornton was disappointed that she and Akers failed to complete all their assignments on that EVA, but she may have been too hard on herself—after all they were attempting to cram two space walks into one. Additionally, she felt they had rushed some of their training exercises in the pool so they could accomplish all their tasks.

> It was difficult. There were some demonstrations as part of our EVA we could do if we got to them and we really wanted to do them. So, in the water tank we would work just as hard as we could to get through all the tasks that we had to do so that we could do these things that we wanted

> to do. And that was a really bad idea. It was real exercise for me—I was really huffing and puffing—and you really don't want to do that. You want to go as slow as you possibly can and then slow it down because that's closer to what you will have to do in orbit. So, I think we set up an unrealistic timeline because of that.

Another factor, the suits they wore while training in the pool, likely led her and Akers to underestimate how difficult it would be to build the truss structure. Thornton compared the EMUs used in training to "an old worn-out tennis shoe" that allowed her to carry out tasks in the pool that were much more difficult in orbit in the less flexible suits used on the actual space walk. Thornton frankly admitted, "We had this long strut to move around. If you're restrained you can take the thing—hold it by one end—and you can swing it around; no problem. But if you look at the moment [of inertia] of that strut relative to the moment of inertia of your body—to move that thing 90 degrees to the left your body has to turn 180 to the right—which you can't do [laughs]. So that sort of thing is basic physics, but it was holy moly." Back inside the orbiter, after having fought the less flexible suit for seven hours and forty-five minutes, she felt like she had done two hundred crunches. "My abs were really sore because I tried so hard to make that suit bend like it had done in the water tank."

Rick Hieb recalled a lesson learned that was well documented during the mission debriefing. "If the task was going to be done from a foot restraint, then the water tank was pretty good training for that. But if the task was going to be done free floating, then the water tank really was not good training. Tom and Kathy particularly experienced that. The tasks ran super long and . . . they were sweating and working—they were both just exhausted." Thornton agreed with Hieb in 2020: "What he said is absolutely true; . . . you have to work really hard to restrain yourself in some way if you're not in a foot restraint. And that's where the difference between the buoyancy of the water and gravity in the water tank and the real thing comes in. You don't have the viscosity of the water to help hold you in place—it's hard." This lesson learned led to an experienced spacewalker being assigned to oversee the training of future crews.

Jerry Ross and Jay Apt made an unscheduled space walk on STS-37 to free a stuck antenna on the Gamma Ray Observatory (GRO), a satellite designed to make astronomical observations in low Earth orbit. Linda Godwin deployed the satellite from the payload bay of *Atlantis* using the robotic arm, but the high-gain antenna failed to deploy. Ground controllers at Goddard Space Flight Center relayed repeated commands to the GRO and the crew made multiple attempts with the robotic arm to dislodge the antenna, all to no avail. Ross and Apt were then sent outside to save the day.

With his left hand holding onto a flight support structure trunnion of the GRO, Ross estimated that he initially applied about thirty to forty pounds of force on the antenna to free it up, with no luck. After about a dozen times pushing and pulling it back and forth, it finally came free. Over twenty-five years after the mission, Ross gave a talk at a meeting—chaired, as it happens, by his daughter Amy, a NASA employee—and an engineer who had worked on the GRO in attendance informed Ross why the antenna failed to deploy as planned. "The technicians had installed the nuts and bolts that were to hold the insulation on to the antenna boom . . . in the wrong direction, so that the threads of the exposed bolts were actually keeping the antenna from deploying." Fortunately, Ross and Apt were able to overcome the oversight. Without EVA capability and astronauts well versed in EVA procedures, the GRO would have required later servicing to bring it to life.

Rick Hieb was very pleased with the EVA training he received for his missions. But there are some aspects of walking in space that cannot be simulated accurately on Earth in a gravity field. His total weight on Earth in the suit was about five hundred pounds. In orbit, he tested his mobility on his initial EVA and intentionally progressed very slowly; he may have been weightless but he was acutely aware that he still had mass. With experience, he felt comfortable moving just a little faster and then even faster. "Then I got going too fast and damn near lost my grip when I wanted to stop because I wasn't yet ready for that inertia, and the handhold that I was using wasn't really a true handhold, so I didn't get a really good grip on it. Then I slowed down again of course to regain that comfort level." Training is essential for a successful space walk, but Hieb needed the weightless environment of orbital space to gauge his movements.

John Herrington was proud of his quick thinking to resolve a problem he was having on one of his three space walks on the STS-113 mission, one that had begun the night before. Experiencing a massive headache that evening, he inventoried the incorrect tool when preparing for the next day's tasks. He was at the worksite before he recognized the oversight, located on the end of the P1 truss of the ISS where his job was to remove a dummy plate so that an antenna could be installed. The job required the use of an ice cream scoop tool, an extension, and a power tool. Herrington planned to insert the extension through a hole in the scoop and use the power tool to remove the nut and bolt. Unfortunately, the extension he packed was too large in diameter to fit through the hole of the scoop, and initially, it appeared that he was up the creek without a paddle. It was a long trek back to the airlock that would have eaten up a significant part of precious EVA time. He felt terrible and embarrassed, because his EVA partner, Michael López-Alegría, was already preparing to head back for the proper extension when Herrington thought to himself, "Given what I have, how can I solve my own problem?" Suddenly, he realized that the way he had been trained to do the task was not the only way it could be accomplished. Logic dictated to him that he should simply remove the problem, so he removed the scoop and used the tools he had with him—which worked to perfection. "When stuff didn't go as planned, I was able to overcome it, one way or the other. You don't like making mistakes and I did and I fixed it on my own without any help. It's like working on a car. You're underneath the car and only have a certain amount of tools and you figure out what you gotta do with what you have. And that is what I'm used to doing. It paid out big dividends."

Linda Godwin agreed with Herrington's approach to conquering the unexpected. "I tried to be smarter; I told myself sometimes you tried to reach something and it's not working, you gotta step back and say, 'Okay, you keep trying the same thing every time and this is not getting any better—you gotta think of another way to do it.'"

Herrington's resourcefulness bailed the NASA planners out of another potential failure. He had been trained to access thermal panels on the back side of the ISS so that he could correct a problem with a series of improperly designed ammonia connectors, about thirty of them as best as he could recall. If the connectors leaked ammonia, they may have become locked in place and rendered inoperable. His job was to correct the problem by adding small pieces of metal—with his hulking gloves—to every single ammonia connector. The plan called for Herrington to be translated to the work site while attached to the robotic arm, but due to a stalled rail car on the Mobile Transporter, the arm could not reach that location. Herrington explained that the work site had not been designed for a spacewalker not secured to the robotic arm—there were no hand holds, which meant he had to search for places to hang on to and tether to. Expediently, he developed an alternative method and proudly boasted, "I did every single one of them in less time than it would have taken to do them with the robotic arm."

The Hubble Space Telescope, NASA's pride and joy and one of the most iconic accomplishments of the space shuttle program, quickly became an embarrassment following its deployment into orbit in April 1990. It was soon discovered that its primary mirror was flawed, and until a fix could be devised and implemented, the images it returned from the deepest realms of the universe would be far less clear than anticipated. The mirror had been fabricated with an aberration two-fiftieths the size of a human hair, all due to equipment that had been calibrated incorrectly. NASA decided that replacing the mirror was not pragmatic and instead designed new instruments to install on Hubble that would correct the defective lens. During the STS-61 mission flown in 1993, spacewalking astronauts installed the Wide Field and Planetary Camera 2 (WFPC2) and the Corrective Optics Space Telescope Axial Replacement (COSTAR), and Hubble was suddenly delivering images with the high resolution expected. Including STS-61, sixteen spacewalkers visited Hubble five times over sixteen years not only to repair the telescope but to upgrade it with the latest technology. An astonishing twenty-four space walks totaling over 180 hours were successfully performed. Among many improvements, spacewalking astronauts added infrared capability so far distant galaxies could be observed as well as the means to capture more detailed images. None of this technology was available in the late 1970s when Hubble was designed. Failed gyroscopes were replaced on several missions so that the eye high in the sky could continue to probe deeper into the vast expanses of our universe. In addition to applying new pieces of technological wizardry to make Hubble better, they also mended equipment that was never intended to be repaired. Solar panels and batteries were also replaced, and thermal insulation was added to further protect the delicate instrument.

Many were concerned with the scope of the STS-61 servicing mission; it was complex, and there was much to accomplish. Jeff Hoffman remembered that during one of the EVA training sessions at the Goddard Space Flight Center, they were instructed to report to NASA headquarters in nearby Washington

DC to meet with NASA administrator Dan Goldin. Hoffman clearly recalls the pressure-packed message that Goldin delivered to the crew: "I hope you realize that the future of NASA's human spaceflight program depends on the success of your mission." Not surprisingly, the crew was fully aware of the importance and risk involved. The success of STS-61 proved that the telescope could be repaired and upgraded, but it would require creative thinking and hard work by those on the ground as well as the astronauts in space to thwart a glut of challenges, many of them unexpected.

By the end of the final servicing mission in 2009, Hubble had become one hundred times more powerful than when it first reached orbit in 1990. And without spacewalkers making repairs and updates as new technologies came to bear, it would never have come close to being the dream maker that it has become. There were numerous "uh oh" moments on every mission that had the potential to render the entire telescope—or some of its instruments—inoperable.

Jeff Hoffman and Story Musgrave were charged with changing out a series of failed gyroscopes on the first Hubble servicing mission in 1993. The two veteran spacewalkers had devised a plan while training in the pool that promised to save precious time. Not surprisingly, all went well as the gyroscopes were successfully replaced. Leaving Hoffman alone to close the two large doors to the telescope gyroscope compartment, a simple task they had accomplished many times during simulations, Musgrave translated to another work site. Each door has several latches that are secured with a large handle. Likely due to thermal stresses as *Endeavour* and the HST passed in and out of day and night passes, Hoffman had difficulty securely latching the doors. He'd fasten the top of the door, only to discover that the bottom wasn't secured. He tried again, and again, but the stubborn doors resisted his every attempt. And this was only the first EVA of five planned to correct Hubble's faulty lens and make other repairs; hopefully the doors were not an omen of what was going to happen on the following space walks. Hoffman was up against a wall: "Now I realized we've got a real problem; the door is warped and it's not going to close properly. Of course, if the door doesn't close properly, we lose the telescope for both thermal control reasons and light leaking in. There were no contingency plans for the door because nobody had anticipated that it wouldn't close. We had closed it dozens of times in the water and no problems. It was a big shock." It was time to call Musgrave back over to troubleshoot this pesky problem. Hoffman explained:

> I just had to do a lot of experiments and I finally figured out what the problem was and realized that we were going to have to push both at the top and bottom simultaneously. Then we'd be able to get it closed. That's when I called Story over; of course, at that time we had not set up a foot restraint and so Story was trying to do it free floating. It took one hand to hold on and stabilize himself so he could push the door closed with his other hand, and with his third hand flip that bolt. Unfortunately, he only had two hands—we kept trying to explain to the ground what the problem was and they clearly didn't quite get it because they kept sending up procedures—try this, try that—which I knew wasn't going to work, but we tried them anyway.

One plan the ground sent up had Musgrave and Hoffman deeply concerned that they might damage the door, hence they worked slowly and carefully not to make the problem worse. Ultimately Hoffman and Musgrave recommended using the payload restraint device—a contingency tool that could apply up to two thousand pounds of force—which according to Hoffman "is a kind of a webbing tool with a ratchet," similar to what movers use to secure furniture when they are moving it. The ground wasn't particularly enamored with this idea. "I think there were people who were worried about collapsing the telescope like when you squeeze an aluminum beer can." Convinced there was a better way, the ground kept sending up more directions—none of which worked. Hoffman and Musgrave held to their guns on using the payload restraint device, and following some lively discussion with the ground, lead flight director Milt Heflin acquiesced, and it worked. It had been a long seven-hour, fifty-four-minute EVA, but Hoffman and Musgrave had bested the obstinate doors allowing Hubble to continue on with its mission.

Kathy Thornton and Tom Akers had the opposite experience when they installed COSTAR on the same mission. She remembered that during training in the pool it would take her and three divers to install COSTAR. In the high-fidelity simulator at Goddard, located inside a large clean room, the first time she and Akers attempted to install COSTAR, they were unsuccessful.

> It turns out there was a bolt head in the way. I think every satellite to that point that we had attempted to repair on orbit there was something different about it from the way we thought it was going to be. There

> was always some kind of issue. I would have bet you a hundred dollars that thing was not going in there—and it just slipped in so slick. I was stunned! I was so happy! Everything just went according to plan. It was nice after we got back in to hear that it was getting data.

Similarly, when COSTAR was replaced with the Cosmic Origins Spectrograph on STS-125—which weighed eight hundred pounds on Earth—it had never come out easily during training in the pool. But in space, it slid out effortlessly, and the new instrument glided right in.

Another potential Hubble showstopper also cropped up on the STS-125 mission. The Space Telescope Imaging Spectrograph (STIS) instrument had failed and needed to be repaired. Mike Massimino and Michael Good had practiced religiously in the pool on everything that anyone could fathom might go wrong. To access STIS, a handrail had to be removed, which required removing four screws, two on top of the handrail and two on the bottom. Murphy's law dictated that the top two and the bottom left came out without a whistle. But when Massimino attempted to remove the lower right bolt with the Pistol Grip Tool, it spun loosely round and round—the bolt was not coming out.

Back on the ground, engineers began brainstorming solutions, and with failure after failure, they began to think about how they would approach a similar problem while working in their garage on the ground. Next, they contacted Goddard and explained the problem. Was brute force an option in space? Fortunately, Goddard had a mockup of the STIS with the handrail attached. In typical engineering fashion, they rigged up a test to see if they could rip off the handle, and if so, was the breaking force low enough for Massimino to repeat the exercise in orbit. An engineer looped a wire underneath the handrail, hooked the ends of the wire to a digital fish scale, and pulled as hard as he could. When the scale reached sixty pounds of force, the handle snapped loose, and it and the bolt careened away. Success! Massimino, a tall, husky astronaut, could surely muster up sixty pounds of muscle. But what if it broke off and there was debris? What if there was a sharp edge? What if the handle careened back and hit him? Fortunately, Massimino needed no wire to pull on the handrail—he could grasp it securely with his gloved hand. He also applied tape to the handrail to help control debris. Massimino steadied himself to ensure that his body would remain in place when he applied force, and he gave it his best. When the handrail snapped, everyone on the ground erupted in joy. There was no floating debris or sharp edges. Another disaster was dispatched and the rest of the repair on STIS went as planned.

Joe Tanner contributed two very successful space walks on STS-82 to make Hubble better, and like most astronauts, his biggest fear was making a mistake: "I got assigned to the Hubble crew and we were worried about doing something wrong on the Hubble and destroying this world scientific asset, and we didn't realize how stressed we were for about a year and a half, until we actually let go of Hubble and we fired away from it—and got our separation and the tension was gone, because we'd passed all of its start-up alignment checks and stuff like that."

Once the final and fifth EVA was completed on the fifth and final servicing mission, John Grunsfeld reached over and gave Hubble a soft pat, as if giving a final goodbye to an old friend. Challenges cropped up regularly during all five Hubble servicing missions, and without the ingenuity of the ground support and astronauts to solve the many unexpected obstacles, it would have been silenced years ago. In spite of occasional glitches, Hubble continues to rain down images that are helping astronomers dig deeper and deeper into the complexities of the universe.

During the shuttle program, despite the difficulty of performing space walks and the toll that working in space placed on many spacewalkers, they successfully accomplished their missions. The EVA experience gained over the years, first-rate training, and tenacity of their astronauts have given NASA the ability to rise above just about any challenge—no matter how difficult—encountered during a space walk.

Not unexpectedly, some astronauts are exhausted after a long EVA, whereas others walk through them with little energy expended. Physical conditioning, body size, the approach to performing tasks in space, and perhaps most importantly getting the right suit fit—especially the gloves—play an integral role in how much energy is expended outside the spacecraft.

EVA was very kind to Jeff Hoffman's body. He's aware of many astronauts who were exhausted after a space walk, their hands shot, but he always felt fine once back inside the orbiter. His height and long arms were an advantage, and his EVAs were all very well designed, requiring less energy. He found it interesting that although all astronauts grow an inch or two taller once in space—two inches for Hoffman—due to his tight suit fit, he lost a quarter to half an inch by the end of each of his space walks.

The EMU was not as kind to Rick Hieb. Like Hoffman, Hieb was tall with long arms, and even selecting the tallest suit, he required assistance from three of his crewmates to squeeze him into it. The suit had been adjusted to provide an additional inch of height, but that was offset by the inch he grew once he was in space. He was literally jammed inside the suit, and the suit soon took its toll. Hieb suffered from large open sores, several inches in diameter, on the top of his shoulders. The ground could offer little help, so he elected to not inform them. "I just put a little gauze over it—yes that part was ugly. I don't remember—I was too busy thinking about what we were doing. I don't remember it hurting. I remember hurting afterwards but I don't remember it being on my mind at all during the space walk."

Linda Godwin admitted that there was always some wear and tear on her body during an EVA but remembered only a few aches and pains, perhaps some sore fingernails. Jerry Ross remembered that on his first space walk on STS-61B, Sherwood "Woody" Spring was tasked with closing and locking the airlock door at the end of the EVA, but his hands were so tired that Ross had to assist him. Ross likens his space walks to a couple of very physical high school football practices—challenging and satisfying:

> But for me, I always came in as tired mentally . . . as I was physically. My brain was always going a million miles an hour thinking about what I was going to next, which tether I was going to use, what I was going to hold onto, how I was going to support myself so I wouldn't float away, or to drift out of work orientation. What was my buddy doing? How was he doing, how were we doing on the timeline? Where were we with respect to the ground? Can I take a sneak peek to see what's out there? All of that kind of stuff. It's just . . . we were on a tight timeline . . . it was always a beat the clock type of situation, but you also wanted to enjoy and take some mental snapshots of what you're seeing and what you're doing so you can remember them forever.

Winston Scott, following the capture of the SPARTAN satellite on STS-87, compared his physical condition to playing a sporting event when you're totally wiped out at the end of the game but had chalked up a victory. "My fingers were like putty . . . before going back inside." He and EVA partner Takao Doi had spent over ten hours in their suits—seven hours and forty-three minutes of EVA time—and Scott had just noticed a warning light on his suit informing him that one of his life sustaining consumables was reading low. He was tired, "But again psychologically [I] felt so good because we had a successful EVA. We captured SPARTAN and got it back into the bay, and everything else was safe and successful—wiped out!"

Scott described the smell after a long and demanding space walk once he was back inside the orbiter: "You're pretty ripe afterwards—you hang that sucker [suit] in the air lock and let it dry out; if you had to do another EVA two days later you have to wear the same equipment, the same underwear, same long johns, same LCVG. You let it dry out in the airlock. But the shuttle purifies the air pretty well—the suit dries out and you're able to wear it again without too much difficulty."

Rick Hieb is convinced that his mental state regulated his physical condition following the long third EVA on his STS-49 mission. He simply felt exhilarated. That space walk lasted eight and a half hours, plus they had to spend an extra hour and a half in the airlock due to a malfunction of the rendezvous software, and Brandenstein had them remain in the airlock while Intelsat was deployed, so they were in the suits for twelve to thirteen hours. "Had we not gotten the satellite on that third try, I'm sure I would have felt terrible, exhausted, and bad. But I don't remember any of those feelings because we were able to accomplish what we went out to do, against all odds. No doubt I was tired, but I don't remember it that way because we were so happy."

Many astronauts refer to Story Musgrave as the Zen master of spacewalking. When he speaks of performing EVAs, the words *choreograph* and *ballet* liberally find their way into the conversation. His every move was planned, and exertion of any kind was forbidden. He claims that he was never tired after any of his four space walks. He felt no compulsion to drink water during his EVAs—he was expending very little energy. On his first EVA, the first of the shuttle program, his heart rate never exceeded sixty beats per minute. "Now how can I be tired if my heart rate is under sixty? And so, it's choreography you see—it's every finger and it's every toe and are things going according to the plan." He agrees that there are some circumstances where an astronaut cannot plan to that level, but they should choreograph to a level where the work is minimal. "I was known to be a total wimp because when I'm planning a space walk if I took one deep breath, I'm going to fix this. I don't tolerate one deep breath—I choreograph around that. You choreograph as a total wimp when you're choreographing and planning the mission so that if you get any surprises, you have the margin to get the job done."

Most astronauts the authors talked to are not surprised that there has never been an astronaut killed during a space walk, which they attribute to their trainers and their own confidence in themselves. Story Musgrave sees it differently. With the number of EVAs that have been carried out since Leonov and White's first space walks in 1965, he is surprised that no one has been lost. Musgrave believes spacewalkers have been statistically fortunate to not have taken a hit from a micrometeoroid. On three of his six shuttle missions, windows on the orbiters had to be replaced once home due to debris that would have taken out a helmet or suit. "You add it all together we're talking three hundred space walks plus—you can't develop a system [suit] that can be worn that much, especially soft goods and such that will hang on. . . . without any significant failures. My God, man!"

Winston Scott is also surprised that a spacewalker has never been lost during an EVA. "I'm very surprised—happily surprised. I tell you what, if you look at the history of EVA we certainly came close a couple of times. Looking at the conditions that those folks [early spacewalkers] encountered—they were true ground breakers. When I went out, we had the benefit of all their knowledge and experience. I tell you what we are very lucky we haven't lost anybody in space. I'm grateful that we've done so well"

John Herrington rationalized that there's about a one in seven thousand chance of being struck by a micrometeoroid while on EVA, whereas there is approximately a one in three hundred possibility of dying on launch. He was much more worried about making a mistake, which is an unrealistic expectation given the complexity of most space walks. "So, you just don't think about that happening to you."

Musgrave once said that the only thing that can be expected on a space walk is the unexpected. Rick Hieb lived that mantra with a number of "suit anomalies" on his three STS-49 space walks. Several hours into his first EVA, he had a suit alarm that he had never seen in training—a wonderful way to welcome a new spacewalker to the vagaries of climbing outside the spacecraft. His first impulse, as trained, was to flip through his cuff checklist that supposedly would tell him how to remedy the situation, only to discover that the anomaly was not covered. "The alarm was 'set to power and switch to spacecraft umbilical position' and you know I'm thinking to myself 'what in the hell is that about?'" Astronauts are trained to follow mission rules, and an unknown suit problem certainly justified a call to Mission Control. But astronauts are smart and trained to logically work through unexpected difficulties and to not overreact or panic.

For several reasons, he decided to not report the anomaly to the ground, at least initially; the crew was extremely busy preparing for the EVA and speaking would have interrupted them. Pierre Thuot was in the middle of securing himself to the robotic arm, and he and Bruce Melnick were talking back and forth on his progress. Dan Brandenstein and Kevin Chilton were discussing the rendezvous they were in the process of executing. Additionally, anything and everything that he would have said to the ground would have been heard by everyone tuned in for the space walk, including his parents. There was no way he was going to announce to them and the world that his life may have been in jeopardy; he was fine in spite of the warning. Besides, he was concerned that someone on the ground would panic and perhaps terminate the EVA. He calmly checked his pressure gauge—his oxygen was good. Hieb was confident that his suit was fine—simply because he had not noted any problems. Still, he had to address the alarm; the best he could figure out was there might be a problem with his suit battery. Quietly, he tried to unobtrusively contact the ground: "Houston this is EVA 2; I've got a suit alarm." Unfortunately, everyone else was talking over him, and there was no response. He tried a second time, again trying not to raise undo alarm. With all good intentions, Thuot suddenly yelled out, "Hey everybody shut up; Rick's got a suit problem!" "Oh my gosh; that's not how I wanted to get that communicated." The ground evaluated the situation and eventually instructed Hieb to not worry about it, with no explanation. Post-flight it was determined that it was likely a solder ball floating around and causing a temporary electrical short. The alarm came up twenty to thirty times during that EVA but never again on his next two space walks on that mission.

Hieb also experienced a problem with his digital display unit on his space suit shortly after climbing out of the airlock during the three-person EVA. Again, he decided to not notify the ground and paused for a moment before sharing the details with the authors. He acknowledged that he had likely confided the incident to Mission Control once back on the ground. However, he informed his crewmates that his suit display had become unreadable and asked them to monitor its status and let him know if anything looked peculiar. Hieb reasoned that his systems were all working and that he felt safe. He knew that the ground would not be happy with that decision, but he wanted to capture Intelsat, and the ground may very well have terminated the EVA had they known his display was inoperative. "So I decided to live with it on my call; they wouldn't have agreed with it probably, therefore that's why I didn't tell them right away. So, you know there were some of those things that happened, but again having had as much time as I did in training—and Pierre and I had a lot of training time—I was pretty comfortable in the suit and doing the things that we were going to do and never really felt like this is not going to work." He completed the eight-hour, twenty-nine-minute EVA without further incident.

Kathy Thornton recalled that on one of her STS-61 space walks, she encountered a communications glitch. She could hear only her EVA partner, Tom Akers; those in the orbiter and on the ground were muted. Like Hieb, she remained calm and remembers feeling quite "peaceful." She was partnered with Akers for all three of her EVAs, one on STS-49 and two on STS-61, and she was extremely comfortable working with him. "Tom—he likes to hum. If everything is going well, Tom is humming. As long as I heard Tom humming—life is good."

Dealing with ammonia connections on the ISS poses a huge risk to spacewalkers. Herrington recollected his work on ammonia connectors on his STS-113 mission and that the possibility of a leak existed on each connector. He and López-Alegría practiced with the 400 psi ammonia lines on the ground to familiarize themselves with the hardware. Herrington bemoaned, "They're stinkers!"

Robert Curbeam knows firsthand that ammonia connectors can be stinkers; he was sprayed with icy ammonia on his STS-98 mission while making a connection. His EVA partner, Tom Jones, heard Curbeam announce that he'd been spewed with ammonia and hurried over to assist him. He was greeted with thousands of ammonia ice crystals shining in the blackness of space. Curbeam compared his predicament to a blizzard. Not wanting to make matters worse, he slowed himself down, knowing he had to stop the leak; in addition to being a hazard, the ISS needed that precious ammonia to function. The solution was to close a lever upstream of the leak—easier said than done. Presumably, the lever required twenty-five pounds of force to close it, but Curbeam, strong and in great physical condition, was certain it took more. After wrestling with the valve and barely maintaining his connection to the station, the valve succumbed to his strength, and the leak was extinguished. He was able to remove most of the ice crystals from his suit with a brush and was then instructed to stay outside for another orbit so the sun could evaporate as much of the remaining ammonia as possible. Further measures were taken once back inside the orbiter to protect the crew from any exposure to the ammonia, which is extremely caustic to the lungs if breathed. Curbeam escaped a potential catastrophe through his cool, calm and measured response—as trained.

Hieb acknowledged that he failed to fully grasp the risk of one of his EVA tasks until the moment of truth had arrived. He'd trained to remove a cover off the solid rocket motor that would launch the Intelsat VI satellite into geosynchronous orbit on STS-49. It was a pretty straight forward job on the ground. But when he approached the real rocket motor, fully cognizant that this time the motor was alive with propellant, he approached the job more cautiously. "At least theoretically it would be possible for an [electrical] arc to ignite that solid rocket motor, and we'd all die, but I'd die first. I was significantly more cautious in peeling that cover off than I had ever been in my training. There are things that are different than on the real day."

Safety near misses in any kind of work environment are worrisome, as they often portend more serious flaws that are being overlooked. There have been several cases where space gloves have been compromised—with the potential for an air leak—but fortunately, none developed into a serious problem requiring immediate mitigation. Jay Apt, on his first space walk on STS-37, developed a blister on his hand when a metal bar inside the glove punctured the inner bladder of the glove. He showed it to Ross once back inside the orbiter, and Ross thinks they may have put a Band-Aid on it prior to going out for the second EVA, as it didn't appear to be a serious problem. Back inside after completion of EVA 2, they observed that Apt's blister had broken, and there was blood on his comfort glove. Comfort gloves, a comfortable fabric internal layer, are optional; they help wick away perspiration and make it easier to don and remove the outer gloves. Back on Earth, the suit personnel informed them that the metal bar had actually broken through the glove bladder, offering a potential leak source. Fortunately, it didn't fail; the congealed blood may have helped maintain a tight seal. Ross realized the potential danger and confessed, "Very interesting thing to see and to find about after you get back on the ground."

Ross believes the riskiest aspect of spacewalking is not having prepared properly in the first place. He made sure that they closely scrutinized every aspect of their planned EVAs. Stored energy, sharp points, radio frequencies, and much more were all heavily inspected for potential problems. "And if you don't have the opportunity to look at that hardware, or if you don't have your head screwed on right when you look at it, you might overlook something that can bite you in the end, and I think that's probably the thing that can hurt you the most."

Perched on the end of the space station robotic manipulator system robotic arm, Scott Parazynski was being translated to the P6-4B solar array blanket where he would attempt to repair several tears for added stability so that the array could be fully deployed. Damaged by a micrometeoroid strike, according to Parazynski, the power generated by the solar array was sorely needed to keep the ISS fully functional. There were no contingency plans for the repair, so the ground had to brainstorm a way to fix the problem, and they settled on threading "cuff links" into the arrays that should add the strength required to fully deploy them. The rub was that the devices would need to be constructed from materials on board. Pieces of aluminum strips and twenty meters (sixty-six feet) of wire did the trick.

The crew received a plan for the repair from the ground about five pages long, and on one of the pages, in large red font, an ominous warning had been left for Parazynski: "Note: some sparking may be expected." Fortunately, Parazynski had two and a half days to prepare for the EVA, and in addition to the written plan, they were sent a virtual reality video that was extremely helpful in figuring out how Parazynski was going to remedy this unexpected challenge. The crew was very confident they could successfully carry out the plan, but still there was some doubt in their minds; this was not going to be a standard well-rehearsed EVA.

Once the cuff links had been completed, Parazynski, armed with a battery of tools, ventured out to the array and first removed a "hair ball," a mass of twisted metal courtesy of the micrometeoroid. Parazynski was secured to the end of the space station robotic arm that had been coupled together with an extension—the orbiter boom sensor system—so that it provided approximately a ninety-foot-long reach; the tall and lanky astronaut would need every inch of it to make the repair. "This is really a springboard to beat all springboards. The deal was I would get onto this contraption and then be taken for the ride of my life. It was a forty-five-minute ride each way. Gave me a lot of time to think about [what] was going to happen. . . . I was taken waaay above the International Space Station. I could see the whole orbiter belly and looking all down onto the planet with the space station in the foreground. It was just extraordinary." Despite the beauty spread out in front of him, Parazynski was consumed with the upcoming repair during his ride to the solar array. The weight of the situation was underscored as the CAPCOM took about fifteen minutes to inform him of all the bad things that could happen during the repair attempt.

Doug "Wheels" Wheelock, Parazynski's EVA partner, was a spotter for him as the crew inside the orbiter could not see the work area, another challenge on this difficult repair. But typical of all crews, this was an integrated team effort requiring contributions from everyone. Wheelock wanted to watch closely to reduce the possibility of Parazynski inadvertently coming in contact with the array which could have ended up in a giant fireball. Unfortunately, Wheelock had to endure excruciating cold during the entire spacewalk—his frigid fingers strained to turn temperature control knobs or take a photo.

It took Parazynski several hours freeing the hair ball and snags before he could execute the repair with the improvised cuff links. Even with the long robotic arm, it took all of Parazynski's height and his extended arms to reach the array as he reached high above his head to begin a task that had never been envisioned. With a plan developed only days earlier, he was ever cognizant that 100 percent oxygen and sparks are a dangerous combination!

As he prepared to begin the repair, Parazynski was warned by IVA crewperson Stephanie Wilson that any activity on his part could cause the arrays to shift and he should be prepared to block the arrays from touching his suit using a quickly designed tool resembling a hockey stick. Or be prepared to back away. Calmly, he replied, "I'm ready."

Parazynski had a bird's-eye view of the worksite high above his head, and from his helmet camera, everyone else could follow along; his lower arms and gloved hands were the only parts of his body that could be seen. Occasionally, his distorted reflection was apparent in the metallic-looking array, whereas at other times, a large gray shadow of his silhouetted space suit on the glistening golden solar arrays provided a glimpse of most of his body.

He had to thread the cuff links—thick pieces of wire—into small holes on the array while working high above his head, a task that would likely have been difficult to a garage tinkerer on good old planet Earth. The EMU had not been designed for overhead work, making it difficult for him to reach the arrays. He had to push up on his toes to extend his body as far up as possible; time was running out as he installed one cuff link after another—a total of five—but he knew he simply had to gut it out. Following the repair, the arrays were fully deployed; NASA had once again accomplished the nearly impossible.

Parazynski returned the hair ball back to Earth, and after NASA had completed their analysis of the train wreck to glean as much information as they could about the events that led to the damage and subsequent repair, they gifted it to Parazynski.

These astonishing accounts are representative of the many sorties outside the space shuttle and ISS during the lifetime of NASA's first reusable spacecraft, highlighting the difficulty and danger of walking in space. They also illustrate the tenacity and creativity of these spacewalkers to successfully carry out their tasks, often under seemingly insurmountable conditions. But along with all the hard work come the rewards. The spacewalkers were gifted with incredible memories that will last a lifetime.

# 10

# A Lifetime of Memories

Favorite People, Favorite Places, Favorite Memories of the past. . . .
These are the joys of a lifetime. Those are the things that last.

—Henry van Dyke

Joe Tanner stressed that often, public perception is that astronauts are having fun during space walks, which is misleading. "It is a lot of work! Work can be fun if you like to do work. You are pretty much totally motivated by getting as much done on the tasks and timelines as best you can without making any mistakes. You're out there working your butt off. For some people it can be very painful and very hard; physically drained." In spite of the hard work, it's nearly impossible for spacewalkers to not marvel at the beauty of space with Earth shimmering miles below and take pride in the feats they accomplished. Being selected for a space walk also brings a sense of thrill to many, and learning that you are going to perform an unexpected contingency or unplanned EVA quickly starts the motor running.

The day after the Syncom IV-3 satellite had failed to ignite following its deployment on STS-51D, the ground began assessing its options to correct the problem, one of which included a space walk by Jeff Hoffman and Dave Griggs. Hoffman's ears perked up when he heard that an EVA was being considered: "EVA; Wow! Naw, they're never going to let us go out. NASA's never done anything like that; sending people out to do a job that had not been planned before." When his wife heard rumors that a space walk was being considered—for her husband—she was concerned and called a friend to find out more information. Were they really going to go outside? Her friend consoled her and told her not to worry, NASA had never done an unplanned space walk. Needless to say, Hoffman and Griggs were elated to carry out their very first EVAs.

The attempt to correct a malfunctioning antenna using tools quickly manufactured by the crew from materials on board *Discovery*—attached to the end of the robotic arm by spacewalkers Hoffman and Griggs—was unsuccessful. Rhea Seddon proceeded to return the robotic arm to its cradle located in the payload bay. With the sun setting, the lighting conditions were not quite right for making a visual inspection of the situation. As much as Hoffman was buzzed when he first learned that he was going to participate in a space walk, his reaction to a query from the ground was even more unforgettable: "'Jeff and Dave, would you mind spending another forty-five minutes outside while we wait for the sun to come up?' Oh man! I had a great time, holding on and watching the world go by. Climbing around. I went from the front all the way to the back of the shuttle. Almost like being a tourist out there. I had no work to do, just enjoy myself."

Winston Scott wasn't overwhelmed on his first space walk. Enigmatically, it wasn't the stunning beauty of the brightly painted canvas that stretched to the horizon that initially struck him: "I thought, 'What if I couldn't get back there?'" Not keen on that possibility, he quickly pushed the thought out of his mind. He was astute enough to take in the complete picture, let his mind wander, and appreciate his view of the universe.

Scott experienced moments during his EVAs that duly impressed him and left him with memories that he recalls perfectly some twenty-five years later. He was selected to test an EMU modified to withstand the colder temperatures spacewalkers would endure when assembling the ISS, which orbits at a 51.6 degree inclination to Earth's equator. The original EMUs were designed to contend with slightly higher temperatures at an approximately 28 degree orbital inclination. One of the objectives of the STS-72 mission was to test the upgraded suit, and Scott was selected to perform a cold soak test on his first EVA. This was going to be the second test of the modified EMU, the first evaluation of the improved suit was evaluated on the STS-63 mission by Bernard Harris and Michael Foale. But while carrying out a mass-handling exercise, the hands of both astronauts became extremely cold and the EVA had to be terminated.

The suit was further modified, and on a night pass, Scott was anchored in the payload bay where the coldest temperatures were expected. The bay pointed away from the sun to ensure he experienced as much cold as possible, and he remained as still he could to avoid generating any excess heat. Every five to ten minutes, he assessed thermal conditions of the suit using the Cooper-Harper rating scale, a technique used by test pilots to assess handling qualities when testing aircraft. Near the end of the night pass, the cold temperatures—minus 104 degrees Fahrenheit—were beginning to seep in but not enough to concern Scott or anyone on the ground. Scott gave the thermal protection capability of the suit a big thumbs up.

Scott enjoyed the test and was particularly proud of his contribution to the future of shuttle and ISS space walks. But another story played out during this EVA that Scott cherishes to this day. The bright and blinding sunlight quickly transitioned to total darkness as he flew into the night pass, and straightway he could feel the gripping cold just as his eyes detected the absence of light. "I'm looking over my shoulder, and I can see the terminator approaching. And then sunlight—feeling comfortable. I go through the terminator and then come out the other side a few seconds later in darkness, and instantly I could feel the chill begin to seep down into my feet." Near the very end of the night pass forty-five minutes later, Scott, again looking over his shoulder, could see the terminator faithfully returning: "I passed the gray area into the sunlight and instantly—a couple seconds—I can feel the heat begin to seep into my suit."

Scott was presented with another unexpected thrill that traced memories into his brain. During one of his three space walks, his commander informed him that he was going to fire the RCS jets; be prepared. On Earth, sound travels through air, and if it reaches the ear, it causes the inner ear to vibrate, which then turns the vibrations into sound. Sound cannot travel through the vacuum of space, but it can be transmitted through a solid structure. Suddenly, Scott glimpsed the bright flash of the firing jets reflecting off the back of the orbiter; naturally, he could not hear them fire. "But just an instant later the sound traveled—I could feel it traveling through the structure of the shuttle, through my boot, inside of my suit, and then rumbling in my ear. I could actually feel the sound—I could see the jets firing and a moment later the sound rumbling through the structure of the vehicle and into my suit. So there are really some cool things in space that don't necessarily happen here on Earth."

It's not possible for astronauts to bring home a piece of space as a child might bring home a rock from a weekend camping trip and then display the trophy beside their bed. Mostly, astronauts bring back memories. But one astronaut was able to retrieve an artifact from Hubble that he had constructed with his own hands. During the STS-82 servicing mission, it was discovered that some of its thermal blankets needed to be replaced—and the sooner the better. NASA opted to temporarily remedy the situation and sent up procedures to construct blankets by hand with materials on board *Discovery*, along with instructions on how to install them during one of the planned space walks. These provisional blankets were subsequently replaced on a later mission with better ones constructed on the ground. But first, the old handmade ones had to come out. Kieth Johnson was working that mission, sitting at his flight console, when one of the STS-82 astronauts came in, quietly took a seat next to him, and asked, "'Hey, what are they going to do with those covers?'" Puzzled, Johnson asked him to clarify his question. Again, the question was the same. Johnson informed the astronaut that the old blankets would be stuffed into bags and returned to the orbiter and then brought back to Earth. Johnson recalled the exchange from the expectant astronaut: "'What are they going to do with them when they get them home?' Wow man; I don't know! They'll come home and they'll unpack them and they'll probably go to Goddard. . . . 'I want the one that's on that door right there because on the back of it I wrote my name, my wife's name, my kid's name, my dog's name, and all of that is on the back of that thing that we stuck on the backside of the Hubble Space Telescope!'"

John Herrington reminisced about his first time exiting the ISS airlock on his STS-113 flight. Unlike the airlock on the shuttle orbiters that open into the protective payload bay, when the spacewalker exits the ISS airlock, their first view is that of Earth directly below them. Laughing, he remembered thinking, "It's a looong way down; you hang on tight for a little bit!" He was pleased with the advice given to him by then veteran spacewalker Jerry Ross to take a mental picture during one of his EVAs and remember it for the rest of his life. He chose a moment when he was on the end of one of the ISS's trusses where he had been tasked with pressing a series of plungers to ensure they were operable for future EVA missions. Don Pettit was working inside with Herrington monitoring his progress on a computer. As Herrington pushed in a plunger, Pettit could see a light confirming a successful test. Herrington remembered: "I'm hanging onto the edge of the space station by a thumb and a forefinger . . . and looking out at the edge of the Earth and thinking Jerry told me to sear something into my mind and never forget. And this is what I'm going to do. I did it, and [I] have a video of it too."

Rick Hieb's three space walks were all out of *Endeavour* on the STS-49 mission; assembly of the ISS had not begun when Hieb retired from NASA, therefore he never had the opportunity to perform a space walk from the ISS. Hieb remembered that when he climbed out of the airlock of *Endeavour*, with the open payload bay pointed toward Earth for warmth, his feet were pointed down toward the bottom of the payload bay with his head facing the bulkhead, which was all he could see. No dramatic views of Earth or deep space—only a white bulkhead. He compared the helmet to wearing a baseball cap, which along with his position restricted his view. Using tethers, he moved over to the bay longeron where payloads could be attached. Eventually, holding on with one hand, he was able to turn his body and be rewarded with his first view of raw space from outside the orbiter. "That was one thing that I was not prepared for. The view was so fundamentally different from looking out the window that I started to open my mouth and say something profound—then I realized I was just going to say, 'Wow! This is awesome.' So, I didn't say anything." He's cognizant that not every space traveler is assigned a space walk, and even today he prefers to not talk about the view from space during his EVAs, ever mindful of the unique opportunity he was blessed with.

It wasn't until he had returned to Earth that Hieb was able to articulate the view and the feeling of floating in space inside the EMU. He compares it to the view seen from the confining window of an airplane to standing atop a mountain. "You stand on a mountain the whole world is spread out and you see it all and that's the way it is in a space suit. That's one of the two or three forever moments for me . . . that first look and just realizing how different it was to be able to see the earth." Hieb chuckled at John Herrington's comment about Earth being a long way down when he exited the ISS crew lock. Hieb admitted with a hint of jealousy that a space walk from the ISS would be wonderful: "Must be quite a stomach clenching moment."

Joe Tanner remembers being elated when he learned that he had been assigned to be one of the chosen few to participate in the STS-82 Hubble servicing mission, his first EVAs; it was a "dream come true." Tanner extolled the views outside the orbiter windows, but once he egressed from the airlock and earned his first glimpse of space on his maiden EVA, he reveled in "a warm feeling of a life goal achieved" and immediately rejoiced with a victorious "Hallelujah!" Once outside the spacecraft, executing the timeline dominated his mindset; there wasn't time for looking at the sights. He soon learned that when he snapped a photograph, he found himself enjoying the view, so naturally he quickly began to find reasons to take more pictures. He was conflicted with the sudden freedom of exploring the entire universe and "the desire to not let go of that handrail!" The experience was visceral. "The Earth was visually breathtaking, of course, but also an emotional experience when I realized I was looking at it with my own eyes from inside a space suit! This isn't a movie and I am really here." He took a few precious moments to relish his surroundings and then went back to work. Comparing that first encounter of raw space with the first time he saw the Grand Canyon, he shared that both seemed too majestic to be real. Tanner made seven space walks during his career and was never pleased when it was time to head back inside. As he was making his way back to the ISS airlock on his final EVA, the sun was setting as if informing him that his spacewalking career had ended.

Like Tanner, Linda Godwin was riveted to her schedule, and looking at the scenery took second place. She was glad that one of her IVA crew members reminded her to occasionally take a look at the planet. Even when she was on the end of the robotic arm being slowly moved to a workstation, she spent more time thinking about upcoming tasks than taking in the view.

Those first moments when astronauts exit the airlock seem to stick with many spacewalkers. Nicole Stott had to remind herself that she was in space to work and not to enjoy the view. As she climbed outside the Quest airlock on the ISS, she saw "what seemed like a glowing, colorful, horizon to horizon view that seemed to just suck me into it—had to remind myself that I wasn't there to just see it." A long journey on the robotic arm, especially when there is little work required, allows ample time to soak in the moment and is another favorite of spacewalkers. Stott was gifted with a twenty-five minute ride on the end of the arm, and even carrying a large box—the European Technology Exposure Facility—that partially obstructed her view, she was awestruck. She still revels in her memory of the ISS slowly drifting out of her view as she was being transported to the payload bay of *Discovery*, and as Earth was spinning beneath her, *Discovery* suddenly came into sight. Deep space, the ISS, Earth, and *Discovery* all in one spacewalk! She described her EVA as being peaceful, but she harbored a concern: "Don't be the one that falls asleep on the end of the arm."

Spacewalkers take much deserved pride in their work. After Jerry Ross and Jay Apt had unstuck the stubborn antenna on STS-37, the ground needed time to check out the Gamma Ray Observatory before releasing it, which allowed time for Ross and Apt to perform some get-ahead tasks for the next day's planned EVA. After those chores were completed, they were instructed to return to the airlock for satellite deployment. Unfortunately, there were no windows in the airlock for them to observe GRO being released, but no one had told them to climb all the way inside the airlock and close the door. "I think the only part of our bodies that were in the airlock were our toes," Ross admitted. "Everything else was outside so we could see everything happen." They were not going to be denied the pride of seeing their recovered patient silently heading off to work where it would fulfill its primary mission to unlock more of the universe's secrets through the analysis of gamma rays.

In his book, *Spacewalker*, Ross described one of his most unforgettable memories, a forty-minute ride while attached to the end of the space station's robotic arm on his very last EVA on STS-110. With darkness approaching, Ellen Ochoa's task was to move Ross past the docked space shuttle and ISS to the station airlock. Ross further illuminated in 2021, "It was a wonderful memory that sticks very vividly with me yet." He had completed all his chores and had time to enjoy the moment. Over the Mediterranean Sea, with Egypt on his left and Europe on his right, his eyes set on the terminator as the sun was setting. Thunderstorms were boiling below, and it was at this time that Ochoa informed him that his ride was about to begin. Continuing to gaze in the same direction, he sensed his body's orientation begin to change as the arm slowly translated him to his destination. "After a while, I'm looking in the same . . . relative direction with respect to my feet, but now I'm seeing stars instead of fires on the ground or thunderstorms." During the night pass, the sky was pitch black, which allowed the stars to blossom, and he noticed that he was now approaching the tail end of *Atlantis*. The payload bay was illuminated by moonlight and by light escaping from the windows of the orbiter. He could see his crewmates inside *Atlantis* as well as its broad delta shaped wings; he's convinced that he could have nearly touched the tail of the spaceship.

> And then as we came around the other side of the space shuttle and down towards the hatch of the International Space Station airlock, the sun started coming up, and I could feel a little bit of warmth on my back as that happened, and then by the time I got back down to the airlock, the sun was pretty well up and it was time for me to re-engage my brain,

other than just looking around and having fun and get on with the rest of the activities. But very, very nice. . . . I think it was about forty minutes long, it was basically an entire night pass that I got to sit there and just enjoy the ride, and certainly did.

An earlier space walk had an even more profound effect on Ross. At night on his second STS-37 EVA, he was completely taken aback by the beauty of space while attached to the end of the shuttle's robotic arm, an experience that led to an epiphany about his place in the cosmos. He had turned off his helmet-mounted lights and arched his body so that he could better see the universe displayed in all its splendor stretching to infinity in all directions. As his eyes adjusted to the darkness, he recalled, "I'm starting to enjoy this incredible view and all of the sudden I had this sensation come over me of being at unity with the universe and that was a pretty incredible experience. Doing exactly what God had designed you to do is a pretty good feeling."

Even with all the meticulous planning and training for what might be a once-in-a-career opportunity, space sometimes has its own ways of disrupting the dreams of astronauts. Ed Rezac's warning that "Murphy is always in the mix" haunted two aspiring spacewalkers, for whom anticipated euphoria was soon replaced with profound disappointment.

On the tenth day of *Columbia*'s twenty-first flight, mission specialists Tom Jones and Tammy Jernigan were set to exit the airlock on the first of two planned EVAs to test space station construction equipment and techniques. But as Jernigan rotated the airlock hatch handle, it came to a hard stop after about thirty degrees of rotation. "Initially, I thought we just had a sticky hatch," Jones later told CNN's John Holomon. "We were both putting out about the maximum force we ever try to put into mechanical systems in our water tank training back in Houston." It was then that he realized it was jammed completely.

Crewmember Story Musgrave had Jernigan open the emergency depressurization valves on the hatch to blow the exterior thermal cover down so the crew could better scrutinize the hatch mechanisms using the RMS wrist camera. They could watch the exterior handle and the push/pull rods move, but not nearly far enough to retract all the latches. "I like these contingency EVAs," Jones radioed. "I didn't think we'd be doing one in the airlock, though," replied Jernigan. "We were in a surreal situation: Just an eighth of an inch of aluminum separated us from the experience of a lifetime," Jones later wrote.

Engineers pored over hatch hardware on the ground and looked at all the possible failure modes. They hoped that heating the hatch in the sun or moving it laterally while cranking the handle would solve the problem. During a media event the following day, Tammy Jernigan described the various ideas that ground controllers had read up to them to try and expressed that "we are optimistic that we will get this hatch open." But it was not to be.

Two days later, after the wakeup-call music "Break on Through to the Other Side" by The Doors, CAPCOM Dom Gorie gave the crew the bad news. The mission management team had just come out of a meeting and decided that "EVAS are not going to be a player for this flight." The media seemed overly concerned that the loss of the two EVAS would impact the construction of the coming space station, but while the crew deftly reassured reporters that the tests would be rescheduled for a future flight and cause no delay, it was a crushing loss to the two would-be spacewalkers. "I hope for better times on future missions while assigned to a space walk," Jones told Holliman.

During a post-flight inspection, an improperly installed screw was found jammed in the gears that the hatch handle was designed to rotate. Like a pebble inside a fine Swiss watch, it had lodged itself in the worst possible place and brought the entire operation of the door to a stop. There were no options to remedy the errant screw. Fortunately, after what had to be a profound disappointment, both Jones and Jernigan would indeed each get their chance on their next shuttle missions during the assembly of the ISS.

John Herrington also believes that seeing Earth from inside a space suit gives spacewalkers "a different perspective on looking back at Earth." It's much different than looking through the orbiter window with three panels of glass. He summed up the difference in the view succinctly: "So now it's ear to ear!" Ever the engineer, he believes it may be possible to mimic the experience with a "Garmin Virb-a 360 camera where you can scroll around and look up and down. God that would be fabulous!"

Years after his STS-120 space walk and repair of the torn solar array, and having retired from NASA, Scott Parazynski cites that adventure as one of his most memorable experiences as an astronaut—perhaps of his life:

> My most vivid life experience probably, but certainly on EVA, was at the end of STS-120 EVA four. I was about ninety feet above the space station on the end of this contraption—cobbled together robotic arm. You have your feet wedged out in this boot plate; you really aren't held on by very much . . . you're just dangling out at the end of the universe. I was way above the shuttle, way above the space station, looking down at my buddy Wheels. Wheels calls out "Hey look down there!" We were flying over Australia at night, and the entire continent was covered in clouds. Thousands of square miles. It's a phenomenon known as mesoscale lightning where one burst of lightning will pop off and then there are these dendritic fingers which will each ignite and so you have ten, one hundred, ten thousand, a million million bursts of lightning, just a fireworks of lightning beneath me. Then I realized I had a pretty good day job.

The accomplishments of the space shuttle program are truly astonishing, and were it not for spacewalking astronauts, much of it would have not been possible. Years following retirement from NASA, many of them—if not all—still take pride in their triumphs.

Jerry Ross looked back fondly on his exploits after his first mission to space on STS-61B—and he still had six more trips to space and eight more space walks prior to leaving NASA. "I had that sensation of total satisfaction of a dream come true, of ya know, busting your butt for years working around things that prevented you from achieving your dream to the point that we actually got there and you had a very successful flight." He was one of the EVA pioneers in testing construction techniques, helped save the Gamma Ray Observatory, and participated in space station assembly—quite a résumé.

Describing the satisfaction he holds for his role in assembling the ISS comes difficult for Joe Tanner, but he certainly feels blessed to be associated with something as historical as the space station. While he's not compelled to step outside and watch every time the ISS passes over, he does "enjoy watching a good flyover. I always feel a sense of accomplishment remembering that I had a hand in installing one half of its solar panels." He celebrates every one of his flight anniversaries and takes particular pride in his three spacewalking missions. "I have another reason to remember Valentine's Day as on that day I emerged from the *Discovery* airlock on my first EVA . . . on HST!" John

Herrington also has a special holiday memory from his STS-113 mission; his parents watched the station pass over during his second space walk on Thanksgiving Day, 28 November 2002. "I have more fun thinking about that than watching it fly over; that's what I'll remember."

Herrington was nearly as elated when he was selected for a space walk. Well before his first flight, his boss, Jerry Ross, asked him what his aspirations were for a mission. Herrington informed Ross that he wanted to be an MS2 flight engineer and to walk in space, which all occurred on his first and only mission to space—STS-113. "So, anything I ever wanted to do on a space mission as a mission specialist I got to do with one mission. I couldn't have asked for more than I did."

Teamwork is essential to successful EVA missions, a collaboration that extends to the non-spacewalking crew and the engineers, scientists, and other people on the ground. Humbled with her role in helping the Hubble reach its lofty expectations on the STS-61 mission, Kathy Thornton believes that many others deserve credit for the telescope reaching its full potential: "Jim Crocker, who figured out how to design the COSTAR that could correct the axial instrument, that was phenomenal. That kind of creativity is far beyond anything I did."

Rick Hieb credits the entire mission-oriented team for developing and implementing a plan to capture Intelsat VI. "I was more happy because we as a crew did the job that we were there to do." Following his retirement from NASA, Hieb remembered speaking with the CEO of Intelsat, Kay Sears, and he inquired how they had fared with the satellite once it had reached its intended orbit. With a hint of pride, he shared that he was pleased when she informed him that the company had netted about $2 billion from it. "So, it's fun to look back at that and kind of have the bigger picture thing. Yes, flying in space is fantastic, I was super fortunate to do it. But the truth is the people are what you remember."

Reminiscing about her spaceflights and EVAs, Linda Godwin, like Rick Hieb, credited those she worked with. "I wouldn't trade anything I did for my career. I really liked being part of that whole team too—[although] I liked being the one that got to go. I worked with some great people and I got to meet people from around the world." Godwin also performed an EVA on the Russian *Mir* space station and is one of only three astronauts to carry out space walks on both *Mir* and the ISS. She often looks for the ISS passing over: "I think about people up there looking down and I think about being a part of that and how fortunate I was to get to do that in my career. And two space stations!"

When asked in 2020 if he still enjoys seeing images captured by Hubble, Jeff Hoffman responded quickly, "Absolutely! The WFPC2; I put it in. And there are still pictures coming down." He's followed the progress of Hubble for decades, gratified by his contributions to what is arguably one of the greatest telescopes ever built.

Hoffman's EVA partner on the STS-61 flight, Story Musgrave, feels no pride in the role he played in repairing and upgrading Hubble. Citing its small size relative to the universe, its pinpoint accuracy, and what it's capable of achieving, he seemed to think of the telescope as an entity, not just a mechanical device. "I don't feel pride, sir. I had a part in the thing. I feel pride for the machine and—the machine is quite a machine."

Winston Scott retired from NASA in 1999 but has never retired from outer space. He lives near the cape and watches just about every launch from his balcony, or he may step outside just a few minutes prior to lift off. He closely follows news of space exploration. He's provided commentary for Blue Origin and Virgin Galactic, and when Bob Behnken and Doug Hurley launched on the first SpaceX flight to carry astronauts to the ISS. He's certain that he left NASA at the right time—it was the correct decision. Still, he underlined, "My life quite frankly is dominated by my space experience. No matter what I do on a nine-to-five basis. It's non-stop. I feel so fortunate to have done this. People who get selected for this, we work hard, study and we persevere, but there is also an element of good luck to it. . . . I feel very blessed to have been selected for this and I'm happy to share it."

Spacewalkers have been gifted memories that will remain with them for the rest of their lives. From dazzling and mind-searing sunsets and sunrises, to bird's-eye views of our planet, to the blackness of deep space, the exploits of spacewalkers were once the making of science fiction. Free time outside the spacecraft led many to the realization that their job ranked up there with the best in the world—or out of this world. They witnessed sights not possible to see on Earth. The first view of open space from their helmet, a ride on the robotic arm, and personal experiences enabled them to see the universe as they had never seen it before. They took pride in their work that many times led to a better understanding of the universe. And their experiences made memories likely to never be equaled.

# 11

# Murphy Is Always in the Mix

Those are our friends out there. Not just somebody in a space suit.

—Tomas Gonzalez-Torres, EVA trainer and flight director

The space shuttle provided access to space for astronauts and foreign visitors for thirty years, by far America's longest running crewed space program. The diversity of missions was truly impressive, from medical and scientific research to satellite capture, repair, and deployment to construction of the International Space Station—and well over 150 space walks. The shuttle enhanced NASA's understanding of space flight, especially EVA, and ushered in what one day may be referred to as the golden age of spacewalking. There were well over a hundred space walks dedicated to the construction of the ISS, nearly twelve per year if discounted for the two years that the shuttle was grounded following the loss of *Columbia* in 2003. Many of these space walks, including the fifty-four prior to the beginning of the space station construction, carried audacious expectations and many lasted for over seven hours. Additionally, numerous EVAs encountered problems that required NASA to pull out their well-conceived contingency plans to remedy the situation. Some required subject-matter experts on the ground—instructors, flight controllers, hardware developers with vast technical expertise—along with astronauts in orbit to come up with new and creative methods on the fly to accomplish the planned objectives.

The astronauts who carried out these space walks deserve acclaim for these astounding accomplishments, but without those on the ground—the instructors, engineers, and trainers—none of it would have been possible. Astronauts don't train themselves to do space walks; experts in EVA plan each space walk extensively and then train the astronauts how to fly the mission. During the shuttle program and operation of the ISS, neutral buoyancy training underwater has been by far the most effective means of training astronauts to do space walks; therefore, the instructors who plan EVAs are highly focused on using the pool to teach EVA skills.

When the Gemini program began, no one had ever trained astronauts to do EVA. The Air Bearing Floor training—short and rudimentary—consisted primarily of using the hand-held maneuvering unit to translate to specific points, remove rotational and translational velocity and intercept a moving target. They also practiced pushing away from a simulated spacecraft and then using the umbilical to pull themselves back to the vehicle. The engineers running this program estimated that about ten hours of training on the floor was adequate to prepare for a mission.

Training in the zero-g aircraft involved practicing specific tasks in about twenty-five-second increments provided by roughly two hundred parabolas for each mission. The first spacewalkers came back from their missions and shared their experiences with engineers, but the program was moving so fast that there was little time to build up a team of dedicated instructors and trainers to integrate all the information into a coherent training program. The weightless EVA training for the lunar missions and Skylab program was a step up from that of the Gemini program, primarily due to the integration of neutral buoyancy training into the syllabus. But the number of weightless EVAs during Apollo and Skylab pale in comparison to what would be accomplished on the shuttle and ISS. With time, NASA had the foresight to employ dedicated EVA trainers and instructors and continued to adapt and improve the training program throughout the life of the shuttle and into the era of the space station.

EVA trainer Kieth Johnson, then a flight controller and instructor at JSC, began his career with NASA learning the shuttle EMU space suit. With that knowledge, he began training astronauts on how the suit functioned, how to respond to a myriad of messages about the health of their suit, and then how to react to potential emergencies, all useful experience to draw on when he later began training astronauts for EVA. Following his stint as a suit instructor, Johnson transferred into the EVA group known as "tasks," where he learned how to plan space walks, about all the tools that might be required during a space walk, contingency EVA planning, and the nitty gritty details of how to capture and repair satellites. "I did both of those [suits and tasks], and then I became a front room flight controller. Then you are responsible for both the suit and the tasks they are doing outside. I'm responsible for the choreography of the EVA and interacting with all the other disciplines in how the EVA affects everybody else."

The JSC EVA training team was divided into "tasks" and "systems" groups. The systems group managed the space suits—how to assemble a suit to fit a specific astronaut or how to respond to emergencies such as leaks or loss of communication. The tasks group was focused on accomplishing the mission EVA objective, including planning the space walks and teaching and explaining all the different types of tools.

Christy Hansen, EVA Task Lead for STS-116 and 125, found herself assigned to the EVA group at JSC when she began her career at NASA in 1999. She felt very fortunate to be in such an exciting and challenging position, one that would lead her to planning space walks, training the crews, and then supporting real-time operations as an EVA task flight controller in the Mission Control Center (MCC), working with a large team to ensure efficient, safe, and successful EVAs. Hansen came onboard during exciting times, at the very beginning of ISS assembly with a plethora of astronauts to be trained for construction of the space station. She always loved human spaceflight and NASA and felt privileged to work both ISS and Hubble Space Telescope (HST) servicing missions. The Mission Operations Directorate (MOD) motto of "plan, train, fly" would become a model for how Hansen approached her future career opportunities at NASA.

Tomas Gonzalez-Torres, lead EVA officer for STS-125 and Hubble Servicing Mission 4, participated in NASA summer internships while still in college, and through NASA pamphlets, he knew that he wanted to work in the EVA branch. Initially, that group was full, but by his third summer placement, he was able to move over to the EVA group, and following graduation from college, he began full-time employment in his preferred career. He started out on the STS-85 mission, working on his certifications as an instructor and a back-room operator. He must have impressed someone, as he was next asked to assist with an upcoming highly coveted Hubble mission. Eventually, he was assigned as the back-up to the lead engineer for the STS-103 Hubble Servicing Mission, which put him in line to become lead for the next mission to repair and upgrade Hubble, STS-125. This method of slowly increasing responsibility and moving people through the program was NASA's way of ensuring continuity on the complex Hubble missions.

Hansen, Johnson, and Gonzalez-Torres worked in the JSC EVA operations group and were responsible for all shuttle, ISS, and Hubble space walks, but they were often supported by specialists at the Goddard Space Flight Center. Ed Rezac, from Goddard, came to NASA with experience under his belt; he began working in 1976 with ILC Dover, the company that built the space suits, and moved over to JSC in 1981. Rezac has been contributing to EVA missions for over forty years, having supported the first shuttle EVA on STS-6, the Manned Maneuvering Unit, repair of the Solar Maximum Mission satellite, and four HST servicing missions. His role early on was to prepare the space suits that the astronauts used in training—real suits but slightly battered from training use. Eventually, Rezac moved to NASA's Goddard Space Flight Center, where he continued to support EVA activity, working alongside the JSC trainers developing procedures and tasks for EVA.

Training an astronaut to do a space walk requires tremendous attention to detail and long hours of planning and training. Johnson, Hansen, Gonzalez-Torres, and Rezac have spent years working in the NASA EVA group responsible for teaching astronauts how to use all the tools they would need in space to complete their assigned tasks. They also planned and choreographed the EVAs, which required the development of detailed procedures, building in as much efficiency as possible so the spacewalkers could accomplish all their tasks in the allotted time. The plan must account for countless minutiae and potential contingencies—how, where, and when to do a task, how to translate and use hundreds of specialized tools—while factoring in the technical specifications and functions for all the hardware. The procedures needed to be written—simply so they could be easily followed—and tested in various environments, including the pool. The NBL and its precursor, the WETF, hosted the lion's share of EVA training, but virtual reality, the Air Bearing Floor, and partial gravity simulator were also employed as needed.

Unfortunately, there aren't any EVA Trainer 101 classes in college. Before NASA allowed new employees to train astronauts for EVA, they needed to be taught how to train the future spacewalkers. Hansen remembers it being a very specialized certification process:

> When I came in as a new person out of grad school and wondered how I was going to train an astronaut (never having flown myself), I was introduced to the overall certification and mentoring process; as new EVA instructors in training, you would be required to observe several classes

> on the technical, functional, and operational aspects of space shuttle, International Space Station components as they pertained to nominal and contingency EVAs. After observing, you would be required to practice (with a certified instructor evaluating you), and then if you did well there, you would be permitted to "go for cert," where you'd teach the class again with the option of passing or failing. If you passed, you would be officially certified to teach that class to astronauts. JSC also employed an "on the job training (OJT)" method to assign new employees and new certified instructor as OJTs to official missions, shadowing an experienced team of EVA instructors before you were assigned as a lead yourself.

The technical training standards were high to ensure new EVA trainers were fully competent before they were given responsibility for training astronauts. New and early career EVA trainers were also paired with experienced teams to understand procedure development, how to work across other technical disciplines (ISS systems, robotics, etc.), chain of command, leadership, teamwork, responsibility, and the overall high performance standards expected of all employees in MOD.

Unlike some of the other groups at JSC that trained astronauts on specific spaceflight systems, the EVA group also supported flight operations during missions, which was particularly interesting to Hansen:

> Some of the other systems groups at Johnson were either an instructor or a flight controller. But for EVA, we were a part of the entire life cycle of the mission EVAs, from early development and planning, to training, and to real-time support of the mission as flight controller. This was an efficient and effective model for optimizing mission success; as the procedure developer and trainer, you are the subject matter expert on all aspects of the spacewalk—so are suited and trained to be spring-loaded for addressing contingencies that may arise on orbit. You also know the crew extremely well and can often anticipate their next steps.

Supporting real-time flight operations requires a whole new set of skills beyond planning and training for EVAs. All new instructors-to-be were required to go through about a year and a half of flight controller certification. Hansen remembers sitting on-console being thoroughly tested by her peers, who were yelling at her while she was trying to sort out "twenty" different conversations at the same time, intentionally pressuring her to see how she responded. "And that's a difficult flow to go through," Hansen admitted. All this training and hard work earned her a seat in the back room; to get into the front room required even more training. During a mission, flight controllers work at computer consoles at NASA's Mission Control Center located at JSC, known as the "front room," and are supported by experts located in nearby less conspicuous accommodations called "back rooms."

Gonzalez-Torres's strength was in the tasks group—tools, techniques, and procedures—but he cross-trained in suits, necessary for him to move into the front room as a flight control room operator. He recalled a four-hour presentation—which he referred to as a dog-and-pony show—he was required to give to certify that he knew the suit inside and out. The audience consisted of engineers and astronauts who grilled him unmercifully until they were confident he knew his stuff. He reminisced with a cynical laugh: "That doesn't sound stressful at all!" This demanding process has since been changed to what is now called "knowledge gates" whereby they are tested as they learn instead of waiting until the very end of their training, which Gonzalez-Torres believes is a much more effective process.

It seems ironic that someone can train an astronaut to do an EVA when they themselves have never done a space walk. Of course, the trainers don't get to go into space, but they can do the next best thing to understand what it's like: put on the space suit and go into the pool. While not required, some of the trainers take that plunge. After passing an air force class 3 physical, Gonzalez-Torres made three runs in the pool wearing the EMU. He admitted that it was an eye-opening event not being able to reach out and grab an object like he normally does; the suit can be viewed as a tool that requires special instructions on how to operate it properly. He had been taught the vagaries of the suit, but experiencing it firsthand made him a better trainer.

Christy Hansen's first suited run in the pool was extremely challenging: "I didn't want to tell anyone that I was extremely exhausted most of the time; but I knew I could do it—because this was the coolest thing I had ever done. I was sweating—and ended up getting several bruises due to a suboptimal suit fit." She stressed the importance of having a good suit fit, but even after having her body carefully measured and a suit built specifically for her, the fit was not optimal:

> The suit was too big for my body, and so there was a lot of extra space that I had to move my arms and chest through before I could physically move my arms and body forward—so a lot of effort for little movement. When I reached my arm out and tried to lean forward, my body and hands were moving around in the space suit—without the space suit actually doing what I wanted it to. When grasping a tool or reaching for a handrail, it felt like my hands and body would move inches inside the space suit before the space suit would move.

Hansen also supported flights on the KC-135 "Vomit Comet" and taught classes in the VR lab, Air Bearing Floor, mass handling trainer (known as Charlotte), and partial gravity simulator. Prior to teaching these classes to crew, Hansen and her EVA colleagues would experience these simulators as subjects; these direct experiences provided context and background to understand the challenges faced by astronauts during training and enabled better instruction.

Kieth Johnson summarized the changes in EVA training from when he first became involved in 1990. "So, we spent most of our time concentrating on the very specific things the crew members needed to do for those [pre-ISS construction] space walks. The crew members would come in and get selected and go through their ASCAN training. They did very few runs in the space suit. Then they would get assigned to a mission and find out they had an EVA and they would get . . . more [task-based] training." He fondly remembered the introduction of the EVA Skills Program and believes it was a great improvement over the previous training curriculum. "I think they recognized that in preparation for space station, there would be EVAs where they may or may not be trained, so we did have to concentrate on more of a skills-based [program]—what's the general things you can do. That's kind of how we prepare now." The skills-based program continues to be fine-tuned to enhance EVA preparation for ISS space walks.

Gonzalez-Torres reflected on the evolution of the EVA training beginning with when he arrived at JSC in the mid-1990s. Space station assembly had not yet begun, and there weren't a lot of space walks carried out in those days, so while JSC had EVA experience, it was limited. Several years into the construction of the ISS, which began in 1998, and with many more space walks, the art of spacewalking had progressed significantly, and by the end of assembly, they had climbed to the top of the learning curve. Gonzalez-Torres exclaimed that "we gained so much experience during the assembly of the space station; I mean it was amazing!" Without the modifications to the EVA training program, it's doubtful the construction of the station would have progressed as smoothly or as quickly as it did.

According to Hansen, following assembly of the ISS and the "wall of EVAs," Expedition crews didn't do as many training runs in the pool as when the shuttle was flying; the training model transitioned to skills-based training, focusing on developing common skills that could be applied to the majority of ISS failures requiring EVA for repair. For more specific and complex failures that may occur in orbit, where the crew didn't receive training specific to that task, a contingency factor would be added to the timeline to allow for an increase in duration due to unfamiliarity with the task and anticipated contingencies that would take time to discuss and address.

Kieth Johnson explained that his role in EVA planning typically began with the initial design of the mission. Shortly thereafter, the crew would be assigned, and often they would jointly begin planning the EVA tasks that needed to be accomplished. He became familiar with the hardware as it was being developed, and he interacted with the crew to better understand what they needed to accomplish. "So, we had insight into the design process and into the operations of whatever the task and hardware was early on so we could influence how the crew was going to be using it."

Choreographing the space walks was an important planning step during the shuttle days and remains a critical phase of the EVA planning process on the ISS. "My group would have to develop the EVA procedures," Hansen explained, "weighing mission priority against performing tasks in the most efficient order possible, while following all flight rules and technical and safety requirements. There are operational constraints and safety implications on how the crew can translate, carry and use tools, and perform assembly and repair tasks, including loading constraints, thermal constraints, power inhibits, general fatigue and human factors issues, life support consumables, and meeting technical requirements—every space walk that is planned must address these critical factors."

During the shuttle days, the instructors employed a method called "training ratios" to help determine how many simulations in the pool were required for a particular EVA. The ratio of training hours to actual EVA hours for a given objective served as an indication of how complex the EVA was expected to be. Straightforward EVAs might have required only five or so suited runs, whereas more complicated EVAs, for example a Hubble servicing mission, might have involved ten or more suited runs.

Gonzalez-Torres expanded on how they approached developing a plan if an EVA was required on the ISS. The EVA tasks group would be given a list of the requirements that had to be accomplished. Then it was his team's job to use their expertise to develop procedures and timelines. This planning is much more complicated than simply estimating how long it would take to do a specific task. Instead, the job would be broken down into its individual components, beginning with how long it would take the crew to translate to the work site. Tether swap locations had to be factored in, along with the amount of time required to do the swap. Next, the additional time required to get from the tether swap location to the worksite had to be estimated, as well as the time it would take to set up a foot restraint, which is another block of time. Tools likely need to be configured. Gonzalez-Torres expanded:

> So now I'm just getting to the worksite and maybe I'm up to twenty, twenty-five minutes, and now I have to determine, okay, they have to drive two bolts, or is it twenty bolts? And then how long is each bolt going to take? You know, I might not cut it down to that detail, but two bolts are going to take a lot less time than twenty bolts. Are they all the same torque? Is it five turns or is it a hundred turns per bolt? And so, as I learned the details of these tasks, then I could estimate how long I thought, based on my experience of training prior crewmembers and actually being involved with the EVAs on orbit, I could then estimate how long will this entire block task take.

There may be additional activities required, with each one getting the same level of scrutiny to come up with an estimate for the length of the entire space walk.

Some objectives have higher priorities than others, explained Gonzalez-Torres, so they try to plan to do those first just in case problems are encountered, and ultimately develop a six-and-a-half-hour space walk, which approaches the limit for an EVA due to astronaut fatigue and the amount of consumables available in the suit. If the estimate to accomplish the required tasks is only five hours, then they can go back to the ISS planners and inform them that they have an hour and a half of time that could be used for other tasks. Once the EVA group had a good timeline estimate, they were ready to test it in the NBL. Often, some of the individual blocks of time are shown to take more time or less time in the pool, so the timeline has to be adjusted accordingly.

Hansen revealed that there are further complications to consider when planning a space walk; the training mockups at JSC are often slightly different from the actual flight hardware. Therefore, the training runs are augmented by trips to the Kennedy Space Center to see the real flight hardware. Learning a task incorrectly on non-flight hardware could mean the difference between a successful space walk and complete failure.

While the EVA group at JSC is in charge of the EVA training according to Gonzalez-Torres, for the HST servicing missions, they were highly reliant on the technical knowledge and expertise of the EVA team at Goddard Space Flight Center located in Greenbelt, Maryland, where Ed Rezac worked. Gonzalez-Torres stressed, "It was definitely a hand-in-hand collaboration. I mean the Goddard folks are incredibly skilled individuals. Very, very hard working. It was just a pleasure to get to work with them." Anyone associated with one of the Hubble missions became part of an unofficial club affectionately known as "Hubble Huggers." There is even an unofficial "Hubble Hugger" patch.

Rezac's role after he transferred to Goddard in EVA training was heavily focused on the HST repair missions, and he noted the value in looking at the history of EVA planning and training. "We looked at those who noodled it out before us and we benefited from those lessons learned—even within our same programs." While planning for the second HST servicing mission, his team closely reviewed the first repair mission back in 1997. They wanted to better understand the problems incurred on that mission, such as doors that wouldn't close and bolts that wouldn't turn. "You build on those past successes, and everyone is a new adventure that provides new successes and lessons learned."

Once a servicing mission had been identified, Rezac clarified that they usually began planning the EVA before the crew had been named. Using life-sized mockups in the NBL, they'd employ available astronauts to perform engineering evaluations. First the trainers would explain the limitations and the capabilities of the telescope—what you can grab and what you can't touch—and then determine if the task could be done as planned. Rezac emphasized that "the astronaut feedback is critical." Once the crew was named, they'd begin developing the detailed procedures necessary to accomplish the job and in parallel identify the required hardware, including special tools. For example, the Pistol Grip Tool (PGT), which Rezac facetiously says every garage tinkerer ought to own, was developed to allow astronauts to tighten or unscrew fasteners despite their bulky pressurized gloves. The PGT turns at a slow speed but high torque and has a computer that can be programmed before the astronaut heads out on the space walk: "Which direction something has to be turned, what the breaking torque is, what the running torque should be. It captures all that, so when you come in at the end of the day you plug it back into your computer and it downloads all that information, which is pretty good to have if you plan on going back there one day." The tool worked so well that it is part of the toolbox on the ISS.

The missions to service Hubble were highly successful and not only corrected a design flaw in the telescope's optics but also made the telescope better than originally intended, and a huge part of that success was providing the right tools for the astronauts to accomplish complicated tasks. The Space Telescope Imaging Spectrograph (STIS) was scheduled for repair on the final servicing mission: STS-125. The STIS was developed to capture data on chemistry, temperature, and movement of planetary bodies by separating light into its various wavelengths. Rezac explained that the panel from the front of the instrument had to be removed to change out an electronics board, and 111 tiny screws had to be removed to get the panel off, each one carrying the threat of escape and the possibility of finding its way inside the delicate optics of the telescope, doing irreparable damage. A specially designed device named the Fastener Capture Plate ensured the screws were not lost.

The PGT, while a great tool, turned slowly and was not capable of backing out 111 screws in a six- to seven-hour space walk, so Mike Massimino, assigned to STS-125, requested a faster turning tool. Due to his limited line of sight at the worksite, he also asked for a smaller tool because he couldn't tell if the bit was on the screw head. Rezac and his team promptly returned to Goddard and developed the Mini Power Tool, which had a slimmer profile and turned at a higher speed, exactly what Massimino requested. Following the next NBL run, Massimino informed Rezac that the tool worked well but that it was dark where he had to work and he couldn't see what he was doing.

Back to Goddard, where they added a light emitting diode (LED) array to the front of the tool. Massimino tested the tool once more in the NBL, and it worked perfectly. Rezac boasted, "We knew that was a good idea because we flew that mission in May and the next Christmas every power tool in Home Depot had an LED light array on the front of it."

Serendipity is often the engineer's best friend. Rezac and an associate were working late one night at the NBL for the next day's training run in the pool in preparation for the repair of the Hubble Advanced Camera for Surveys (ACS) on STS-125. The ACS, which had developed an electrical short, utilized the visible and red wavelengths to survey large areas of the universe. To accomplish this task, the astronauts would have to remove four small electronic boards. Rezac described the boards as being smooth on one side, but the other side had soldering points and sharp edges that could easily damage a glove, so a tool was needed that would allow the astronauts to remove the boards without placing their gloves at risk. They had already developed some elaborate tools that did the trick, but that evening, while replacing all the parts to the mockup that had been moved around during training earlier in the day, they discovered an existing tool that served the same purpose and allowed them to remove the electronic boards more quickly and with greater ease. Rezac was so excited that he immediately called his good friend astronaut John Grunsfeld at home to inform him of their discovery. Grunsfeld made his way to the NBL that night to test out the tool, and it worked superbly. Rezac acknowledged that "sometimes, we just over-engineer."

What if? That's the question that Gonzalez-Torres believes keeps spacewalkers safe and makes missions successful. He wanted to know what the consequences might be for every potential scenario that may occur during a space walk. These range from life threatening catastrophes to inconveniences that might lengthen the EVA.

The "what if" questions lead to the development of written contingency plans just in case problems arise during the space walk. Hansen recalled that for one of the Hubble servicing missions, the contingency document had grown to thirty pages! Still, it's not unusual to encounter problems that had never been anticipated and that were therefore not in the contingency plan. Rezac explained that the lead trainers were usually "flying a console" during the mission somewhere in the Mission Control Center at JSC located in

Building 30. Not only are they following the procedures; they're also looking ahead, anticipating what might go awry:

> This is going on right now but over here on the next page I see that they are going to move to this position and this is the challenge they are going to have to face there, and this is the problem we may have to solve. Every time we tell an astronaut, "I want you to turn that bolt counterclockwise five times," we have to think what if that bolt doesn't turn, what if that bolt only turns two times, what if the bolt snaps off? We have to be thinking of everything that can possibly go wrong and then have a contingency response prepared and ready to put into action. Murphy is always in the mix!

Having the right tool makes any job go faster and much easier, whether you make a living using tools or simply tinker with them around the house and garage. The same is true for spacewalkers working in space. JSC and Goddard have developed a set of space-age tools—generic and specialized—catered to the needs of EVA astronauts working in a weightless environment wearing a bulky space suit. The NASA engineers have been designing tools for years and are very good at building devices that fit the astronaut's needs, for example, a large handle or big buttons to accommodate the gloves.

Tools developed for generic activities would initially get what Gonzalez-Torres referred to as a crew assessment; they'd bring in a small group of astronauts early during the development phase and ask for their feedback. Their input gave the engineering team a broad range of ideas that helped them build a tool that just about any astronaut could use, and it also prevented every astronaut's attempt to cater the tool to their own specific needs. "What this does is it tries to prevent every astronaut coming in and saying, 'Hey I like it blue; I think this would be great,' and the next person comes in and says, 'Oh, my favorite color is red; let's make it red,' you know, and everybody has an opinion. Oh yeah that's great but we're not going to design these fifty different times because fifty different people are going to use it." The tool may not be perfect for every astronaut, but it will perform adequately for almost everyone. It's not uncommon for astronauts headed to the ISS to request a change be made to a tool, but unless the job cannot be done with the tool in its current design, those changes are usually resisted by the engineers.

The tool design for Hubble was done differently according to Gonzalez-Torres due to the uniqueness of the servicing missions. There were going to be only a handful of repair missions to Hubble, which meant that there would be a limited number of spacewalkers—often only a single astronaut, perhaps two, who would use a specific tool—so if the astronaut really wanted the tool design changed to meet their requirements, they had the flexibility to make changes. The Mini Power Tool is a good example. Gonzalez-Torres agreed that there were a lot of good changes made to tools, but on occasion, even on Hubble missions, he'd pull the crew aside and ask them if the change they were recommending was really needed, because changes usually require engineers to work overtime and weekends at additional cost. He was willing to have his engineers put in the extra time and hard work, but only if it was absolutely necessary.

It's clear from talking to Hansen, Rezac, Johnson, and Gonzalez-Torres that developing the training plan and then training the astronauts for EVA was more than an eight-hour workday for them. They worked long hours and put their hearts and souls into their jobs. Hansen recalled that fourteen-hour days were not uncommon. Each of them realized—and accepted—that the success of the mission might depend on how well they did their jobs just as much as the EVA astronauts who did the real work in space. If they failed to train the astronauts properly and a mission EVA objective couldn't be accomplished, the cost to fly that mission again could be staggering. Hansen summarized her attitude:

> As an instructor, I took my job very seriously. You work hard, and you think of everything. On paper, we did everything we could; we checked all the boxes off. But in my head, every night it was like, "What if we forgot something?" I felt confident that we had done everything we were supposed to do, but in the back of my mind, all the time I knew there were things out there that maybe we did not anticipate, which is the nature of space flight. The crew were not only your professional work colleagues but often became your close friends and felt like family.

Like Hansen, Johnson was always confident that he had adequately trained the astronauts to carry out their assigned space walks safely. But he knows that there is always the risk that something will not go as planned: "When you ask me if I'm confident when we send crew members out, I say yes with a caveat. There are always challenges that could come up." Success is predicated not only in the training flow and contingency plans but also on how well Johnson and others on the ground are prepared to get the right information up to the astronauts in real time to mitigate problems that inevitably arise during a space walk.

While filming for the PBS NOVA documentary *Hubble's Amazing Rescue*, Rezac had joined Neil Degrass Tyson at the KSC press site when a reporter asked Rezac if he was worried about the upcoming servicing mission on STS-125. "Well yeah," Rezac said. "Because there are so many things that are just out of our control. Look at all the horrible disasters we've had—the two space shuttle tragedies. Anything can happen."

The disciplined and well-organized EVA training program and its cadre of effective and dedicated trainers give astronauts confidence to exit the spacecraft without fearing for their lives. Tom Akers, with four space walks under his belt on STS-49 and STS-61, summarized his position, saying that he never felt any danger during his EVAs, citing the trainers who prepared him for his missions: "Never worried about safety, given the hours in the suit training underwater and in vacuum. . . . Took all my brain power to be sure to get the job done without too many mistakes! If you knew the time and effort plus the expertise of all the suit folks on the ground designing and maintaining the suits and equipment, you'd not be surprised either." Additionally, he had absolute confidence in the highly trained staff who monitored his every move while he was outside the spacecraft.

Joe Tanner also had complete faith in everyone involved in preparing him for his space walks. When he climbed outside the orbiter, he knew that the suit was as good as it could be. "The EVA flight controller, tasks lead, and systems lead are also following every step and constantly anticipating what might go wrong. The back rooms are filled with flight controllers monitoring the vehicle and payload systems." The flight surgeon was scrutinizing his heart rate and other life systems. The EVA trainers had prepared him to carry out his mission, and Tanner was highly confident of his training and pleased that someone was always keeping an eye on his safety.

Winston Scott echoed the confidence that Akers and Tanner had in their trainers:

> Trainers were very experienced people. They knew what we needed to do. They were good coaches—good knowledgeable people from the text-book standpoint. They were very encouraging. A good group of trainers and support people make all the difference in the world, even though they may not have worn the EMU—maybe once in the water—[and] they've never flown in space. They are very knowledgeable, and they understand what needs to be done from a technical standpoint, from an operational standpoint. They can guide you along and coach you. I don't recall anybody ever getting impatient with an EVA person in the water no matter how much difficulty they may have been having. Just an outstanding group of support people on the EVA staff as well as all over NASA.

Astronauts have been performing space walks on the space shuttle and ISS now for almost forty years. The success of these highly complex excursions outside the spacecraft and NASA's spotless EVA safety record are testimony to those on the ground that prepared them.

Astronauts are highly intelligent and motivated individuals dedicated to getting the job accomplished as planned, safely and efficiently. They come in many sizes and shapes—tall, short, thin, husky, petite—an almost infinite variety. They also have diverse personalities, just like the people in every workplace in the world. When asked if he had ever encountered any problems while training astronauts, Ed Rezac quickly retorted, "Yes, just about all of them." He rationalized that when they are outside the spacecraft in the dangerous void of outer space, even though they have support from their intravehicular astronaut and the ground, they are still a long way from the safe confines of the spacecraft. "When you get down to it, it's their life on the line, so you don't mind them being type A—or I never did."

Johnson recalled working with veteran astronaut Story Musgrave preparing for the EVAs on the STS-61 mission. He was already familiar with Musgrave's outgoing and eclectic personality and often marveled at some of his antics: "Hah! He can get away with that?" Johnson realized that training for such a complex mission was going to be a challenge, even with four experienced spacewalkers, but with Musgrave's long history with the suit, including development of the EMU, he figured there wasn't much he could teach him.

During training, Johnson often presented a suit failure scenario to Musgrave and reviewed the accepted malfunction procedures with him, only to hear Musgrave pronounce, "You know what I would do? I would turn this off. I'd see what this sounds like under this condition," which was completely contrary to what Johnson had been taught. Johnson was always on the lookout for Musgrave to deviate from the accepted path and sometimes wondered if he should inform his management about some of Musgrave's wild ideas. "From my perspective, it was as much a learning for me. I don't know what he learned or what he took out of it. But it was nice interaction, my time with Story."

Kieth Johnson had known Canadian astronaut Chris Hadfield for years but had not worked with him directly. Then he was assigned to train Hadfield for a space walk for the upcoming Expedition 34 mission aboard the ISS. Johnson had worked successfully with many crews training them for EVA missions but admitted that he shared some trepidation about working with Hadfield based on his many years of knowing him. Hadfield was a very talented and confident astronaut and an experienced spacewalker—Mr. EVA—and that is what worried Johnson. "I . . . had some apprehension about working with him. I love the guy. . . . But I will say there is an air of—he knows what he's doing and don't question it." Fortunately, the training went extremely well, and the entire team had a great time. During a debriefing session following one of the NBL training runs, Johnson made a statement to the debriefing group, which included astronauts and trainers: he wanted the junior trainers to understand that while Hadfield was assertive and sure of himself, they shouldn't let Hadfield lull them into thinking he knew everything. Johnson wanted them to give Hadfield advice if they felt it was warranted. That counsel was obviously directed toward Hadfield too, but Johnson wasn't sure that his sage advice was appreciated at the time.

Hadfield's space walk on that long-duration mission was highly successful, and shortly after his return to Earth, he retired from the Canadian Space Agency and wrote a very popular astronaut memoir, *An Astronaut's Guide to Life on Earth*. Hadfield refers to advice given to him during an EVA debriefing session—very similar to what Johnson described—and confides that he was initially insulted by the remark. However, he soon realized that the trainer had been correct.

Johnson had purchased a copy of Hadfield's book but had not read it. One of his coworkers approached him at work and informed him that he, Johnson, was in the book. Naturally, he couldn't wait to find out for himself, and although his name wasn't in the book, a story about the comment that he had made during the training debriefing session was. Johnson was ecstatic! Sometime later at a flight directors' Christmas party, Johnson overheard one of his friends tell a fellow reveler, "I'm in his book." Piqued, Johnson wandered over and asked his cohort which book he was referring to. To his surprise, it was Hadfield's book, and his friend was adamant that Hadfield was referring to him, as he'd made the same recommendation to Hadfield prior to the mission. A friendly argument quickly ensued between the two over which one of them Hadfield was referring to, with neither giving in. With time, Johnson had a revelation. "Wait a minute, what if it took the two of us to get Hadfield to learn this lesson about himself and how we should approach things.... Maybe the two of us together are in the book! We work with all kinds of those various personalities that, you know, love them, hate them, deal with them—it's all part of the job, but it also makes it exciting."

Long hours of training together can create treasured relationships that often last years or longer. Kathy Thornton shared her impressions from her days training for shuttle EVAs:

> You get really close to them, especially on the missions with lots of space walks. They're great people. I guess all the shuttle trainers are in a similar position—they are training people to do something they themselves have never done, which is extremely hard. They maybe get some suit time in the water tank but very little, compared to the amount of time we get per mission. Given those constraints, I think they do an awesome job of working with crews, training them, debriefing them, and finding out what went right and what went wrong and synthesizing that across a number of crews. I might have my opinion on how things should be done—and of course I'm right [laughs]—but other people's opinions might not necessarily agree with mine in a lot of ways.

Gonzalez-Torres believes that the EVA group is unique at NASA. Other groups also spend inordinate hours training astronauts for space flight, but for some reason, the members of the EVA training group seem to become close friends, not only with the astronauts but with their families. They attend parties together, kids get to know kids, and spouses talk to spouses. "Those are our friends out there that we want to make sure are safe. It's not just somebody in a space suit." As his first mission as lead approached, Gonzalez-Torres began having vivid, terrifying nightmares about an astronaut's tether breaking and the astronaut then floating away from the station.

Hansen felt like she became a member of the crews she trained and thought of them as her family. There were hours on the phone with them, weekends and evenings, to ensure that nothing had slipped through a crack, checking and cross-checking technical specs with operational preferences. Procedural changes to improve efficiency were common, with more runs in the NBL, and procedures had to be updated—everyone needed to communicate and to ensure they always had the latest information captured accurately. As launch day neared, the communication hastened.

Support divers wearing SCUBA gear accompany the astronauts when they enter the water to ensure their safety, take video of them, and move them to new worksites. They are constantly looking for any kind of problem that may crop up, observing up close the facial expressions of the astronauts as they descend into the pool just in case a problem arises that the astronauts are not able to communicate. They are also responsible for ensuring that the astronauts are as near perfectly neutrally buoyant as possible, critical to the comfort of the astronauts. Rezac praised the divers: "I have never seen a more professional group of people than the divers and the team working at the NBL." The training operation is quite complex, with two astronauts and a cadre of divers amid the large mockups—in water depths up to forty feet—focused on learning how to complete tasks, and as Rezac makes perfectly clear, "It's dangerous as hell!"

Kieth Johnson went out of his way to show his respect for one highly experienced EVA crew by letting them decide how they would carry out their assigned objectives—which had likely never been done and may never happen again—to route cables on the ISS. According to Johnson, the crew was extremely appreciative of his decision to let them figure out how to get the task done, but he informed them afterward, "Yeah, don't get used to it."

Many of the trainers would trade places with the astronauts in a heartbeat for a trip to outer space, but they understand that that privilege is reserved for very few. They are proudly content with their role in preparing the spacewalkers, knowing that without their expertise the astronauts could not succeed. They also treasure the relationships they develop with them and are very appreciative when the astronauts recognize their efforts.

Hansen believes that succeeding in the EVA training group requires a combination of strong technical skills, interpersonal skills, and extreme perseverance and stamina. For her, keeping her knowledge base sharp and her proficiency high required constant studying and frequent technical interchange and discussions with colleagues; information changes so quickly in EVA that you could never assume you had the latest information. She always felt a huge weight on her shoulders. As part of the EVA and flight crew family, she would not be the weak link who could potentially lead to mission failure. "I felt like I carried that burden with me every day before and after our training runs."

Hansen has moved on from JSC to Goddard Space Flight Center, yet she constantly reflects on and applies the knowledge and experience she gained there to her work at Goddard. From working ISS technology payloads to project management for airborne science missions, she implements that full lifecycle "train as you fly, and fly as you train" philosophy. One of the most long-lasting lessons Hansen learned and carries forward is anticipating failures, knowing that every undertaking and mission will encounter both known and unknown failures and that careful forethought and planning can help prepare for those.

Christy Hansen carries fond memories of long fulfilling days working in the JSC EVA group. She was so busy training spacewalkers in those years that she didn't have time to grasp the impact of her work. Reflecting on what she calls the "best job I ever had," she was simply too busy to appreciate the magnitude of what she and her team were doing every day: "We were building ISS and upgrading and repairing the Hubble Space Telescope, two national assets that were both advancing science and technology. We faced deadlines and issues every day, and it took full bandwidth to address those and remain ahead of the curve preparing for our space missions." Today she is amazed that she was an integral part of such an important endeavor. She often attempts to describe her EVA work to her coworkers at Goddard but simple words fall short of conveying it clearly. "Most people are not aware that the job I had even existed. The public is generally aware of astronauts in space, but not often of the many ground teams who are preparing them for flight. I laugh thinking of my attempts to explain this job to folks outside of JSC. They look at me like I'm nuts."

Tomas Gonzalez-Torres retired after nineteen years at NASA—four as a flight director—and returned to his alma mater Iowa State as a senior design lecturer but is now back at NASA. Ed Rezac recently retired after more than forty-five years in the business to progress space flight.

Kieth Johnson, still in the JSC EVA group, enjoys looking back but is focused on the future. He's proud of his role in the Hubble servicing missions and construction of the ISS, but like Hansen, he is too busy with an overloaded work schedule to applaud his own accolades. Perhaps he too will look back one day after he has moved on and marvel at what he helped accomplish. For now, he's looking forward to NASA returning to the moon—with fingers crossed that the project maintains its momentum.

# 12

# A Hole in the Ice

Development of the space station is as inevitable as the rising of the sun; man has already poked his nose into space and he is not likely to pull it back. . . . There can be no thought of finishing, for aiming at the stars—both literally and figuratively—is the work of generations, and no matter how much progress one makes, there is always the thrill of just beginning.

—Wernher von Braun

Sutton's Bay, on the northwest side of the upper peninsula of Michigan, was gouged out to 130 feet deep by the ancient sheets of glacial ice advancing from the Arctic thousands of years ago. Today, fishermen and adventurers from the nearby town brave the brutally cold winters, venturing out on the hard-frozen bay to pitch small shanties, where they bore through the ice and fish for days. Dr. Jerry Linenger, decades removed from his days as an astronaut, has occasionally even skated on the thick ice with a hockey stick and puck, across "water so clear below that you almost feel suspended . . . like flying as you glide."

On occasion, Linenger plunges into the frigid waters, even during the coldest months. Suited up in a full kit of dry suit and SCUBA gear, he leaves behind the relative safety of the world above and slips through the drilled circular porthole in the ice to explore the inhospitable world below. Down among the tangled webs of seaweed waving gently in the wake of this passing intruder, Linenger is intently aware of his environment. As on his single venture into open space so many years ago, his gear is critical to his survival, and he keeps a constant check on his air supply and the time. Most importantly, he is careful to keep track of his location in relation to the open hatchway back to the life-sustaining environment on the other side of the ice above. "You must eventually find your way back to that drilled hole," he emphasized.

The single highest priority of conducting an extravehicular activity in open space is to simply survive the experience. The protective suit that envelops the astronaut and the engineered wizardry of its life support system must work flawlessly in an extremely hazardous environment. If either the suit or its systems fail in a non-catastrophic manner, there are procedures to be followed, and they all ultimately lead one back to the safety of the airlock hatch. "If you lose suit pressure . . . you had [a] little card on your sleeve that said the pressure is falling at this rate, you've got . . . twelve minutes to get back, or you've got six minutes to get back," explained Linenger. Even in normal operations, as a spacewalker gradually depletes the suit's consumable supply of oxygen and battery power, an ever-tightening radius of safety shrinks closer and closer to the open hatch. "So there's a little bit of—closer to the hatch, you feel . . . like [you've] got some better options if something does go wrong," he said.

With his SCUBA tank nearing empty and the cold of Sutton's Bay sinking into his bones, Linenger would eventually head back to his hole in the ice. Like the rickety old hatch of the Russian *Mir* space station he floated through in 1997, he hoped this natural airlock would still function for him—in this case, not be frozen over—allowing his return to the realm of air-breathing humans above. His careful planning, care for his equipment, and cautious approach to diving are long-honed skills learned over a lifetime of training in the U.S. Navy, at NASA, and with the Russian Space Agency. This same mindset that preserved his life in open space in pursuit of exploration now preserves it as he pushes his own limits today.

Linenger, clad in a training version of the Russian Orlan space suit, gingerly maneuvered around a strangely crude mockup of the space station *Mir*'s airlock at the bottom of the Hydro Lab in Star City, the Russian training facility just northeast of Moscow. He had been immersed in the Russian culture and language for many weeks at this point, in preparation for his mission to the aging *Mir*, where he would be the first U.S. astronaut to perform an EVA from the facility. With a steady stream of bubbles racing toward the surface of the water around him, he heard the thickly accented voice of his Russian Space Agency (RSA) instructor in his earphones. "Jare-dee, Jare-dee . . . open thee hatch *very* carefully . . . *veeery* carefully, swing it open," he was warned.

"Russian is not finesse, and so they overbuild everything," Linenger explained. Examining the mockup he was working with, he noted, "It's got these big, solid, iron hinges on it that look like you can't break them with a sledgehammer, and I'm thinking, 'Okay, open the door very carefully.'" In a culture that was still coming to grips with the transition toward capitalism after the fall of the Soviet Union, institutional knowledge was worth its weight in gold. Many aspects of Linenger's training were not learned from textbooks and diagrams but were rather shared verbally from the instructors and cosmonauts of Star City. As long as they held the knowledge, their paychecks would keep coming.

He had no idea why the instructor was so worried about him damaging the crude, heavy hinges of the mockup hatch, and the voice in his helmet felt no impulse to explain it to him. But after being admonished so many times over the course of his training in RSA's version of NASA's weightless environment training facility, Linenger could hear the gruff Russian voice in his mind before he even spoke. "*Veeery* carefully, Jare-dee . . ." In a few short months, crammed into the real airlock aboard *Mir* preparing for his historic space walk, he would learn why for himself.

Following the space walk demonstration between *Soyuz 4* and *5* in January 1969, the Soviet Union would not stage another EVA for nearly nine years. The Zvezda design bureau had developed a semirigid lunar suit design for use by the commander on the moon's surface, dubbed Orlan. With the demise of the lunar program and a shift in priorities toward orbital space station operations, this design was adopted for EVA use aboard the Soviet's *Almaz* military space station. As the U.S. *Skylab* station neared launch, the Soviets also decided in 1970 to use various elements from their orbital spy station to build a "long-term orbiting outpost" (abbreviated DOS in Russian) known as *Salyut*.

Like the Americans did in developing the shuttle EMU, the Russians left the Orlan as a dedicated suit for use outside and utilized a soft pressure suit for critical flight phases such as launch and reentry. Various modifications of the Orlan design were required for it to meet the requirements of long-duration storage aboard the space station and to be recharged and maintained between the three to four EVAs expected for each suit's service life. These improvements resulted in the new Orlan-D suit (for DOS) to be used aboard the upcoming series of *Salyut* stations.

As with NASA, in the acronym-heavy language of Soviet spaceflight, a space walk was referred to as a VKD, an abbreviation loosely translated as "activity outside of the spacecraft." The first planned VKD of the Soviet space station era came somewhat unexpectedly when *Soyuz 25* was unable to dock with the *Salyut 6* and had to return to Earth.

The *Soyuz 26* cosmonauts Yuri Romanenko and Georgi Grechko were then assigned a stand-up EVA to inspect the docking port to see if any debris or such blocked it. "In this case, we were prepared to dock to *Salyut 6*—not at the front docking port, but at the aft docking port," Grechko explained to author Bert Vis in a 1992 interview. "It was unusual, because this aft docking port was made for automatic cargo spacecraft and we docked successfully and immediately prepared for a space walk through the forward docking device, to check this device from the outside of the station."

On 10 December 1977 Grechko opened the hatch of what he referred to as the station's "torpedo tube," simply floated about halfway out, and gave the docking cone and latches a quick visual inspection. "I went out and checked the forward docking port. Everything was alright—no damage from the spacecraft when they failed to dock," he recalled. "Before this checking, nobody knew whether this docking port was still good for docking, and this was the reason for us to dock at the aft docking port." Satisfied with its condition, Grechko prepared to seal up the compartment, but not before his commander took the opportunity to see the earth from open space himself.

Romanenko pushed a little too eagerly for the open hatchway, and Grechko grabbed him by the legs. The cosmonaut boasted for years after about how he saved his partner from propelling himself into oblivion, but it was well understood by all involved that Romanenko had, at the least, a communication umbilical that would have arrested his departure from the space station.

Two more space walks were conducted from *Salyut 6* in July 1978 and August 1979 by subsequent crews. But it wasn't until the launch of the *Salyut 7* space station in April 1982 that Russia really got any momentum going on routine extravehicular work.

On 28 July 1982 cosmonaut Valentin Lebedev, flight engineer aboard the *Salyut 7* space station, huddled in the connecting compartment between the two humanoid Orlan-D space suits he and his commander, Anatoly "Tolia" Berezovoy, would wear on their first VKD in less than two days' time.

"I remember how, during my first flight on *Soyuz 13*, I wanted so much to get out into open space, spread my limbs apart, and soar above the Earth," he wrote in a small notebook that would later be published under the title *Diary of a Cosmonaut.* "I've come a long way to accomplish my dream; nine years . . . I have shed so much blood sweat and tears to make my dream come true. Today is the day it will happen. This will make all those years worthwhile."

He and Berezovoy had spent days preparing for the space walk, rehearsing their procedures, installing new carbon dioxide scrubbers in the suits, and checking all their photography gear. Lebedev wrote with the honesty entrusted only to his diary of his nervousness and the resulting weight loss over the past week. "Tolia and I realize that we won't be able to joke anymore. We are facing a very serious job ahead," he confided. "The most important thing is to calm down, to take it easy."

But it was to no avail. The night before the VKD, he didn't sleep for a moment. Around 10 p.m., Berezovoy floated by with a sarcastic "good morning" as they gave into the sleeplessness and got on with preparation for the work ahead. Lebedev was anxious to get started. As he looked at a photo of his son on the compartment wall, he reflected on the difficult discussion he had with him before launch. Asked what would happen if he became loose on the space walk, Lebedev was honest, telling him no one would be able to save him. Now, with his son's face staring back at him with a slight smile, he gathered his courage and said aloud, "Onward, my boy!"

In preparation for the space walk, the cosmonauts deactivated the station's gyroscopes and prepared their Soyuz vehicle for an emergency departure, turning on all of its systems before closing the hatches between the various modules and their ship. With their suits thoroughly inspected and leak-checked, the moment they had both waited for so long had finally come.

They did a pressure check of their suits by bleeding some air out of the connecting compartment, and once back in communication range, they were given permission to open the hatch. "I turned the handle of the worm-lock and immediately a shaft of bright sunlight shone in." His pulse spiked to 140 beats per minute as his excitement peaked. "Space, the giant vacuum cleaner, began to suck everything out of the station," Lebedev noted. Glittering dust, small bolts and screws drifted past him as they tumbled off into the emptiness.

As Lebedev floated about waist high out of the hatch, he suddenly felt no fear. His heart rate slowed, and he began to relax. But as he looked around the outside of *Salyut*, he had difficulty orienting himself. As he installed a movie camera and floodlight, he recognized the solar panels and antennae, and after a period of time, he became more comfortable in his new, alien surroundings. This was a unique feature of working on the outside of a cylindrical structure in space that would continue to plague spacewalkers for years to come.

He went about recovering the Etalon experiment, a materials exposure panel that had been left outside for several months, and replaced it with a new panel of samples to be recovered later. Still floating in the hatchway, he then unlocked and deployed the Yakor foot restraint. He worked himself out of the hatch and wiggled his boots into the anchors. Berezovoy would remain in the hatchway monitoring him.

"Space is very beautiful with the dark velvet of the sky, the Earth's blue halo, big lakes and fast rivers, the masses of clouds and complete silence all around," he later wrote. "The panoramic scene is very peaceful and majestic." The station itself appeared to be immobile to the cosmonauts, "a huge mountain frozen in space set against the beautiful background of the spinning Earth."

During one period of contact with the ground, the familiar voice of the world's first spacewalker, Alexei Leonov, broke through the static. "Valentin, do you see any yellow spots in your eyes after you look at the clouds?" He did not. "What do you see now on Earth?" The Russians had a habit of pestering their cosmonauts with trivial questions during times of intense work that required great concentration. Lebedev jokingly replied, "I see our hotel in Baikonur, and our instructors sunbathing near the pool!" The cosmonauts and everyone listening in laughed.

As they passed into darkness, the moon was not bright enough to illuminate their work area, so they turned on a floodlight. They took turns working on a test panel of tightened bolts that they were to loosen and retighten using a torque wrench. They found the work surprisingly tiring, "because our hand and forearm muscles hurt clear up to the elbow . . . the metal rings on the space suit sleeves press against our hands, making them go numb."

The hard work generated a heat load in the suit and a buildup of moisture that caused the helmet visor to fog over. This would be an ongoing problem with the Orlan design for years to come. "It becomes quite difficult to see the cooling regulator dial. When we set the cooling regulator on three to four mark, our legs begin to freeze, even our knees ache, but the rest of our bodies remain in complete comfort," Lebedev reported.

The spacewalkers also conducted some initial feasibility assessments on installing additional solar panels on the station. Based on their observations, the work was later carried out during two space walks by the subsequent crew, Vladimir Lyakhov and Aleksandr Aleksandrov, adding significantly to the station's power supply.

Being outside in darkness reminded him of being on a street corner of a small village. The light from the open hatch was like the welcome glow under a home's front door, and their floodlight like a streetlamp illuminating their seemingly stationary neighborhood. But when the sun quickly rose and pierced the darkness, his serene, comforting environment was immediately wiped away. He could feel the blistering heat of the sunlight on the upper side of his gloves and the hot handrails of the station in his palms.

Back in the hatchway, Lebedev took a brief moment to sit on the edge, "like on a porch step, and admired the starry sky." With the hatch closed and the airlock repressurized, they reversed the process of suiting up, opening the rear entry backpack door of their tiny spacecraft and crawling out backward. "We were immediately struck by a smell in the connecting compartment, exactly like a room which had been irradiated with UV light. The smell came from our space suits after being irradiated by the sun." *The smell of space.*

After the space walk, Lebedev found a dent in his helmet about 20 mm in diameter. More shockingly, the metal was split. "I probably hit it on one of the protruding objects in the connecting compartment. Thank God that helmet is built with double layers of metal," Lebedev wrote.

From that first *Salyut 7* EVA until the *Challenger* disaster in January 1986, a true space walk renaissance took place in both the Soviet and American programs. Twenty-four EVAs were successfully completed, eleven from *Salyut 7* and thirteen from the U.S. space shuttle. In April 1984, when the Americans were performing the first in-space satellite repair on the Solar Max satellite, the Russians were also carrying out maintenance work on *Salyut*, with a series of five complex space walks to fix a ruptured oxidizer line. Vladimir Solovyev and Leonid Kizim installed bypass lines around the area of the leak and wrapped them in insulation in a first-of-its-kind orbital repair job.

As of this writing, a handful of spacewalkers in history have completed ten individual EVAs. Alexandr Serbrov did all his from the new space station *Mir* between two missions, as did Sergei Avdeyev over the course of four flights. Gennady Padalka accomplished this feat from both *Mir* and ISS during five long-duration Expeditions. Americans Michael López-Alegría and Peggy Whitson joined the fraternity with half of López-Alegría's coming aboard space shuttles and the rest during ISS missions. Bob Behnken reached the milestone in 2020 during a short stay aboard ISS. A space shuttle veteran, he performed four EVAs from the station after having launched on the first SpaceX commercial crew mission. Chris Cassidy notched three shuttle excursions, then seven more during his two ISS Expeditions.

And then there is Anatoly Yakovlevich Solovyev. Selected as a cosmonaut in 1976 to pilot the ill-fated *Buran* space shuttle, Solovyev eventually performed his first VKD from *Mir* in July 1990. Over the course of the next seven years, he would accrue a total of *sixteen* space walks totaling some eighty-two hours in open space, a record that stands to this day.

On his first foray into orbital space aboard *Mir*, Solovyev and Aleksandr Balandin were to repair torn insulation on their *Soyuz TM-9* spacecraft. But as they released the outward opening hatch, the crew suddenly realized the airlock section of the Kvant 2 module was not fully depressurized. The hatch slammed open violently, and after their repair work on the spacecraft, they were unable to reseal it. Fortunately, Kvant was a three-section module, and they were able to use the center area as a makeshift airlock, leaving the outer section, with its inoperative hatch, unpressurized.

A week later, Solovyev and Balandin exited *Mir* again to collect equipment left behind on their first excursion and got a better look at the damaged hatch. They took some photos and removed debris and were eventually able to wrestle it closed and repressurize the compartment. However, in October 1990, the new crew of Gennadi Manakov and Gennady Strekalov found the hatch to be more severely damaged than was previously revealed. Over the course of two space walks, they were able to install hardware to further brace the structure, and the repair was considered to be successfully complete—at least by Russian standards.

In the midst of Solovyev's string of VKDs, the visiting crew of *Soyuz T-12*, arrived for an eleven-day stay aboard *Mir*. Among the crew was Svetlana Savitskaya, who had earlier become the Soviet Union's second woman in space. Savitskaya was to now become the world's first female spacewalker, narrowly beating out U.S. astronaut Kathryn Sullivan by less than three months. In the secrecy behind the Iron Curtain of the time, Savitskaya had been given the assignment just a month after NASA announced that Sullivan would be conducting an EVA.

Two other women among the ranks of cosmonauts, Irina Pronina and Natalya Kuleshova, revealed in an April 2000 interview that Savitskaya had great ambitions and had proposed the mission herself. As Kuleshova recalled, "It was her mission.... She joined NPO Energiya and left her test group and started working as the deputy head of the department for training of cosmonauts, to convince the ... authorities of the firm that she must participate in that flight and to perform an EVA."

Once the mission was approved, controversy immediately ensued when Irina Pronina was proposed as the backup crewmember. "The head of the firm told me that I would be backup," Pronina related. "But when Savitskaya heard that, she did her best to ruin that initiative." According to the two cosmonauts, Savitskaya wanted to avoid competition that might threaten her chance to conduct the space walk. She instead insisted on Yekaterina Ivanova as her backup.

"It was Savitskaya's initiative to avoid all competitors because I was quite suitable and in a very good state. I was trained and healthy, and Savitskaya was afraid that in the end, I would be chosen to perform the flight." To avoid the threat of being replaced, she invited Ivanova, who "anthropologically wasn't suited for EVA. She had short arms and short legs."

Kuleshova described Ivanova as "a rather intelligent woman. She was clever, technically." But she was never considered a contender for EVA because "she was a candidate of technical sciences." But Savitskaya got her wish—Pronina would not be her backup. Unfortunately for Ivanova, she had apparently been set up to fail. Of her time training in the Hydro Lab in Star City, Pronina recalled, "It was torture. She was the worst of all." Bert Vis addressed the controversies with Savitskaya—the gamesmanship surrounding the selection and the sketchy reports that she was given the space walk only after Pronina had fallen ill. She responded firmly, "The first crew was Savitskaya, [Vladimir] Dzhanibekov, and [Igor] Volk. This was the *only* first crew, from the beginning.... She couldn't replace me, because I was the cosmonaut-engineer.... Vladimir Dzhanebekov, and I as an engineer, had to decide the main problems during the flight. The woman in the second crew was [a] researcher, and she couldn't decide technical problems as an engineer, in spite of the fact she was ready to walk in space."

Savitskaya, Dzhanibekov, and Volk launched aboard *Soyuz T-12* on 17 July 1984 and docked to *Salyut 7* a day later. On 25 July Savitskaya and Dzhanibekov opened the VKD hatch and left the airlock. While the Soviets released news of the visiting crew flying to their orbiting space station, no mention was made of the space walk until after it was successfully completed, and no live television was shown, even to their countrymen.

Savitskaya had trained extensively to perform a welding experiment, using a hand tool dubbed *universalny rabochy instrument* (URI). While there was some concern that the equipment itself could burn through a space suit, and an unclear practical use for performing welding outside the spacecraft in the future, Savitskaya carried out a variety of tests cutting and welding several different types of metals.

The Soviets had their "first," with Savitskaya beating Sullivan by a narrow margin. It was a propaganda coup for the Cold War adversaries of the United States but would not open any doors for female cosmonauts to come. To this day, Savitskaya remains the only woman of the former Soviet Union to have performed a space walk.

When Sergei Krikalev left the planet for his second trip to *Mir* on 19 May 1991, his country was in disarray. As the Soviet Union began to collapse over the previous several years, intense infighting had ensued, and financial resources for spaceflight were increasingly scarce. When two upcoming visiting flights were reduced to one, Krikalev agreed to remain aboard the station for far longer than he ever expected.

In the ensuing ten months he conducted six consecutive space walks with Anatoly Artsebarsky and one with Alexander A. Volkov. On 25 December 1991, at 7:32 p.m. Moscow time, the bright red hammer-and-sickle-adorned flag of the USSR was lowered for the last time, and a short time later, the red, blue, and white banner of the new Russia was raised. The following day, the Soviet Union was officially dissolved. Krikalev returned to what was now a newly independent Kazakhstan, after 311 days in orbit. The future of his space program was shrouded in much uncertainty, as was that of his home country.

But the collapse of the old communist regime opened the possibility of some further cooperation between the world's two spacefaring nations. Building on the groundwork laid by the Apollo-Soyuz Test Program of 1975, which saw the first docking of an American and Soviet spacecraft in orbit, the United States and the new Russian Federation renewed a cooperation agreement first signed in 1987 and pledged to work together in space going forward.

It was a logical step geopolitically and rapidly gained momentum throughout the early 1990s. The United States would help stabilize the Russian space program financially with critically needed funds, and space engineers and scientists would remain working on peaceful endeavors to explore space rather than seeking other opportunities with less desirable regimes outside of Russia.

Early efforts saw Russian cosmonauts, including Krikalev, flying aboard several space shuttle missions. Even by that time, agreements were made to not only fly the space shuttle to dock with Russia's *Mir* station but to have American astronauts spend months-long missions aboard the complex to gain long-duration flight experience not available since Skylab.

Russia was formally invited into the International Space Station program in September 1993, and phase 1 of the program would be known as Shuttle/*Mir*. A major area of interest between the United States and Russia was what form cooperative EVAs might take, and *Mir* would serve as the perfect platform to expose American astronauts to their counterpart's equipment and techniques.

In 1992 Hamilton Standard, the builder of the shuttle EMU suit, began evaluating Russian space suit technology using its own funds. The following year, the company convinced the Zvezda Research, Development, and Production Enterprise to lease them an Orlan-DMA space suit for assessment, the first time Russia shared their EVA equipment with the United States and a major sign of cooperation between the two countries. When the ISS agreement was signed, it was envisioned that crewmembers from any country would be able to perform EVA work utilizing either the U.S. or Russian space suit, and work was redirected to develop a joint airlock that could service and store both models. It was during this evaluation period, in late 1994, that *Aviation Week & Space Technology* editor James Asker provided global readers with his firsthand account of what it was like to wear and work in the two very different space suits.

Asker found that the suits "reflect the differences in the engineering philosophies and the activities of the world's two manned space programs," and he surmised further that "they even reveal a bit of the ideologies of the two nations that produced them."

Each suit had its strengths and limitations, he found. The most radical and obvious difference was the rear-entry backpack "door" that allowed for easy self-donning of the Orlan suit. It was a suit one *entered* as opposed to put on. While the life support systems were comparable in design, the Orlan featured a higher operating pressure, which reduced the pre-breathe time required. But the tradeoff was that the EMU offered more mobility, with the ability to pivot at the waist and to twist the shoulders, albeit with some effort.

Where the EMU had a clear advantage was in glove design. All the early input from the astronauts had paid off, and Asker found that "the tactile sensitivity of the NASA gloves is outstanding; I could easily pick up a coin." The Russian-designed glove, in contrast, was somewhat hampered by the suit's higher operating pressure.

Another area of different design philosophy was that the EMU was robustly designed to be fail-safe, while the Russians approached Orlan with greater redundancy in all life-critical systems. With the exception of the gloves, it had dual pressure bladders should one be punctured or torn.

While most preferred the rear-entry design of the Orlan and the higher operating pressure offered some operational advantages, it was considered by some to be overbuilt. But the redundancy it offered resulted in a simpler design that worked well, even if it wasn't as elegant as the American space suit.

At the time, three EVAs were planned for docked missions, one from the shuttle on STS-76, one from *Mir* using Orlan suits on STS-81, and one on the seventh Shuttle/*Mir* docking in May 1997, when two cosmonauts from *Mir* and two astronauts from the shuttle would work together to deploy and attach a solar dynamic power module onto the station.

Program managers stressed that the interoperability of EVA systems be advanced to the maximum extent for Shuttle/*Mir* and the coming ISS. With the initial space walks serving as rehearsals for the assembly of ISS, the focus was on "making existing suits, tools, and hardware more compatible." But as time went on, this noble goal became more elusive than anyone could have envisioned.

Astronaut Jerry Linenger, fully kitted up in his new Orlan-M model space suit, reached for the outer hatch that led to the open space outside of *Mir*. He chuckled to himself and grinned slightly as the voice of his Russian space walk instructor echoed from his memory: "Open it *veeery* carefully, Jare-dee." As he swung the bulky hatch open and the sunlight flooded in, he glanced over at the hinges and was shocked at what he saw:

> I look over, and [laughing] instead of these big solid hinges, they've got two C-clamps holding the hinge in place because during a space walk

sometime earlier, someone . . . had not depressurized totally and they blew the hatch out. And so it looks like these little C-clamps you'd buy at Walmart or something holding on this big heavy hatch! And no one told me that, because . . . they don't want to report any bad things . . . because they need the U.S. engaged. And when I saw that, I'm thinking, "Okay, thank God I am being careful with this and I hope that thing closes when we're done because if it doesn't seal . . . we're dead."

Linenger had learned over his months of training in Star City that institutional knowledge was the new currency of the Russian space program, and on the rare occasion when "you're out in the woods talking to someone . . . because that's where you got the real story away from any potential microphones hidden in the walls or whatever . . . people would tell you of space walks that happened in the past," including harrowing stories of things that had gone wrong.

Reflecting on the cagey admonition from his Star City instructor about the hinges, Linenger confided that he never told him about damage to the airlock hatch from the 1990 space walk by Solovyev and Balandin. "Oh, he absolutely knew it! The Russians absolutely knew it. They wouldn't tell me . . . they just, ya know, *open veeery carefully*."

The Shuttle/*Mir* program would see seven Americans spend a combined 907 days aboard the *Mir* outpost, 815 days of which were consecutive. With the exception of Norm Thagard, the first U.S. astronaut to launch aboard a Soyuz spacecraft in March 1995, all were ferried to and from the station aboard the space shuttle fleet. For Linenger, his selection for a long-duration stay aboard *Mir* came very early in his astronaut career.

Linenger was assigned to be the fourth NASA astronaut to represent the United States aboard *Mir*, on a mission that was to launch in January 1997. Training for a flight with the Russians was full of challenges—long separations from family during trips to Star City, intense language study, and a training culture far removed from the relatively hyper-accurate simulators of the Johnson Space Center. As fate would have it, Linenger's assignment came along at the time that the first joint EVA was being planned, using the latest Orlan-M suit.

Training for a space walk in Russia was similar in many ways to how it was done in the United States. RSA used vacuum chambers, with crewmembers suspended by cables to offset the weight of the space suit, to familiarize them with the operation of its systems and the limited mobility of the rigid, pressurized suit. The Russian's version of neutral buoyancy training was carried out in the Hydro Lab, very similar to NASA's various pools, except that it featured an open-grid platform that could be raised out of the water, which held all the mockup equipment the cosmonauts trained on.

"The U.S. is just kind of sleek and modern and methodical on how we do things," Linenger recalled. "The Russians were more . . . an individual just kind of chatting with you . . . and almost winging it a lot of times it seemed, not as planned out." This style of training was only more foreign to the career navy man when it came to the EVA:

> The space walk in particular, in the U.S. side, it was set up like you would military training . . . the very methodical way of doing things. On the Russian side, it was a little different, because I was preparing for a specific task but, you show up, "Well we're not quite ready." Maybe forty-five minutes later you're getting in your suit, you get lowered down in there, they pull you back out. "We don't have it set up down there yet." Then, they do get it set up, then, you're plunged back into the pool again. Working down there, "Well this is not actually what it's going to look like. It's going to be a little different but this is an analogous thing." So it was a lot of just using what they had available and let you work with that and not so high fidelity.

Linenger viewed his space walk as a test flight, of sorts, of the new Orlan-M suit: "It was the first time it was used, and of course, I was the first U.S. astronaut in a foreign space suit, so it was a lot of firsts there which puts a little pressure on you."

The Orlan-M suits were too bulky to launch with Linenger's crewmates Vasily Tsibliev and Sasha Lazutkin aboard their Soyuz spacecraft. Instead, they were delivered by a Progress resupply vessel after the three had joined each other in orbit, with Linenger flying up on space shuttle *Atlantis*. "That was another good day because I was looking forward to that space walk," the astronaut recalled. While they were unpacking the suits, Linenger found a surprise that was indicative of the cultural differences between the two space programs. "They had some 'special medicine,' as the Russians called it, stuffed in the sleeve of that Orlan-M suit . . . and [I] gave that to my crewmates." Alcohol would never be officially permitted, but the unauthorized contraband was a welcome gift for the cosmonauts.

The Orlan series of suits historically tended to collect moisture within them that could lead to fogging of the visor if not properly addressed. Frenchman Jean-Loup Crétien encountered this on 9 December 1988 while he was attempting to close the hatch at the end of his first EVA aboard *Mir*: "I had a ventilation problem in my EVA suit at the end so I had some condensation . . . I could see only half of it at the end of the EVA. Unhappily, I don't know why, when coming back in the air lock, it was totally—more than fog, it's water, because in zero-g, that water stays there. So it's very hard to see through it."

It was a known problem, Linenger recalled, one of the warnings he received from his Star City mentors when out of earshot of others. "Hey, be careful when you get in that suit that you go quickly from when you unhook that umbilical. . . . Get your pumps going to get the ventilator going fast, and the sublimator going even before you step outside the airlock."

"Vasily, during the space walk, very quickly did his and immediately came to me to double check to make sure I had mine going," Linenger remembers. He had to cram himself into the suit, his spine having stretched several inches in the microgravity of space. Within all the layers of undergarments, liquid-cooling long johns, and the suit itself, he felt strangely secure. "I was definitely not floating [within the suit]. I felt tucked in to the suit with all these different layers, and it gives you sort of the feeling of protection and puts you at ease."

With just the gentle hum of the suit's fans and a breeze through the helmet keeping his visor clear, Linenger basked in the silence: "I've been sitting inside this factory, basically, and I didn't even realize it. . . . When it got quiet, it was the most beautiful sound I've ever heard. And so inside the suit, yes you hear [sounds], but . . . probably less than the churn and the machinery on board that space station."

Linenger's primary objectives on the EVA were to attach an Optical Properties Monitor (OPM) experiment to the outside of the station, recover two cosmic dust/debris collector panels, and attach a radiation dosimeter. Once outside in the blinding sunlight, he translated across a short ladder and attached himself and the dresser-sized OPM to one of *Mir*'s Strela telescoping booms.

Looking around the exterior of the aging station brought another shock to the rookie spacewalker. Much like the interior of the ship, with jury-rigged gear and old equipment stashed in random places, the exterior was a maze of nearby solar panels and obstructions sticking out at odd angles. "*Mir* is basically an obstacle course with lots of sharp edges, lots of potential ways to cut your suits and to get rapid decompression," Linenger said. "Trying to translate across the surface would be almost impossible." The boom he had mounted would be manually hand cranked and swiveled by his spacewalking crewmate to avoid the hazards, but it was not without its own limitations.

"It was a lot of 'Stop. Go this direction,' which is hard to describe because he's down at the base and I'm out there," Linenger explains. "It was more like, 'Okay the direction you just moved me, go back the other direction'... and then he cranks me back. Then, 'Nope! Not that way, the other way three clicks'... it was just a trial and error thing where he's trying to move me because I can see that I'm about to ram into a sharp solar panel if he keeps continuing in a certain direction on that arm."

The Strela boom flexed unnervingly as it swung and stopped abruptly with the mass of the American spacewalker and his OPM package. At times, Linenger even saw it whip into an S-curve until the motion fully dampened out. "I'm out there dangling kind of like the fish on the end of a fly rod is what it felt like, and it was a pretty wild ride out there to be that far removed." Once he reached his destination on one of *Mir*'s modules where the experiment would be mounted, he attached his end of the boom, and Tsibliev shimmied up the obstacle-avoiding bridge to the worksite.

Linenger experienced two memorably unexpected sensations during his only EVA, the first being plunged into complete darkness while out at the end of the Strela and simply having to wait for sunrise again to get back to work. Once the twilight of orbital dusk was gone, there was about a twenty minute period of total blackness. "Man, you are just blacked out," he explained. "I had a light on my suit and... if I'm hooking cables, I can keep working during that time for close work," but swinging through sharp thicket of solar panels and old gear on the end of the boom was out of the question. "Eventually, your eyes adjust and now you're seeing the stars all around you and it's wonderful." With nothing to do but wait, Sasha Lasutkin, from inside the station, offered to put some music on over the radio to pass the time: "We say, 'Yeah, pipe it out!' So for [twenty] minutes we listen to rock 'n' roll music whizzin' through space. I'm on the end of a pole in the blackness listening to rock 'n' roll.... It's like... where am I?

This is incredible! Looking down . . . lights of the earth . . . so that's a unique thing for twenty minutes. We'd get a break every time."

Without the blazing-hot sun, the deep cold of space soon seeped into the spacewalker's suit. Linenger dialed his suit cooling to a higher temperature, but it was not enough. "You would definitely be shivering by the end of those twenty minutes," he recalled. Fortunately, with the help of Sasha's music, Linenger was able to bounce up and down on the end of his boom in an orbital dance of sorts to keep his body temperature up. But with the welcome orbital sunrise a few minutes later came the shock of his astronaut career.

Every space walk, in Linenger's interpretation, "is a unique, different experience, and I think a lot of that depends on orientation of whatever the space structure is, lighting, day/night . . . where you are in those cycles, what you're doing, how close to the structure you are versus how far away from the structure you are."

In his case, everything was going well for him. He was doing meaningful work and felt no disorientation or apprehension, until something suddenly hit him—an overwhelming, primal fear as old as human consciousness itself. He was *falling*. "All of the sudden I've got a different interpretation and the interpretation is we are whizzing through space. So I felt the velocity at gut level. And obviously . . . grab on and try to say, 'Okay,' same sort of mental process, 'Okay I have to overcome this.' It was very strange because it was like the whole structure is falling with me and yet I'm attached to it but we're all just falling."

Trying to analyze the sensation years on, Linenger supposed that up to that point, "I faked myself out sort of to think that I'm not flying at 17,500 mph, now I am." He tried desperately to fix his eyes on a different scene—something more stable on the station to orient himself. Despite the overwhelming attack on his senses that might have paralyzed a less experienced man, he found that his "rational being takes over and you say, 'Okay . . . I'm fine, I'm not going to crash, I'm not going to hit the bottom . . . so just tuck it away.'" *What an experience! This is fantastic! Never felt this before and what a thrill!*

It was time to get back to work, but Linenger is adamant that it wasn't just a simple case of mind over matter. He couldn't simply just "tuck it away":

> It's not like I can just pretend that I'm not whizzing through space. When that was happening I was stuck with that sensation. And the only thing

> I could do is say, "Don't fear this" because your genetics over . . . hundreds of thousands of years are telling you you are falling and that's a good instinct we have to grab onto something. It's an instinct saying, "I'm falling and I'm going to crash." The only thing I could tell myself is I'm not going to crash, and so I'm able to take the worst end result and rationalize, I guess then just . . . say, "Okay, I don't care about that. That's in the background, I can still perform right now."

And as quick as it came, *snap!* The gut-wrenching turmoil stopped. "When I get in a different spot, different lighting, closer to the station, whatever it was . . . you're back to just being in space going normal again." The human brain, in all of its yet-to-be-understood intricacies, once again allowed the astronaut to accept that he was back in the comfort of his new environment, with the perception of endless falling and blinding speed left behind, never to return.

Linenger and Tsibliev finished installing the OPM and moved on to retrieving the sample panels and installing the radiation monitor. Although the visor restricted their peripheral vision a bit, Linenger found that "it's got a reasonable amount of coverage. You can see fairly well, but . . . you can't see totally up, [and] you can't see one hundred percent down or left or right to the extremes." So while working within arm's reach was perfectly functional, he had no context of what was around him. *Mir*'s hull was a "convex surface so it's kind of falling away from you . . . so you can't really see around the bend." When translating along a handrail, or loosely hanging by a short tether, backing into a solar panel or other equipment was a constant worry.

This limited visibility and challenging workspace led to one of the funnier episodes of Linenger's stay aboard *Mir*. After removing the cosmic dust collector panels, he handed them off to Tsibliev, who announced he was headed back to the airlock with the experiments. Linenger watched as his crewmate headed off *in the wrong direction*. He hadn't gotten far before he turned around and looked at his spacewalking partner, confused. Not wanting to say anything over the radio that might embarrass him, Linenger recalled they resorted to hand signals:

> He puts his hands in the air like, "Which way's back? Can you tell from where you are?" I can't tell from where I am, because I am again in the middle of a clutter of things, but I see a hatch above me and I float up to . . . the porthole basically, and I look inside the space station through

> the porthole. Now I see familiar things.... I see the treadmill and I say, "Okay, well let's see we're upside down so we must be on the bottom side of this module and therefore if I want to get to the airlock I have to go to the left from this position." And so I push myself way back up and I told Vasily, "It's not the way you were going, it's back that way." So by looking inside the space station of familiar ground I am able to orient myself on the outside of the space station.

Linenger would pass along his improvised method of navigating around the exterior of *Mir* to all of his colleagues who would follow.

Linenger had smelled space before, when opening the hatches between spacecraft. But as he crawled in reverse out of the turtle shell-like door of the Orlan suit's backpack, the overwhelming assault on his senses came not from space but from the spacecraft he now called home. "It smelled like I was back like in a gas station where someone had spilled some gasoline. So I realized the whole space station was kind of full of these volatile chemicals—probably ethylene glycol leaks that we are having. So the suit was almost an escape. It's almost like going from the factory with, say, a steel foundry or something like that and the air a bit smoky and everything else. Now you get inside that suit and I just got my own little ecosystem inside there."

Mentally exhausted, Linenger stowed his gear and had a moment to reflect on the day's adventures. They had performed admirably, but the Russians were scant on compliments. He recalled:

> That was a lot of work up there! And you're not getting [any] feedback whatsoever. We're all human, and it's nice when someone gives you that pat on the back and says, "Job well done." ... That was a two-year effort getting ready for this thing—five months pretty intensely, and it all worked out. You're executing not only on *Mir*, you're going from a dangerous situation, and you're putting yourself in a more dangerous situation. You want to know that what you're doing is worth your life. When I look back on it, it was worth my life and it's a heck of an experience in that five hours of doing that space walk. Put that in the top ten of my life.

Not every one of the U.S. cadre of astronauts who visited *Mir* had the opportunity to perform an EVA. Before Linenger's flight, Shannon Lucid spent her time onboard watching with some envy her Russian colleagues walking in space. As they began preparations for an upcoming EVA, Lucid fantasized that "maybe this time the guys will invite me to go out with them. Yes, the stars are always brighter on the other side of the hatch."

Unfortunately, during Michael Foale's stay aboard the station, a Russian Progress cargo ship lost control and careened into *Mir*'s solar panels and damaged the U.S.-funded Spektr scientific module. Considerable effort was required to assess and potentially repair the damage, and the RSA demanded that all crewmembers henceforth—Russian and American—be capable of performing an EVA.

With the damage to *Mir*, any consideration of the planned four-person EVA to install the solar dynamic power module was shelved. Foale conducted an EVA with Anatoly Solovyev to examine the exterior of Spektr, making him the first astronaut to have conducted EVAs in both U.S. and Russian space suits. David Wolfe arrived aboard STS-86, and it was during this docked period that cosmonaut Vladimir Titov and Scott Parazynski performed their space walk from the shuttle, a first for a Russian in a U.S. EMU, and had the undetected SAFER failure. It was eventually determined that the damaged, depressurized Spektr could not be saved, and the module was sealed off for the remaining lifetime of the station. But Wolfe still had the chance to conduct an EVA of his own. He was just nine years old when he saw the film of Ed White on America's first space walk, and at that moment, he decided that he, too, would do that someday: "It was thirty-one years later that I did it. It was worth every minute of the wait. But I never dreamed it would be from a Russian spacecraft, in a Russian space suit, speaking Russian with a Russian who had been out sixteen times, the most experienced spacewalker in the world, and that's what Anatoly Solovyev is. So it was a real first-hand lesson from the number-one guy in the field, and that was a privilege."

The lessons from the Shuttle/*Mir* program, particularly in the EVA realm, would have great implications for the coming International Space Station. It was imagined that routine operations would occur with U.S., Russian, ESA, Canadian, and Japanese spacefarers using both of the latest generations of two proven space suit designs, from an airlock capable of supporting all EVA operations. Over the entire course of *Mir*'s lifetime, twenty-nine Russian cosmonauts, three American astronauts, two French, and one German representing ESA conducted over three hundred hours of station-based EVA in the course of seventy-five space walks.

Based on *Mir* EVA experience, "We learned that you want to have some well-marked pathways that are obstruction free," Linenger explained. Today's ISS includes these accessibility features along with clearly legible "road signs" that direct a spacewalker, with the limited visibility offered by the helmet visor, to the desired location on the immense structure. "I think in the design we were very careful, and probably that's a good iteration, partly from our experience looking at the *Mir* space station."

Phase 1 of the ISS program was complete, and with the unexpected gain of EVA experience offered by NASA's participation aboard *Mir*, it was time for the world to go build a space station.

The ISS became permanently occupied in November 2000 while it was still in an embryonic stage of its development. During this fast-paced period of long-duration ISS Expedition missions and the visiting construction crews of the space shuttles, astronauts and cosmonauts had to be prepared to perform EVAs in both countries' suits. The Hydro Lab in Star City was modified with equipment to support training in NASA's EMU, while the NBL in Houston was configured to accommodate the Orlan. Yuri Usachev and James Voss of Expedition 2 conducted the first space walk to be staged from the ISS itself in June of 2001, both wearing the Orlan space suit. It was the only EVA ever conducted from the transfer compartment of the Russian Zvesda module, as other dedicated airlock modules would soon be delivered.

The Quest joint airlock, which resembled a high-tech genie bottle of sorts with its large circumference equipment lock topped with the smaller, cylindrical crew lock, arrived the following month on the STS-104 mission. After spending two EVAs making connections and installing exterior oxygen tanks to Quest, Mike Gernhardt and James Reilly staged the first excursion from the new airlock module on 21 July 2001. But even though it was initially dubbed the joint airlock, it was never used to conduct an EVA using Russian hardware.

The Russians instead provided their own airlock module, Pirs, which was launched to an automated docking to ISS, and later outfitted by Expedition 3 cosmonauts Vladimir Dezhurov and Mikhail Tyurin. All three EVAs by this increment's crew, including one by American Frank Culbertson were conducted using Orlan suits.

Following the loss of space shuttle *Columbia* on 1 February 2003, the ISS, its complex assembly far from complete, was severely hampered for the next several years. Ken Bowersox and Don Pettit performed an EVA on 8 April, the last from the station for over ten months. With just a minimal crew of two aboard during subsequent rotations, there simply wasn't enough crew time available to maintain the station, perform scientific and medical experiments, and have any time left for EVA work.

The two crewmembers of Expedition 9, Gennady Padalka and Mike Fincke, were to conduct an EVA in June 2004 from the Quest airlock utilizing the U.S. suits to replace a critical component that controlled the orientation of the ISS in space. But one of the EMUs stored aboard the station developed a problem with its cooling system, which led NASA to develop an alternate plan to get the unit replaced. An opportunity to test international EVA procedures had presented itself, and American and Russian teams got to work planning the excursion from the Russian segment.

The plan was not without its challenges. The Pirs airlock was a long distance from the worksite at the S-0 truss atop the Destiny laboratory module. The crew, in Orlan suits, would have to traverse along the Strela boom to the forward end of the Zarya module. From there, they would cross over to the U.S. segment, around the Quest airlock they would have preferred to use, then up to the S-0 truss. Russian communication antennas would be blocked by structure between the two segments, so the crew would have to occasionally float to the top of the truss to get a clear line of sight and allow suit telemetry to be observed by the ground.

Operational control of the EVA was complex as well. Roscosmos would be in control for the egress and translation until the crew crossed over to the U.S. segment, where NASA Mission Control would take over. With the work completed, Russia would retake control once again as the spacewalkers returned to their side.

The first attempt to carry out this unprecedented EVA on 24 June 2004 ended abruptly after the primary oxygen tank in Fincke's suit showed signs of a leak. The pair was back inside with the hatch closed just fourteen minutes after they had opened it. "I could have done without the malfunction, but that's okay," Fincke radioed to MCC. "I've got experience now. They can't call me rookie." Six days later, they ventured out again, this time successfully completing the repair, and would go on to perform three more EVAs during the mission.

On occasion, old equipment would be discarded from the ISS during an EVA. In February 2006 a decommissioned Russian Orlan space suit was fitted with a radio transmitter and taken outside with Valeri Tokarev and Bill McArthur. Dubbed SuitSat, the old Orlan had been stuffed with dirty clothes and towels by the crew. Just outside the hatch, Tokarev unfastened a tether and pushed the suit, referred to as "Mr. Smith" by ground controllers, into a tumbling orbit of its own with an unceremonious "*do svidaniya!*" In TSUP—RSA's Mission Control—U.S. flight surgeon Shannan Moynihan was told they were going to jettison something, but she was not aware of what that something was. She looked up at the large screens just in time to see the ghastly image of what appeared to be a doomed astronaut spinning away against the black sky.

"Evidently, she found it quite alarming," recalled McArthur. "When I give talks today, the story I tell the audience is that the ground didn't know. We just did this all to play a joke on them. I have a lot of fun with that."

Although its batteries died prematurely, SuitSat continued to transit space until its orbit decayed enough to be dragged down by the upper atmosphere on 7 September 2006.

Early on, the ISS program substantially reduced EVA time requirements through the use of robotics. A rail line of sorts operates outside the station, with a system of motorized and manually propelled carts that translate astronauts, equipment, and even the robotic arm to several locations along the length of the truss. The mobile transporter (MT) and mobile base system (MBS) feature a grapple and power fixture for Canadarm 2, allowing it to be moved to otherwise unreachable areas of the structure. Dextre, the special purpose dexterous manipulator, is a robotic "handyman" that can also be carried by the arm to assist in tasks that might otherwise require an EVA. On either side of the MT are identical crew and equipment transfer aids, or CETA carts, advanced versions of a smaller unit Jerry Ross had tested in space years before. These can be unlatched from the MT and moved manually by EVA astronauts and locked in place with a hand brake. All this equipment rides along rails stretching along the truss, attached by rollers both above and below the rail similar to the cars of a modern roller coaster.

If there were any mundane EVA work, lubricating the wire snares of each latching end effector, or "hand" of the robotic arm, might be considered as such. The large bearings of the solar array rotary joints, which allow the gigantic solar wings to track the sun throughout the station's orbital flight, also need periodic lubrication. But just like maintaining any mechanical gear or vehicle, be it on Earth or 220 miles above it, the job has to be periodically done to keep things in working order.

With a completed keel stretching 310 feet, an internal pressurized volume of its modules equal to that of a Boeing 747, and an acre of solar arrays, the ISS is easily visible as a smoothly gliding star in the heavens as it passes over nearly 90 percent of the earth's population. Often, during an EVA, a ground observer would have no idea that two human beings were floating gently along with that star as they went about their work outside of its life-sustaining modules. But sharp-eyed amateur photographers, utilizing specialized equipment, have managed to capture images of the ISS from the ground in which spacewalkers are discernably visible.

While astronauts and cosmonauts still occasionally work together in either U.S. or Russian space suits, it has become increasingly rare, and the days of crossing over from one segment to another have come to an end. While the astronauts can't say exactly why, the differences in equipment, safety protocols, and operational control all seem to have played a role, and the 2004 EVA by Fincke and Padalka certainly highlighted some of these issues. Jerry Linenger believes that "letting each iterate on their own is probably good from the redundancy viewpoint," but he goes on to speculate that "I think there's a little . . . 'my turf, your turf' kind of thing."

The final space walk conducted at ISS during docked shuttle operations occurred on 27 May 2011, during a mission that was added late in the program to deliver the alpha magnetic spectrometer. This massive particle physics detector promised to unlock the mysteries of antimatter and dark matter in the universe, giving insight to the origin of the cosmos itself. Mike Fincke and Greg Chamitoff attached the orbiter boom sensor system of the shuttle to the starboard truss to be left behind. NASA figured that if another hard-to-reach area of the station needed human hands to tend to it, like Scott Parazynski with his solar array repair, the long boom would be invaluable.

Also left aboard the station were three "garages" full of spare parts. Attached to the Columbus module, Quest airlock, and portside truss, these external stowage platforms contain ORU's such as spare ammonia pumps, electrical control boxes, control moment gyros, and even a spare latching end effector—the "hand" of the station's robotic arm. This stock could be tapped by spacewalkers to keep critical systems functioning aboard the ISS.

With the retirement of the space shuttle after its 135th flight in July 2011, the ISS began a far more self-sufficient era of operations. ISS would rely on Russian Soyuz for crew transport, Progress and ESA's automated transfer vehicle for logistics, and a burgeoning U.S. commercial resupply program leveraging new space companies like SpaceX and Orbital Sciences Corporation. EVA from that point on became the sole responsibility of the Expedition crews onboard ISS, as the winged spaceplanes headed for museums.

On 27 September 2008 a new player entered the field of EVA when China became only the third country in history to conduct its own space walk. Zhai Zhigang exited the orbital module of the *Shenzhou 7* spacecraft, with Liu Boming assisting him from inside the depressurized section of the ship. The earth's limb "above" him and deep black space as a backdrop, Zhigang attached two tethers to a handrail, faced the camera, and said, "I have been out of the hatch and I am feeling good. To all of the people in my country and the world, my greetings! My country, please have faith in me, and my team will finish this mission."

If Zhigang's space suit looked familiar, it was by design. Russia, under an agreement signed in April 2004, provided three Orlan-M suits to China, among other space equipment. China's space engineers used the Orlan design as a starting point, with its decades of use and improvements, to produce their own updated model dubbed Feitian, roughly translated as "flying in the sky." In fact, since this was the first test of the suit in space, Boming wore one of the procured Orlan-M suits as a backup should anything go wrong with Zhigang.

The Feitian featured some improvements over the Orlan—a larger visor, digital electronics, and softer knee and elbow joints—yet the initial version lacked the "moonroof" window in the top of the helmet. It was assumed by Western analysts that the fluid and air connections matched Orlan, since both suits were used on this first EVA. Even the tethers, wrist mirrors, and other gear appeared so similar to Russian equipment; the scene lent itself to being mistaken for a Russian space walk.

Boming poked his helmet up out of the hatch and handed Zhigang a Chinese flag on a short pole, which he waved around enthusiastically in the vacuum of space, to the cheers of ground controllers. The EVA lasted a short twenty minutes, and the pair retrieved a materials exposure experiment for return to Earth. The main objective of the excursion, the first use of the newly designed Feitian suit, had been successfully carried out.

Like Orlan, the Chinese suit was not intended to be returned to Earth for reuse. With no room in the descent module for such bulky gear, Feitian's demise came with the destructive reentry of the orbital module as it was discarded by the crew. As was custom for Russian cosmonauts, Zhigang and Boming were able to keep only their EVA gloves as mementos of their historic first walk in space.

There are two environments—water and the vacuum of space—that could not be more different, yet both are unforgivingly hostile to the alien visitor whose body was not designed through evolution to survive there. And while spacewalkers spend enormous amounts of time training for EVA underwater in a rough simulation of weightlessness, no one ever considered having an emergency involving water intrusion into a suit in space, let alone actually drowning in it. That was until the twenty-third U.S.-based EVA aboard the ISS in July 2013, when an Italian astronaut ventured out for some routine work on the orbiting complex. As fate would have it, his spacewalking partner had come to NASA from the elite U.S. Navy SEAL teams and had already survived the crucible of their grueling qualification course, the infamous Hell Week, and years of underwater training and operations.

*I really wish Chris could come with me.*

Luca Parmitano's path along the station's modules back to the airlock was different from Chris Cassidy's due to their tether configurations. The pair hadn't been far out on the truss of the ISS when his helmet had inexplicably begun filling with water, but relatively close to the airlock where all the silver, cylindrical sections were joined. Cassidy recalled watching his Italian partner heading off to the safety of the airlock just as the sun began to set. As the space station raced around the planet, the next forty minutes or so would be spent on the night side of Earth.

In the waning twilight, a brief moment of orange-gold light brilliantly illuminated the gargantuan complex before plunging the two spacewalkers into total blackness. Only the pair of helmet-mounted lights provided some visibility immediately before them. Parmitano's movement along the structure sent the metastasized sheet of water behind his head sliding forward over his face, covering his eyes and nose. "In that split second I knew I was in trouble," Luca later recalled. He was blinded by the water and the darkness, and his ability to communicate was only worsened when the water, which had already soaked his earphones, began interfering with his dual microphones.

*I don't know which way to go.* "I couldn't even see the hand holds that I was supposed to hold onto to go back. I didn't know how much time I had before the water reached my mouth and completely drown me." Here he was, floating high above the planet, lost in the tangled mechanical seaweed of tethers and handholds in frigid darkness, blindly searching for the airlock's open hatch, the hole in the ice he had left behind over an hour before.

The "really awful sensation" of having the water in his nose and having to breathe only through his mouth was as close as one could come to drowning inside the fishbowl helmet of the suit. Parmitano only made the situation worse when he tried in vain to shake his head to clear some water away. Even in this debilitated condition, he remained astonishingly calm. Cassidy was nowhere nearby, and now no one was answering his transmissions. Then, like a life ring thrown to an overboard sailor in the ocean, he felt a slight tug on his suit—the inertial reel mechanism on his body tether was retracting, pulling him ever so slightly back to the tether's anchor point in the direction of the airlock. "It's not much, but it's the best idea I have," he recalled.

It wasn't until many spacewalkers were on their first EVA that they discovered this peculiar difference between reality and training. "I found, as others did, that even though the reel was meant to auto spool, it didn't always do so," Nicole Stott recalled. "When it did, it actually felt like it was always pulling me back in the direction of the tether point location. [I] never remember experiencing this in the pool."

Parmitano loosely held onto whatever structure he could feel through his pressurized gloves as he let the tether reel him gently back toward the hatchway and safety. Ground controllers, assuming he was doing fine on his own and more concerned with directing Cassidy's clean-up tasks, only then realized that it had been ten minutes since anyone had heard Luca's voice. When he finally reported he was back at the hatch, it was clear he was struggling; through broken static, he managed a garbled transmission, ". . . a lot of water . . ."

"I'm on my way back," Cassidy replied. "Hey Chris?" Parmitano gulped out. "I hear ya, Luca, go ahead," answered his spacewalking partner. *Static.* The original plan was for Cassidy to get into the airlock first, and then Parmitano would close the hatch. But Mission Control told Luca to get in quickly and have Chris follow to do hatch operations. Parmitano dove in headfirst. "Okay I'm in," he managed to transmit. "I'm going slow and smooth here," Cassidy explained to the ground, adhering to the old Navy SEAL adage. *Slow is smooth, and smooth is fast.* A mindset equally applicable to clearing a room of enemy combatants or saving your buddy's life in outer space, the last thing he wanted to do was rush, miss a step, and make the bad situation even worse.

Inside the equipment lock of Quest, NASA's Karen Nyberg and her Russians colleagues could only peer through the tiny porthole on the large, square hatch separating the two compartments until the outer hatch was sealed and the air pressure equalized with the station's interior.

Once Cassidy reported that the outer hatch was closed and locked, Nyberg took over the IVA role from Shane Kimbrough on the ground. Cassidy's helmet-mounted TV cameras still transmitted images of the twisted crowd of two space-suited astronauts in the airlock as Nyberg began the re-pressurization. With the loud, throaty hiss of air rushing into the chamber, Mission Control asked Luca for his status. *Nothing.*

There was no procedure for the situation the two spacewalkers now found themselves in. Cassidy calmly improvised and reached for Luca's gloved hand. "Luca, squeeze my hand if you can hear me," he suggested. This got Nyberg's attention; she placed her hand on the emergency re-pressurization valve on the large inner hatch. If Parmitano was in extremis, she would not hesitate to open the manual valve and rapidly flood the airlock with the space station's air and get him out. Cassidy couldn't reach his partner's hand but got the OK sign from Luca after a few scary seconds. "Shane, I don't know if his voice is going out or if he can't hear you. I'm trying to see him," he reported. "He looks fine. He looks miserable but okay."

As the pressure continued to slowly rise, Luca was unable to clear his ears. The Valsalva device inside the helmet, which pinches the nose closed while the astronaut exhales and pops the ears, had become detached and was bouncing around inside what had become a watery snow globe around his head. "Luca, squeeze my hand. Everything okay? You okay?" Cassidy kept up a reassuring pep talk, even though Parmitano couldn't hear any of it through the soaked headset and blocked ears. He yelled as loud as he could for Nyberg to slow down the flow of air into the airlock.

Finally, with the pressure between the two chambers equal, Nyberg was able to announce, "Hatch coming open." She slid the hatch aside on its rails and reached for Parmitano, floating him into the equipment lock like an astronaut balloon in the Macy's Thanksgiving Day parade. The three crewmates floated around busily trying to assist him. He still had the SAFER rescue jetpack and his chest-mounted tool carrier attached, so the voluminous space quickly became crowded. As anxious as everyone was to remove his helmet, Luca couldn't just reach up and take it off. There was still a slight amount of pressure in the suit that had to be released to avoid the helmet popping off the neck ring and potentially injuring the astronaut.

Cosmonaut Pavel Vinogradov, in the urgency of the moment, tried to reach for the helmet, and Nyberg stopped him. It was a catch not fully appreciated until the later analysis of the entire event. She then unlocked one of Luca's gloves, venting the residual pressure, and immediately began to get the helmet removed. Luca's crewmates stuffed towels in his face to wipe it dry, as cables, his Snoopy cap, and blobs of water floated crazily around the crowded compartment. Nyberg's long blonde ponytail flailed behind her, adding to the visual chaos of the scene.

Finally, Luca's bald head could be seen poking out of the neck ring. He looked spent, taking in big gulps of air as his crewmates disassembled the white suit of armor encasing him. With water still in his nose and his ears blocked, Parmitano could hear none of the reassuring words of his friends. But the intense looks on their faces made it clear to the Italian spacewalker just how concerned they had been for his survival. Cassidy floated himself along the ceiling of the equipment lock, looking down on a scene of incredible relief.

The nearly catastrophic event on EVA 23 led to a months-long stand down of extravehicular work using the EMU while engineers assessed the problem. As EVA office manager Chris Hansen explains, "What actually happened was a complex chain of scientific events, combined with human error and the lack of knowledge of this particular failure." A 2017 NASA Inspector General Report summarized the findings of the Mishap Investigation Board following what in NASA parlance was tagged a "high-visibility close call":

"This investigation concluded the water had accumulated because inorganic materials had blocked holes in the EMU's water separator, allowing water to spill into the ventilation loop that circulates air into the helmet. The investigation also found the buildup of inorganic matter occurred because a water filtering facility at Johnson had not been managed to control for silica. As a result, silica-laden water was used in the processing of flight hardware filters that later were used in four on-orbit spacesuits."

Among some 210 recommendations of the board, a water absorption pad was added to the helmet, as well as a makeshift "snorkel" tube that ran from the neck ring to the lower part of the suit, enabling the wearer to breathe if water did present itself. As importantly, the subject of "cultural complacency" was addressed. "We should have ended this EVA the moment Luca called down to Mission Control that he had water in his helmet," Hansen admits unequivocally. "But it took us an additional twenty-three minutes to end this EVA for Luca's safety." Lessons learned to be passed down to the next generation of EVA controllers.

The typical EVA today is thankfully far more routine than Parmitano and Cassidy's, and the sight of a spacewalking astronaut on ISS is no longer headline news. White-suited figures, floating with their bulky silver Pistol Grip Tools at the hip, look as much like space-age gunslingers as they do orbital repairmen. Helmet-mounted cameras provide views of the work being done for the engineers on the ground, but the constant stir of floating tethers and equipment in front of the astronaut doesn't capture the media's attention. As of this writing, the number of space walks performed by astronauts and cosmonauts in the construction and maintenance of ISS is nearing 270. Of those, almost two hundred have been carried out using the U.S. EMU, while the remainder utilized the Russian Orlan.

During a six-month Expedition to the orbiting complex, extended periods of time may have elapsed since a crewman was last trained for a space walk. Often, a new objective requiring an EVA may present itself that the astronauts never trained for at all. Virtual reality (VR) training that was commonplace on the ground now provides a critical capability onboard the ISS. The same dynamic onboard ubiquitous graphics (DOUG) software is used in conjunction with a virtual reality trainer, complete with 3D goggles, to generate a complete, three-dimensional recreation of the required tasks.

Among the routine space walks, there are the EVAs that stand out. The extraordinarily complex (some thought even impossible) repair of the alpha magnetic spectrometer by Luca Parmitano and Andrew Morgan after one of its cooling pumps had failed was completed over four EVAs beginning in November 2019. The fluid lines had not been designed for repair in orbit and had to be physically cut, one by one, to remove the failed pump. One wrong cut, and the entire device would be rendered useless.

With this hurdle behind them, they would later install a new pump, with intricately designed jumpers and couplings to bypass the old fluid circuit by reconnecting the severed tubing. To the great relief and satisfaction of a large team of engineers and scientists on the ground, when the pump was started up and fluid began to flow, it all worked as designed. AMS, an instrument important enough to warrant delivery by an additional space shuttle mission, was back in action.

In February 2021 a long campaign of EVA began aboard ISS to install six new roll-out solar arrays to supplement the eight original arrays, which have now begun to degrade after providing station power for more than twenty years. And in September a series of eleven space walks was initiated on the Russian segment—utilizing the latest model of their venerable space suit, the Orlan-MKS—to outfit the new Nauka research module, its attached Prichal compartment, and the latest European robotic arm (ERA), giving the Russians a new capability to use a robotic arm to perform work far superior to the old, manual Strela cranes.

Eighteen PLSS/HUT assemblies have been constructed since the inception of the space shuttle program. Generally, four units are aboard the ISS at any given time. One was lost during the pivotal test chamber fire in 1980. Four more were destroyed along with the vehicles *Challenger* and *Columbia* and another during the SpaceX CRS-7 resupply mission failure. Several others have been retired to testing status. But while outdated, and still revealing its secrets after more than forty years of use, the venerable EMU continues to fill the need for EVA until the next generation of suit comes along.

China has continued its advance in EVA experience as well. Thirteen years after the first space walk of Zhai Zhigang, he returned to open space in November 2021 from the Tianhe core module of China's first space station, *Tiangong*, along with Wang Yaping. The following month, Zhigang and Ye Guangfu conducted the second EVA of their six-month mission, installing a video camera and other equipment. Each of these space walks lasted just over six hours.

Liu Boming, who had accompanied Zhigang in 2008, performed two more space walks from *Tiangong* in 2021, testing the latest generation of the Feitian space suit. These later versions feature the overhead window in the helmet like Orlan. This began a regular series of space walks from China's orbital station that continue today, although the Chinese space agency has become more secretive about their EVA work as of late. The crew of *Shenzhou 15* conducted four space walks in the first quarter of 2023, but almost no details of the excursions were shared with the world. Although shrouded in mystery, China continues to claim that in addition to maintaining their space station, the ultimate goal of their EVA development is a manned lunar landing by 2030.

In addition to American and Russian spacewalkers, EVAs have been conducted from ISS by a growing number of international partners—Canada, France, Japan, Germany, Sweden, Italy, Great Britain, and now even the United Arab Emirates. A 12 December 2021 EVA by Japan's Akihiko Hoshide and France's Thomas Pesquet marked the first time a space walk was conducted from the ISS by two astronauts that were neither Russian nor American. EVA has truly become a global international endeavor.

ISS continues to expand its capabilities as it heads into many more years of operations by the international partners, and none of it would be possible without the ability for astronauts and cosmonauts to perform work outside the vehicle. But NASA now has two paths of exploration to pursue—the continued scientific research aboard the ISS, and another that will take humankind farther from its cradle among the stars than ever before.

# 13

# The Promise of Tomorrow

It would be a shame to not be able to send the world's best geologist to Mars because they couldn't fit into the space suit.

—Astronaut Scott Parazynski

EVA in the foreseeable future is highly dependent on humankind's vision for upcoming crewed spaceflight and exploration of the solar system. Assuming future space exploration will continue to include human beings, EVA will surely be required for these missions. NASA plans to continue its presence on the ISS but will likely eventually turn the station over to the private sector, or deorbit it, while concurrently pursuing crewed missions to the moon, Mars, and perhaps destinations beyond. The Russians and Chinese are considering a partnership to construct a lunar base later this decade or next—the International Lunar Research Station—that would evolve from a robotic endeavor to human habitation. The Chinese have also recently announced that sending humans to the Martian surface is in their plans, though few details have been released.

Not to be overlooked, commercial partners are making plans for space tourism, which could lead to space walks for pleasure by private citizens, perhaps becoming commonplace. Space tourism company Space Adventures has signed a contract with the S. P. Korolev Rocket and Space Corporation Energia to send one lucky space tourist outside the ISS for a spacewalk, initially planned for as early as 2023 but clearly delayed. Following specialized training, the sightseer, wearing a Russian Orlan space suit, will be accompanied by a cosmonaut during the space walk.

SpaceX leads the pack in commercial space travel, ferrying astronauts and supplies to the ISS and carrying private citizens into low Earth orbit. Building on their tremendous record of success over the past fifteen years, on 12 September 2024, SpaceX conducted the first commercial/private EVA in history. Billionaire Jared Isaacman emerged from the SpaceX Dragon spacecraft on the

Polaris Dawn mission for about fifteen minutes in a newly designed SpaceX suit capable of withstanding the harsh vacuum of space. He was attached to the spacecraft by an umbilical providing life-sustaining oxygen. Isaacman never entirely left the spacecraft or floated freely, maintaining contact with rails and footholds just inside the open hatch—similar to what NASA calls a stand-up EVA. Sarah Gillis followed suit after Isaacman completed his space walk. Dragon, with no airlock, was depressurized before the hatch was opened, requiring two other crewmembers—who did not participate in the space walks—to also wear pressurized space suits. The EVA training regime was quite rudimentary compared to NASA's well-honed procedures and incorporated simulations, scuba diving, and skydiving but reportedly no neutral buoyancy preparation. The hatch had to be opened manually instead of remotely as planned, and Gillis noticed "bulges" in the hatch seals as she climbed through the open hatch, a concern that will surely be investigated once back on the ground. Fortunately, there were no problems repressurizing the spacecraft once the hatch was closed.

Prior to the Polaris Dawn space walks, spacewalker Bob Stewart declared that EVA space suits were likely too expensive for the private sector in LEO. The development of the new SpaceX EVA suit was shared by magnate Isaacman, and without Isaacman's infusion of cash, it's unclear if SpaceX would have funded the new suit at this time. Regardless, Stewart concedes that some kind of miniature spacecraft or self-contained pod may offer views and experiences like a space walk—and cost less. He also believes that a private EVA adds unnecessary risk: "I just don't see any reason for a non-professional to do an EVA."

SpaceX also plans to land humans on Mars in the near future, and if successful, venturing outside of their habitat will be necessary for maintenance, repairs, and perhaps pleasure. The first missions to Mars will be rife with danger, and astronauts must be expertly trained to carry out EVAs. Perhaps SpaceX could leverage NASA's EVA training regimen; superficially trained astronauts walking on the surface of Mars could lead to disaster. Whether in Earth orbit or on a rocky surface, Stewart is circumspect: "I still don't favor someone who is not exquisitely trained going out and doing an EVA—there's just too much that can go wrong."

Space walks will continue to be carried out on the ISS and the Chinese space station *Tiangong* throughout their existence; maintenance and repairs will be required. Following forty years of human activities in LEO, there are numerous opportunities to venture out into the cosmos—Earth's moon, Mars and its moons, asteroids, and perhaps the moons of several other planets, although many of those suitable for human exploration orbit around the outer gaseous planets of our solar system, which will require inordinate travel times with present propulsion technology. EVA will likely be a part of these explorations, many on celestial bodies with a gravity field.

Deep space weightless EVAs will likely not be as common as space walks have been in LEO. In *Requirements and Potential for Enhanced EVA Information Interfaces*, astronaut Jeff Hoffman, along with nine additional authors, states that although space walks were common during the first seventy years of space exploration, with time, gravity-assisted EVAs will become the norm. Scott Parazynski shares similar thoughts; advances in robotics, automation, and artificial intelligence along with lessons learned from the shuttle and ISS will reduce the need for microgravity repairs and maintenance: "I think future space stations will be designed with robotic servicing in mind, where possible (as long as weight and cost are not prohibitive), making EVA installation and servicing a solution of last resort, rather than the plan." Parazynski believes that gravity-assisted space walks will also benefit from advances in technology: "While I think that future planetary rovers will allow for easier donning and doffing of suits and make getting outside much less of a hassle . . . controlling sensors and deployable rovers from the crewed vehicle will place less emphasis on actual EVA." A suitport that connects the EVA suit to an outer wall of a habitat or rover could be entered from the inside via an airtight hatch and preserve precious consumables by eliminating the need for an airlock.

Astronaut Clayton Anderson's position, elucidated in *This Is the Future of Robot-Assisted Space Walks, According to a Veteran Astronaut*, is consistent with Parazynski's; spacewalking will probably not be eliminated by robots, but the number of EVAs should be reduced by performing some of the routine work presently done by astronauts. Anderson notes that Robonaut, a spindly looking humanoid-shaped robot, was designed by NASA to assist spacewalkers on the ISS but so far has not ventured outside. Another robot, the special-purpose dexterous manipulator (SPDM), known as Dextre, developed by the Canadian Space Agency has shown promise in performing tasks outside the ISS without sending out spacewalkers. But according to Anderson, Dextre is slow and methodical and is dependent on humans on the ground to manipulate it. Even so, Dextre may be a harbinger of fewer EVAs down the road.

Missions beyond LEO where deadly radiation is blunted by Earth's magnetic field add risk. Galactic cosmic rays will expose spacefarers to a low dose of radiation throughout the mission, and it's highly probable that solar flares and coronal mass ejections, although relatively short-term events, will occur randomly during lengthy missions and expose the crew to solar energetic particles. Spacecraft can be constructed to protect astronauts from radiation on their journey to distant locations and the return trip home, including heavily armored recesses inside the spacecraft where the astronauts can retreat to ride out the dangerous solar flares. But it will be difficult to fully shield the spacefarers from radiation. Once on the surface of a rocky world, the crew can construct living habitats covered by thick accumulations of soil to armor their home, providing effective protection. China's *Chang'e 4* lander carried a dosimeter to the far side of the moon in 2019 to accurately measure the amount of radiation astronauts can expect to receive on a lengthy voyage. Fortuitously, it's only about two hundred times more than they would be exposed to on Earth, which can be managed with proper habitat construction.

EVAs will expose astronauts to additional dangerous levels of radiation, thus future space suits will require improved protection, but it's doubtful that exposure can be fully eliminated. Until recently, NASA's limits for the amount of radiation astronauts are allowed to receive during their lifetime while in space varied considerably depending on age and gender. Much has been learned since guidelines were developed in the 1980s, and in August 2021 a proposal was made to eliminate the age and gender criteria and instead limit the amount of radiation received in the lifetime of an individual astronaut to six hundred millisieverts—less than they'll receive on a typical voyage to Mars and back. NASA may be forced to limit the amount of time that astronauts spend outside the living habitat or make exceptions for extended deep space destinations, accepting the increased health risk.

Planetary EVA is much different from spacewalking in LEO, which is not the same as an asteroid EVA. Exploring asteroids, which have negligible gravity fields, presents challenges for the spacewalker; there will be no handholds or rails to grab hold of to stabilize the astronaut, and dust and rocks could be easily dislodged from the asteroid by the movement of the spacewalkers, presenting potential hazards to themselves and the spacecraft.

Spacewalker Tom Jones and his colleagues, in a 2010 paper, *Astronauts beyond the Moon: Mission Operations at a Near-Earth Object*, explore the potential of near-Earth asteroids as attractive targets for NASA to pursue. Following cancellation of the Constellation Program and the Ares I and V rockets in 2010, NASA's future seemed in question. Instead, a project which many called a fill-in mission, was developed to visit an asteroid. Beyond Earth's moon, asteroids offer the easiest accessible bodies in the solar system, although they still present many challenges, including up to year-long missions and exposure to deadly radiation. But Jones and colleagues believe that they could provide "attractive steppingstones offering exciting science return, space resources, deep-space experience, and key knowledge needed for planetary defense . . . and can pave the way for more extensive lunar exploration and eventual missions to Mars." Due to asteroids' small size, ranging from mere meters up to more than thirty kilometers in diameter, and safety concerns to spacewalkers, robots likely provide the best method to explore them.

NASA's return to the moon includes the lunar Gateway, a component of their plan to send astronauts back to the lunar surface. More an outpost than a space station that will orbit the moon, the Gateway is a key element of NASA's Artemis program and may also become a staging point for journeys beyond the moon into deep space. The first two sections of the planned Gateway—the power and propulsion element and the habitation and logistics outpost—are planned to be launched fully assembled on a single heavy lift rocket; hence, there will be no need for space walk construction as was required to build the ISS. Additional components are planned, possibly an airlock that would provide astronauts the capability to carry out EVAs. While the Gateway may never become a hotbed for space walks, it will provide a base for astronauts to return to the lunar surface to carry out scientific research.

Stephen J. Hoffman, in *Advanced EVA Capabilities: A Study for NASA's Revolutionary Aerospace Systems Concept Program* (2004), examined the future technology requirements for EVA and developed a comprehensive report which identified EVA design concepts for missions beyond 2020. Although at the time, NASA had yet to develop what that future entailed, Hoffman assumed that it would include both space walks and gravity-assisted EVAs on the moon, Mars, and beyond.

NASA's return to the moon during the initial phases of the Artemis program will likely have more of an engineering focus, such as mining ice to develop rocket fuel. Stephan Hoffman believes that when NASA missions transition to scientific investigation, astronauts should be extensively trained in geological concepts or be professional geologists. Many of the tasks required of them will be like those carried out by the Apollo astronauts, but expectations will be considerably higher. With only six landings during the Apollo era, most of the lunar surface remains unexplored. Therefore, NASA will need to develop the capability to reconnoiter much larger areas and travel greater distances to collect samples that reflect the detailed geology of the moon. Astronauts should also be expected to observe and catalog geological relationships, the way one rock body relates to another. For example, observation of cross-cutting relationships in the rocks can determine which rock unit is younger and which one is older. Space suit face plates will need to be developed that allow astronauts to detect subtle changes in color, which often reflect important geological relationships. Hoffman astutely points out that the more rocks the field astronaut-geologist inspects, the better the resulting geologic maps.

Documentation of the locations sampled is essential to the construction of accurate geologic maps. Onsite preliminary analysis of selected rocks and regolith is expected, while some samples will be sent back to Earth for more detailed examinations using higher resolution equipment. Robotics will also be beneficial for sampling difficult to reach locations.

Much was learned about the moon from the Apollo missions during their short forays over its surface, and the next generation of spacewalkers will need to build on the lessons learned from Apollo. Keep it simple. Suits and gloves need to be improved. Increased automation will be extremely beneficial and crews will profit from operational flexibility—and less control from the ground. Equipment should be easy to maintain and repair. Longer stays on the surface will be essential to unraveling the complete story of how the moon came to be and what benefits it offers to humankind. Priorities will need to be established.

Mars will present new challenges to future spacewalkers in addition to those encountered on the moon that must be overcome prior to attempting such an audacious mission. The long travel time to Mars, likely in excess of two hundred days depending on the path taken, will challenge even the hardiest well-trained astronauts. Traveling for months on end with a small crew in a tiny spacecraft, isolated, sustained by a diet with limited variety, and with few of the niceties that one is accustomed to on Earth will be a challenge. However, NASA's astronaut selection process and excellent training regimen will allow them to choose crew members who are compatible and fully capable of making this long voyage.

W. W. Mendell and A. D. Griffith, in *Lunar Precursor Missions for Human Exploration of Mars*, cite some of the challenges for excursions on Mars. The long travel time adds logistical difficulties to the entire mission, including training for EVA due to the months that will pass before stepping onto the Martian surface. Refresher EVA training will be required, but it will not be as beneficial as the training on Earth under the watchful eyes of EVA experts. Jerry Linenger acknowledges that while the bodies of astronauts will likely weaken on long-duration missions, there is a bright side: humans are highly adaptable. "If I were . . . going to Mars or going to do space walks in other places, the longer you can adapt inside that environment, say a module on Mars before you actually go do a spacewalk, probably the better off you are."

The substantial delay in communication from Mars to Earth and back is another challenge that must be reckoned with. Depending on the position of the two planets, the time for a radio signal to reach Mars from Earth ranges from five to twenty minutes, much too long for effective communication during an EVA. The astronauts will need to rely on their fellow crewmembers in the living habitat or rover for real-time support during ventures outside, whereas ISS spacewalkers are in near constant communication with Mission Control at JSC.

In addition to venturing outside for scientific observation, regardless of the celestial body, the astronauts will also be responsible for servicing and maintaining equipment external to the living habitat, such as rovers, landers, the habitat, robots, and numerous other pieces of equipment. They'll also need to be extensively trained in rescue operations in case their fellow astronauts encounter difficulties which require assistance or rescue during an EVA.

There are many unknowns associated with working on the Martian surface, such as productivity and metabolic rate of the astronauts. Mars's gravitational field is a mere 37.8 percent that of Earth's. Until astronauts have spent considerable time on the surface, they'll have limited data for daily schedule planning, including the amount of time allocated to sleeping, working, eating, exercising, and recreating. Schedules will need to be adjusted as lessons are learned.

Dust management must also be considered, another lesson learned by the Apollo moonwalkers. Razor sharp edges are common, which can easily damage a space suit. Lunar dust seemed to penetrate every conceivable seam or opening and adhere vigorously to their space suits, and it's doubtful that the lunar space suits would have remained operational if more than three space walks had been attempted on a single mission. To complicate matters, Mars and lunar dust are different, and NASA has yet to return to Earth any samples from the Martian surface to analyze. Both are very fine grained, but the lunar dust is angular in shape due to numerous meteor impacts over billions of years coupled with the lack of erosion and weathering; Mars dust is likely less angular and more rounded due to abrasion from windstorms in its albeit thin atmosphere. Additionally, regular suit maintenance will need to be factored in, including a stock of readily available spare parts, or a way to manufacture them.

Mendell and Griffith stress that the EVA space suit of the future must be customized for the mission and incorporate the latest technological advancements. Simple tasks such as walking, bending over, and picking up tools must be accomplished with minimal effort, thus it's imperative to involve the operator early in the design process. For example, boots for space walks need no treads, whereas they will be crucial for gravity-assisted EVAs. Ease of use and minimization of specialized training should be designed into the suit. Planetary EVA considerations also include the astronaut's center of mass and ground reaction force—mass is needed to create friction—which will impact the design of the space suit and the ability to traverse the surface. Jeff Hoffman believes that in some aspects, planetary EVAs may be more complicated than space walks, primarily due to the weight of the suit: "I don't want to say that one is harder than the other, but it is a different set of problems that you have to solve."

NASA's Amy Ross has extensive experience working on advanced space suit designs—and problem solving—including an improved EVA glove to replace the EMU 4000 series glove introduced in 1985 in preparation for a surge in EVA activity to build the ISS. The new Phase VI glove provided spacewalkers a more comfortable fit with better mobility and was the first EVA glove to be developed utilizing computer aided design. Her father Jerry recalled that she worked on a team "that helped to do the final testing and certification to get them ready to go fly in space." Serendipitously, the elder Ross, on the STS-88 mission, was the first to test the new glove. The plan was for him to perform his first EVA with the older gloves, switch to the new gloves on the second

EVA, and then choose the ones he liked best for his third EVA. Whether it was his familiarity with the older gloves during previous space walks or his confidence in his daughter, he wore the Phase VI gloves on all three EVAs and gave them high marks.

Unsurprisingly, advanced computer technology has vastly improved the development and testing of new space suit prototypes. Amy Ross explained to NASA public affairs officer Josh Byerly in a 2014 interview that prototype suits are outfitted with reflectors for testing that can be photographed using multiple infrared cameras; those images are then input into a computer program to enhance suit mobility. Previously, Ross had to take standard photographs and make measurements manually, a much slower and cumbersome process.

NASA's Z-2 prototype EVA suit, delivered in 2016—the latest design—is based on decades of testing since the Apollo program. The suit, designed for gravity-assisted EVAs, provides vastly improved mobility. According to Amy Ross and Richard Rhodes in a 2014 paper titled *Z-2 Prototype Space Suit Development*, an earlier Z-1 prototype contributed to decisions that NASA made with respect to mobility joints that were specified in the Z-2 contract. Ross has carried out demonstration runs in the pool in both the EMU and Z-2 prototype to evaluate firsthand the products she designs.

The exploration Extravehicular Mobility Unit (xEMU), a near flight-ready Z-2 suit with a few modifications, was designed to support NASA's Artemis program to return to the moon. More dexterous than the Apollo space suits, it can tolerate temperatures ranging from minus 250 degrees up to 250 degrees Fahrenheit, manage the fine-grained angular lunar regolith, and facilitate EVAs up to eight hours. The Portable Life Support System keeps the suit functioning properly and provides a livable atmosphere and temperature control. Miniaturization of electronics and plumbing allow duplication of many features, adding a layer of redundancy. The suit incorporates a rear-entry design making it less prone to shoulder injuries, which Bob Stewart believes is superior to the EMU presently in use on the ISS. "I think the [rear entry] Russian suit is a damn good design."

The Z-2 also incorporates the Phase VI gloves that Ross worked on in the 1990s, but she stressed that eventually, they'll need to be upgraded or replaced. Gloves can be adequately heated to keep the hands warm in the coldest environment, such as lunar craters that never receive direct sunlight, but working with cold tools in these extreme environments will be demanding.

The EMUs are more than forty-five years old—only a handful of functioning suits remain—and need to be replaced to support the ISS for the remainder of its life. A modified Z-2 prototype—the Z-2.5—was developed and tested in the NBL for a planned microgravity test on the ISS. The modified Z-2.5 upper torso assembly can be mated to the EMU lower torso assembly creating a hybrid suit for ISS space walks. However, Amy Ross shared that the entire Z-2 suit appears to be suitable for ISS space walks, even though it was designed for gravity-assisted EVAs.

The xEMU is also made to accommodate astronauts of all sizes, unlike the EMU. Tall astronauts often had to be stuffed inside the EMU, causing much discomfort. Smaller women astronauts struggled inside a space suit designed when there were no female astronauts, and more than a few of them were denied the opportunity to perform space walks simply because the suit was too large for them to function inside it.

The Snoopy communications cap, long a staple of space suits, soon becomes hot, sweaty, and uncomfortable during a long EVA. Velcro used on the caps tends to rub—and irritate—right behind the ears. The xEMU incorporates embedded microphones that are voice-activated, eliminating the Snoopy cap.

Once NASA moved to long-duration stints on the lunar surface, plans were to include a heads-up helmet display, a checklist of readily available procedures, schematics, imagery, navigational data, suit status and other helpful information, intended to save precious communication time and make EVAs more efficient. For example, readily available oxygen levels may help rookie spacewalkers slow down and better manage their consumables.

Tests have demonstrated the xEMU's increased mobility and improved field of view over the EMU, while providing a more comfortable fit. Initially NASA expected to have two flight-ready xEMUs delivered by November 2024. Unfortunately, lack of funding, the COVID-19 pandemic, plus technical setbacks have plagued progress and pushed that date back to at least April 2025.

NASA has already spent nearly half a billion dollars on this vastly improved next-generation space suit, and it's anticipated that delivery of the first two flight-ready suits will total over a billion dollars. The suit was intended to support the ISS, Gateway, and landers that will transport astronauts to the lunar surface from the Gateway outpost.

Due to delays in the xEMU program, NASA awarded contracts to Axiom Space and Collins Aerospace companies in 2022 for a new Artemis space suit. Prompted in part by the then optimistic goal of a lunar landing in 2024, NASA decided to seek commercial expertise to develop new suits. Axiom, in partnership with the David Clark Company, KBR, and Paragon Space Development Corporation, introduced their new suit—the AxEMU—in March 2023. Still not fully developed, the suit is entered from the rear, features newly developed gloves, and incorporates many of the improvements deigned into the xEMU. Collins, jointly with ILC Dover and Oceaneering Space Systems, is also leveraging NASA's expertise to develop their new suit along with their experience in the shuttle EMU, which they developed and built in partnership with ILC Dover. NASA expects the new suit designs to support both the ISS and Artemis EVA missions.

Bob Stewart would like to see a hard suit developed for lunar EVAs and has creative and intriguing ideas for space suits of the future: "What I would really like to see is a spray-on space suit. Because all I have to do is have a body restraint so that my gases don't percolate out and blow me up. Am I going to get it? Probably not. But a hard suit would eliminate the need for a prebreathe." He also envisions an alternative system for weightless space walks. Acknowledging that human legs are not necessary in space, he explains, "So what if you had a little flying machine where you had both of your legs in a little bucket and an anchor on the bottom of that bucket that you could put into . . . an EVA workstation lock? Then you could run that thing using remote manipulators at any pressure you wanted to; you could run it at 14.7 [psi] because [you] would be using a little space craft yourself to fly around. . . . I don't see any reason why we can't do that."

Once Mars has been explored, there are abundant opportunities throughout the rest of the solar system to investigate that will require boots on the ground. The moons of Mars—Phobos and Deimos—make attractive exploration targets, but their small size and negligible gravity fields would be like exploring an asteroid, perhaps easier and more safely done robotically.

The gaseous giant planets in the solar system have numerous moons, each with their own secrets and mysteries, ripe for astronauts to explore up close. NASA will certainly continue to explore these bodies remotely by sending satellites to capture more data. One such voyager is the Europa Clipper satellite planned to settle into orbit around Jupiter in 2030 to search for life on one of its many moons, Europa. But eventually, technology will be developed to land humans on these celestial bodies to further assess the remotely captured data and collect additional information necessary to unravel their scientific mysteries.

Most of these moons have surfaces of ice due to their extreme distance from the sun, and many of them are suspected to hold vast oceans beneath their icy crusts, possibly containing liquid water which could contain life. Traversing icy surfaces will be much different from walking on the rocky landscapes of the moon or Mars, so new space suits will need to address these terrains with reduced gravity fields and much colder environments.

Europa is one of these worlds. It retains a thin atmosphere of oxygen, nitrogen, and maybe water vapor, but it doesn't generate sufficient pressure at the surface to venture outside without a pressurized space suit. The gravity is nearly that of Earth's, but walking on its surface would be an eerie experience; due to its distance from the sun, it receives twenty-five times less sunlight than Earth, requiring artificial light. And Europa is cold; space suits would need to be adequately robust to protect astronauts from extreme conditions. The surface is believed to be quite young, perhaps less than a hundred million years old, with scattered impact craters and long fractures in the frozen crust that offer ample objectives for exploration.

Miranda, which orbits Uranus, may be geologically active due to the tidal forces caused by its mother planet, but it has no atmosphere. Large surface scars, the result of enormous faults breaking its surface, invite investigation. Imagine chasms twelve times deeper than Earth's Grand Canyon, yet with less than 1 percent of Earth's gravity, an astronaut could fall from a cliff thousands of feet high and descend so slowly that they would likely avoid injury if properly trained.

Saturn's moon Enceladus also has a gravitational field so low that it will be very difficult for astronauts to negotiate its surface—they'll float away if not anchored to the planet. Saturn's moon Titan has about half as much of an atmosphere as Earth, providing enough pressure to eliminate the need for a space suit. But surface temperatures average nearly minus 300 degrees Fahrenheit, well below shirt-sleeve comfort.

Outposts will certainly spring up on the moon and Mars that evolve into settlements where EVA will be the norm. With time, improved rocket propulsion will allow routine visits to the strange worlds of the outer solar system. Fairytale landscapes await future spacewalkers on the icy moons of the gaseous planets—towering mountain ranges, seemingly bottomless chasms, geysers spewing water and methane hundreds of miles high. Discoveries beyond imagination await eyes to comprehend. The Apollo missions discovered no life on the moon—as expected—but some scientists remain optimistic that life will be found on Mars, although the probability is low. But the soups beneath the icy crusts of the moons of the outer planets may just hold the holy grail of space exploration—life, just toiling along until a future astronaut happens by and finds evidence of it. There is simply no robotic substitute for human eyes to observe and the human mind to analyze those observations. For this reason, it's certain that spacewalkers, standing on the shoulders of giants, will play an integral role in the continued and future exploration of the solar system.

# Sources

## Books

Anderson, Clayton C. *The Ordinary Spaceman: From Boyhood Dreams to Astronaut.* Lincoln: University of Nebraska Press, 2015.

Bean, Alan. *Painting Apollo: First Artist on Another World.* Washington DC: Smithsonian Books, 2009.

Burgess, Colin. *Footprints in the Dust: The Epic Voyages of Apollo, 1969–1975.* Lincoln: University of Nebraska Press, 2005.

Chaiken, Andrew. *A Man on the Moon: The Voyages of the Apollo Astronauts.* London: Viking, 1994.

Collins, Michael. *Carrying the Fire: An Astronaut's Journey.* New York: Farrar, Straus and Giroux, 1974.

Hacker, Barton C., and James M. Grimwood. *On the Shoulders of Titans: A History of Project Gemini.* NASA Special Publication 4203, in the NASA History Series, 1977.

Hadfield, Chris. *An Astronaut's Guide to Life on Earth.* Toronto: Random House, 2013.

Harwood, William B. *Raise Heaven and Earth: The Story of Martin Marietta People and Their Pioneering Achievements.* New York: Simon and Schuster, 1993.

Irwin, James B., with William A. Emerson. *To Rule the Night: The Discovery Voyage of Astronaut Jim Irwin.* New York: A. J. Holman, 1973.

Kelly, Scott. *Endurance: A Year in Space, a Lifetime of Discovery.* New York: Knopf, Borzoi Books, 2017.

Lebedev, Valentin. *Diary of a Cosmonaut: 211 Days in Space.* New York: Bantam Books, 1990. Originally published in 1983 by Nauka/Zhin.

Massimino, Mike. *Spaceman: An Astronaut's Unlikely Journey to Unlock the Secrets of the Universe.* New York: Crown Archetype, 2016.

Melvin, Leland. *Chasing Space: An Astronaut's Story of Grit, Grace, and Second Chances.* New York: Amistad, 2018.

Millbrooke, Anne. "More Favored Than the Birds—The Manned Maneuvering Unit in Space." In *From Engineering Science to Big Science: The NACA and NASA*

*Collier Trophy Research Project Winners*, edited by Pamela E. Mack, 299–318. Washington DC: NASA, 1998. Open access at https://history.nasa.gov/SP-4219/Chapter13.html.

Parazynski, Scott, with Suzy Flory. *The Sky Below*. New York: Little A, 2017.

Phelan, Dominic. *Cold War Space Sleuths: The Untold Secrets of the Soviet Space Program*. New York: Springer Praxis, 2013.

Pogue, William R. *But for the Grace of God: An Autobiography of an Aviator and Astronaut*. Self-published, CreateSpace, 2011.

Ross, Jerry L., with John Norberg. *Spacewalker: My Journey in Space and Faith as NASA's Record-West Setting Frequent Flyer*. West Lafayette IN: Purdue University Press, 2013.

Scott, David, and Leonov Alexei. *Two Sides of the Moon*. New York: St. Martin's, 2004.

Shayler, David. *Gemini 4: An Astronaut Steps into the Void*. New York: Springer Praxis, 2018.

———. *Walking in Space*. New York: Springer Praxis, 2004.

Slayton, Donald K., with Michael Cassutt. *Deke! U.S. Manned Space: From Mercury to Shuttle*. New York: Forge, 1994.

Stewart, Robert. *To Fly*. Published by the author, 2008.

Virts, Terry. *How to Astronaut: An Insider's Guide to Leaving Planet Earth*. New York: Workman, 2020.

### Periodicals and Online Articles

Anderson, Clayton C. "This Is the Future of Robot-Assisted Spacewalks, According to a Veteran Astronaut." *Forbes*, 29 April 2016. https://www.forbes.com/sites/quora/2016/04/29/this-is-the-future-of-robot-assisted-spacewalks-according-to-a-veteran-astronaut/?sh=5fedf2e02c71.

Ares67. "Challenger STS-6—A Walk into History." NASA Spaceflight, forum post, 7 April 1983. https://forum.nasaspaceflight.com/index.php?topic=33194.msg1117285#msg1117285.

Asker, James R. "U.S. Russian Spacesuits: Aviation Week Comparison." *Aviation Week & Space Technology* 142, no. 3 (16 January 1995).

Azriel, Merryl. "Jim LeBlanc Survives Early Spacesuit Vacuum Test Gone Wrong." *Space Safety Magazine* 28 (November 2012). http://www.spacesafetymagazine.com.aerospace-engineering/space-suit-design/early-spacesuit-vacuum-test-wrong/.

Dunn, Terry. "The Spacesuit Fire That NASA Refuses to Forget." *Space Safety Magazine*, 9 December 2016. http://www.spacesafetymagazine.com/aerospace-engineering/space-suit-design/spacesuit-fire-nasa-refuses-forget/.

Evans, Ben. "A Cog in a Political Machine: The Career of Svetlana Savitskaya." AmericaSpace, 10 February 2012. https://www.americaspace.com/2012/02/10/a-cog-in-a-political-machine-the-career-of-svetlana-savitskaya/.

———. "Four-Time Russian Cosmonaut Aleksandr Serebrov Dies at Age 69." AmericaSpace, 12 November 2013. https://www.americaspace.com/2013/11/12/four-time-russian-cosmonaut-aleksandr-serebrov-dies-at-age-70/.

———. "Is My Hair Gray? The Story of Soyuz 4 and 5 (Part 2)." AmericaSpace, 5 January 2014. https://www.americaspace.com/2014/01/05/is-my-hair-gray-the-story-of-soyuz-4-and-5-part-2/.

Fédération Aéronautique Internationale. "Leonov Successfully Made the First Spacewalk: 18 March 1965." 18 March 2015. https://www.fai.org/news/18-march-1965-leonov-successfully-made-first-spacewalk.

Fossil Hunters. "Interesting Things Attract Me." Updated 6 April 2022. https://www.fossilhunters.xyz/salyut-in-space/interesting-things-attract-me.html.

Germani, Clara. "Space Heroine as Hard-Line Hopeful Russian Election: Svetlana Savitskaya, the First Woman to Walk in Space, Is a Top Communist Candidate in the Once All-Powerful Party's Comeback Bid in the Duma." *Baltimore Sun*, 14 November 1995.

Graf, Gary R. "Manned Maneuvering Unit: Taking a Look Before the Leap." *Space World*, June 1985.

Harding, Pete. "EVA Successfully Frees Stuck CETA Cart outside ISS." NASA Spaceflight, 21 December 2015. https://www.nasaspaceflight.com/2015/12/nasa-eva-free-stuck-ceta-cart-iss/.

Harwood, William B. "Suit Problem Ends Station Spacewalk." SpaceflightNow.com, 24 June 2004. https://spaceflightnow.com/station/exp9/040624spacewalk.html.

Jain, Supriya. "40 Best Buzz Aldrin Quotes." Kidadl, 12 December 2023. https://kidadl.com/facts/quotes/best-buzz-aldrin-quotes-from-the-apollo-11-astronaut.

Jones, Andrew. "Astronauts Complete First Chinese Space Station Spacewalk." SpaceNews.com, 4 July 2021. https://spacenews.com/astronauts-complete-first-chinese-space-station-spacewalk/.

———. "China's Shenzhou 15 Astronauts Complete Record-Breaking 4th Spacewalk." Space.com, 19 April 2023. https://www.space.com/china-shenzhou-15-astronauts-fourth-spacewalk-tiangong.

Jones, Thomas D. "Above and Beyond: No Way Out." *Smithsonian Magazine*, July 2002. https://www.smithsonianmag.com/air-space-magazine/above-amp-beyond-no-way-out-30027019/.

Jones, Thomas D., Rob R. Landis, David, J. Korsmeyer, Paul A. Abell, and Daniel R. Adamo. "Strengthening U.S. Exploration Policy via Human Expeditions to

Near-Earth Objects." *Academia*, 15 September 2009. https://www.academia.edu/20717500/Strengthening_US_Exploration_Policy_via_Human_Expeditions_to_Near_Earth_Objects.

Kennedy, Gregory P. "Jet Shoes and Rocket Packs: The Development of Astronaut Maneuvering Units." *Space World*, October 1984.

Kross, John F. "Enter the Dragon: China and the Next Era in Space." *ad Astra* 33, no. 4 (Fall 2021): 35–37.

Kurkowski, Seth. "NASA Artemis Spacesuit Contract Given to Axiom Space and Collins Aerospace." Space Explored, 1 June 2022. https://spaceexplored.com/2022/06/01/nasa-artemis-spacesuit-contract-winners/.

Lowth, Marcus. "10 Moons Humans Could Colonize." Listverse, 3 July 2016. https://listverse.com/2016/07/03/10-moons-humans-could-colonize/#:~:text=%2010%20Moons%20Humans%20Could%20Colonize%20%201,moons%2C%20Miranda%2C%20the%20smallest%20of%20the . . . %20More%20.

Lucid, Shannon. "The Other Side of the Hatch." 1996. https://www.nasa.gov/history/SP-4225/documentation/lucid-letters/letter2.htm.

Mann, Adam. "Radiation Levels on Moon 200 Times Greater Than on Earth, China's Chang'e-4 Lunar Lander Finds." *Science*, 25 September 2020. https://www.science.org/content/article/moon-safe-long-term-human-exploration-first-surface-radiation-measurements-show.

Marwaha, Nikita. "Space Wardrobe Design: Chinese Spacesuit Analysis and Inspiration, in Space Suit Design." *Space Safety Magazine*, 28 November 2013. http://www.spacesafetymagazine.com/aerospace-engineering/space-suit-design/space-wardrobe-design-chinese-spacesuit-analysis-inspiration/.

Mattingly, Sam G., and John B. Charles. "A Personal History of Underwater Neutral Buoyancy Simulation." *Space Review*, 4 February 2013. https://www.thespacereview.com/article/2231/1.

Mohr, Ed. "Our Patch." NASA EISD Newsletter, August 2016. https://www.nasa.gov/sites/default/files/atoms/files/august_2016_eisd_newsletter_outreach.pdf.

NASA. "Steppin' out with Flash and Buck." *Space News Roundup* 23, no. 4 (24 February 1984). https://historycollection.jsc.nasa.gov/JSCHistoryPortal/history/roundups/issues/84-02-.

Navin, Joseph, Pete Harding, and Chris Gebhardt. "Three Spacewalks Completed across Two Days on Two Space Stations." NASA Spaceflight, 17 November 2022. https://www.nasaspaceflight.com/2022/11/three-spacewalks/.

Neufeld, Michael. "Inventing Underwater Training for Walking in Space." National Air and Space Museum, 17 February 2016. https://airandspace.si.edu/stories/editorial/inventing-underwater-training-walking-space.

O'Toole, Thomas. "Astronaut Sold NASA on Jet-Pack Flight." *Washington Post*, 7 February 1984. https://www.washingtonpost.com/archive/politics/1984/02/07/astronaut-sold-nasa-on-jet-pack-flight/752b4696-e389-42fd-b4e7-be42e5eeea9f/.

Parmitano, Luca. "EVA 23: Exploring the Frontier." ESA blog post, 20 August 2013. https://blogs.esa.int/luca-parmitano/2013/08/20/eva-23-exploring-the-frontier/.

Price, Jenny. "Shared Space." *On Wisconsin*, Spring 2012. https://onwisconsin.uwalumni.com/shared-space/.

Ross, Lillian. "Interest." *New Yorker*, "The Talk of the Town," 17 January 1969. https://www.newyorker.com/magazine/1969/01/25/interest-2.

Russian Space Web. "Final Decision to Launch Soyuz-1." Updated 7 September 2018. http://russianspaceweb.com/soyuz1-decision.html.

———. "Gagarin Cosmonaut Training Center in Star City." Updated 18 May 2021. https://russianspaceweb.com/star_city.html.

Scheuring, Richard A., Charles H. Mathers, Jeffrey A. Jones, and Mary L. Wear. "Musculoskeletal Injuries and Minor Trauma in Space: Incidence and Injury Mechanisms in U.S. Astronauts." *Aviation, Space, and Environmental Medicine* 80, no. 2 (1 February 2009). https://ntrs.nasa.gov/citations/20110020735.

Skipper, James M. "Women Space Firsts: A Footnote to History: Tereshkova, Savitskaya, Ride, Sullivan. First EVA by a Woman of the United States." Updated 30 July 2003. https://jamesmskipper.tripod.com/jamesmskipper/SpaceWomen.html.

Thorbecke, Catherine. "China Unveils Ambitious 'Roadmap of Human Mars Exploration.'" ABC News, 24 June 2021. https://abcnews.go.com/Technology/china-unveils-ambitious-roadmap-human-mars-exploration/story?id=78463069.

Tomorrow's World Today. "NASA Recommends New Radiation Limits for Astronauts." 15 August 2021. https://www.tomorrowsworldtoday.com/2021/08/15/nasa-recommends-new-radiation-limits-for-astronauts/.

Vinogradov, Sergei. "Cosmonaut Anatoly Solovyev: Studying Space Is Costly, but It Has to Be Done." Russkiy Mir Foundation, 24 March 2017. https://russkiymir.ru/en/publications/223915/.

Zak, Anatoly. "Alexei Leonov's First Spacewalk Wasn't Quite as Dramatic as We Thought." *Smithsonian Magazine*, 26 March 2020. https://www.smithsonianmag.com/air-space-magazine/turns-out-alexei-leonovs-first-spacewalk-wasnt-quite-dramatic-we-thought-180974522/.

———. "Cosmonauts Transfer from Soyuz-5 to Soyuz-4." Russian Space Web, updated 17 January 2019. http://russianspaceweb.com/soyuz4-soyuz5-eva.html.

———. "Leonov Performs the World's First Spacewalk." Russian Space Web, updated 20 March 2020. https://russianspaceweb.com/voskhod2-eva.html.

———. "Mir's Kvant-2 Module." Russian Space Web, updated 26 September 2018. https://russianspaceweb.com/mir_kvant-2.html.
———. "Mission of Kosmos-57." Russian Space Web, 25 March 2021. https://russianspaceweb.com/voskhod2-kosmos57.html.
———. "Planning Joint Mission of Soyuz-4 and -5." Russian Space Web, updated 15 January 2019. http://russianspaceweb.com/soyuz4-soyuz5-scenario.html.

## Interviews and Personal Communications

Akers, Tom. Email correspondence with Melvin Croft, 4 February 2020.
Allen, Joseph P. JSC oral history interview by Jennifer Ross-Nazzal, Washington DC, 18 March 2004.
———. JSC oral history interview by Jennifer Ross-Nazzal, Washington DC, 18 November 2004.
Armstrong, Neil A. JSC oral history interview by Dr. Stephen E. Ambrose and Dr. Douglas Brinkley, Houston TX, 19 September 2001.
Cernan, Eugene. JSC oral history interview by Rebecca Wright, Houston TX, 11 December 2007.
Chretien, John-Loup. JSC oral history interview by Carol Butler, Houston TX, 2 May 2002.
Collins, Michael. JSC oral history interview by Michelle Kelly, Oakville ON, 8 October 1997.
Duke, Charles. JSC oral history interview by Doug Ward, Houston TX, 12 March 1999.
Foale, Michael. JSC oral history interview by Rebecca Wright, Houston TX, 7 July 1998.
Fullerton, Richard K. JSC oral history interview by Mark Davison, Houston TX, 21 May 1998. https://historycollection.jsc.nasa.gov/JSCHistoryPortal/history/oral_histories/Shuttle-Mir/FullertonRK/FillertonRK_5-21-98.htm.
Gibson, Ed. Telephone interview with John Youskauskas, 1 January 2021.
Gibson, Robert L. JSC oral history interview by Jennifer Ross-Nazzal, Houston TX, 22 January 2016.
Godwin, Linda. Telephone interview with Melvin Croft, 5 August 2020.
Gonzalez-Torres, Tomas. Telephone interview with John Youskauskas, 30 April 2020.
Grechko, Georgi. Association of Space Explorers, Eighth Planetary Congress interview by Bert Vis, Washington DC, 30 August 1992. Used with permission.
Hansen, Christy. Telephone interview with Melvin Croft, 11 April 2020.
Herrington, John. Telephone interview with Melvin Croft, 12 March 2020.
Hieb, Rick. Telephone interview with Melvin Croft, 19 May 2020.
Hoffman, Jeffrey. Skype interview with Melvin Croft, 23 October 2020.
Johnson, Kieth. Telephone interview with Melvin Croft, 5 May 2020.

Jones, Thomas. Email correspondence with John Youskauskas, 1–2 February 2024.

Kerwin, Joseph P. JSC oral history interview by Kevin M. Rusnak, Houston TX, 12 May 2000.

Kleinknecht, Kenneth. JSC oral history interview by Carol Butler, Littleton CO, 10 September 1998.

Kranz, Eugene F. JSC oral history interview by Jennifer Ross-Nazzal, Dickinson TX, 7 December 2011.

Leonov, Alexei. “The First Man to Walk in Outer Space.” Interview with FAI, 18 March 1965. https://www.fai.org/sites/default/files/article/document/en-18-march-1965-leonov-first-spacewalk-interview.pdf.

———. Interview. YouTube, posted by NASA, 18 March 2015. https://www.youtube.com/watch?v=HnREkhEOKBE.

Linenger, Jerry. Telephone interview with John Youskauskas, 7 August 2020.

———. Email correspondence with John Youskauskas, 8 August 2020.

Lousma, Jack R. JSC oral history interview by Carol Butler, Houston TX, 7 March 2001.

———. Telephone interview with John Youskauskas, 16 October 2020.

Mattingly, Thomas K., II. JSC oral history interview by Rebecca Wright, Costa Mesa CA, 6 November 2001.

McArthur, William S., Jr. JSC oral history interview by Jennifer Ross-Nazzal, Houston TX, 17 April 2018.

McBarron, James W., II. JSC oral history interview by Rebecca Wright, Houston TX, 28 September 2012.

McMann, Harold Joseph. JSC oral history interview by Kevin Rusnak, Houston TX, 25 January 2002.

Mitchell, Edgar D. JSC oral history interview by Sheree Scarborough, Houston TX, 3 September 1997.

Musgrave, Story. Telephone interview with Melvin Croft, 1 June 2020.

Parazynski, Scott. Email correspondence with Melvin Croft, 23 January 2021.

———. Email correspondence with Melvin Croft, 21 May 2021.

———. Email correspondence with Melvin Croft, 11 August 2021.

———. Email correspondence with Melvin Croft, 3 November 2021.

Peterson, Don. JSC oral history interview by Jennifer Ross-Nazzal, Houston TX, 14 November 2002.

Pronina, Irina, and Natalya Kuleshova. Interview by Bert Vis, Star City, Moscow, 14 April 2000. Used with permission.

Reightler, Kenneth, Jr. Telephone interviews with Melvin Croft, 1 April 2020; 3 and 4 April 2021; 20, 21, 26 March 2021.

Rezac, Ed. Telephone interview with Melvin Croft, 1 April 2020.

Ross, Jerry. JSC oral history interview by Jennifer Ross-Nazzal, 5 February 2004.
———. Telephone interview with John Youskauskas, 26 May 2020.
Savitskaya, Svetlana. Interview by Bert Vis and Gordon Hooper, Groningen, Netherlands, 3 July 1995. Used with permission.
Schmitt, Harrison. JSC oral history interview by Carol Butler, Houston TX, 16 March 2000.
Schweickart, Russell. Telephone interview with Melvin Croft, 11 August 2020.
Scott, Winston. Telephone interview with Melvin Croft, 27 July 2021.
Shepard, Alan B., Jr. JSC oral history interview by Roy Neal, Pebble Beach FL, 20 February 1998.
Stewart, Robert. Telephone interview with John Youskauskas, 27 September 2021.
Stott, Nicole. Email correspondence with John Youskauskas, 4 October 2021.
Tanner, Joseph. Email correspondence with Melvin Croft, March 2020.
———. Telephone interview with Melvin Croft, 23 March 2020.
Thornton, Kathryn. Telephone interview with Melvin Croft, 28 July 2020.
van Hoften, James D. A. JSC oral history interview by Jennifer Ross-Nazzal, Lafayette CA, 5 December 2007.
Volynov, Boris. Interview by Bert Vis, London, England, 16 May 2001. Used with permission.
Weitz, Paul J. JSC oral history interview by Rebecca Wright, Flagstaff AZ, 26 March 2000.
Wolf, David J. JSC oral history interview by Rebecca Wright, Houston TX, 23 June 1998.

## Other Sources

Ayrey, William. "ILC Spacesuits and Related Products, 0000-712731 Rev. A." NASA, 28 November 2007. https://www.nasa.gov/wp-content/uploads/static/history/alsj/ILC-SpaceSuits-RevA.pdf.
B., Scott, Sandy Guthrie, Jill Shinefield, and Gail Willumsen, dirs. *Black Sky: The Race for Space*. Discovery Channel and Vulcan Productions, 2004. https://mubi.com/films/black-sky-the-race-for-space.
Bland, Daniel L., Jr. "Space Shuttle EVA Opportunities." JSC-11391. Johnson Space Center, 1 January 1976.
Brown, Travis J. "Skylab Astronaut Life Support Assembly." NASA Manned Spacecraft Center. https://core.ac.uk/reader/10297709.
"Bruce McCandless STS-41B and STS-31." UK Space Lectures. YouTube, posted by Leo Bakker Flight and Space, 4 November 2017. https://www.youtube.com/watch?v=c8zBkqYMNTo.
"Chinese Astronaut Makes Nation's First Spacewalk." YouTube, posted by LunarTV1, 27 September 2008. https://www.youtube.com/watch?v=9PAkbT4kTpI.

Committee on Government Operations, House of Representatives. *Space Station: Improving NASA's Planning for External Maintenance*. Report to the Chair, Government Activities and Transportation Subcommittee, July 1992.

Ellis, James L. *Shuttle EVA Description and Design Criteria*. JSC-10615. Johnson Space Center, May 1976.

"EVA 23: Chris Cassidy and Luca Parmitano Spacesuit Water Leak." YouTube, posted by Eric Pruden, 20 March 2017. https://youtu.be/0KQ3-thZDUo.

"EVA 23: The Most Dangerous EVA in U.S. History." YouTube, posted by NASA Video, 25 June 2020. https://www.youtube.com/watch?v=9iE_69aeVZ4.

"First Spacewalk [real speed, 2 cameras]—Alexey Leonov—VOSKHOD 2 (1965)." YouTube, posted by Retro Space HD, 4 June 2018. https://www.youtube.com/watch?v=7mUO1TDr7fQ.

Gaier, James R. "The Effects of Lunar Dust on EVA Systems during the Apollo Missions." NASA, March 2005. https://www.nasa.gov/wp-content/uploads/static/history/alsj/tm-2005-213610.pdf.

Gray, Mark, prod. *Gemini: Complete Download Edition*. Atlanta GA: Spacecraft Films, 2003. DVD set.

———. STS-41b TV Highlights and Post-Flight Crew News Conference. Shuttle Mission Series. Atlanta GA: Spacecraft Films, 2011.

Guy, W. W. "Shuttle EVA Training Options, Crew Systems Division." Johnson Space Center, 12 April 1982.

Hansen, Christopher P., et al. "International Space Station (ISS) EVA Suit Water Intrusion—High Visibility Close Call." NASA, 20 December 2013.

Hewes, Donald E., and Kenneth E. Glover. "Development of Skylab Experiment T020 Employing a Foot Controlled Maneuvering Unit." NASA Langley Research Center, March 1972.

Hill, Terry. "Lessons Learned from the EMU Fire and How It Impacts CxP Suit Element Development & Testing." PowerPoint presentation, 15 September 2008.

Hodgson, Edward, Ronald Sidgreaves, Stephen Braham, Jeffrey Hoffman, Christopher Carr, Pascal Lee, Jose Marmolejo, Jonathan Miller, Ilia Rosenberg, and Steven Schwartz. "Requirements and Potential for Enhanced EVA Information Interfaces." Thirty-Third International Conference on Environmental Systems (ICES), Vancouver BC, 7–10 July 2003. https://www.sae.org/publications/technical-papers/content/2003-01-2413/.

Hoffman, Stephen J. "Advanced EVA Capabilities: A Study for NASA's Revolutionary Aerospace Systems Concept Program." NASA, April 2004. https://ntrs.nasa.gov/citations/20040200983.

Hubbell Power Systems. *Saving Skylab: America's First Space Station*. 18 April 2020. Film. https://www.hubbell.com/hubbellpowersystems/en/savingskylab.

"ISS Expedition 36—US Spacewalk EVA #23 Is Aborted Early Due to a Suit Water Leak." YouTube, posted by Matthew Travis, 16 July 2013. https://www.youtube.com/watch?v=9vk_UdxAH_g.

Jones, Eric M., and Ken Glover, eds. *Apollo Lunar Surface Journal (Apollo's 12, 15, and 16)*. NASA, last revised 5 June 2018. https://www.hq.nasa.gov/alsj/index.html.

Jones, Thomas D., Rob R. Landis, Paul A. Abell, Daniel R. Adamo, Ronald D. Mazanek, Timothy P. Kennedy, David, J. Korsmeyer, and Donald K. Yeomans. "Astronauts beyond the Moon: Mission Operations at a Near-Earth Object." SpaceOps 2021 Conference: Delivering the Dream, Huntsville AL, 25–30 April 2010.

Leonov, Alexei. "The Spacewalker." YouTube, posted by Vladimir Kozlov, 26 September 2017. https://www.youtube.com/watch?v=C-FPhZHzLxE.

"Let's Fly to the Moon." YouTube, posted by CommericalThyme, 17 December 2019. https://youtu.be/ajFLUvpenw8.

"Lost in Space." British Broadcasting Corporation, 1984. YouTube, posted by lunarmodule5, retrieved 22 June 2021. https://www.youtube.com/watch?v=qW5Qa9QOpZY.

McMann, Harold J., and James W. McBarron II. "Challenges in the Development of the Shuttle Extravehicular Mobility Unit." NASA, 1 January 1985. https://ntrs.nasa.gov/citations/19850008613.

Mendell, W. W., and A. D. Griffith. "Lunar Precursor Missions for Human Exploration of Mars: Studies of Mission Operations." Fifty-Third International Astronautical Congress, World Space Congress, NASA, Houston TX, 10–19 October 2002. https://www.researchgate.net/publication/234346426_Lunar_Precursor_Missions_for_Human_Exploration_of_Mars_-_II_Studies_of_Mission_Operations.

NASA. "Advancing Automation and Robotics Technology for the Space Station Freedom and for the U.S. Economy." Progress Report 11. Advanced Technology Advisory Committee. Submitted to the United States Congress, November 1990. https://ntrs.nasa.gov/api/citations/19910006379/downloads/19910006379.pdf.

———. "Apollo 15 Statistics and Performance." Retrieved 16 January 2022. https://history.nasa.gov/afj/simbaycam/a15-stat-perf.html.

———. "Apollo 16 Air-to-Ground Voice Transcription." June 1972. https://www.hq.nasa.gov/alsj/a16/AS16_TEC.PDF.

———. "Apollo 16 Command Module Onboard Voice Transcription." June 1972. https://www.hq.nasa.gov/alsj/a16/AS16_CM.PDF.

———. "Apollo 16 Statistics and Performance." Retrieved 16 January 2022. https://www.hq.nasa.gov/office/pao/History/afj/simbaycam/a16-stat-perf.html.

———. "Apollo 17 Statistics and Performance." https://history.nasa.gov/afj/simbaycam/a17-stat-perf.html.

———. "Apollo 17 Technical Air-to-Ground Voice Transcription." Manned Spacecraft Center, December 1972. https://historycollection.jsc.nasa.gov/JSCHistoryPortal/history/mission_trans/AS17_TEC.pdf.

———. "Astronaut Rips Handle off Hubble Space Telescope." YouTube, posted by VideoFromSpace, 27 July 2015. https://www.youtube.com/watch?v=rUL38xXd7js.

———. "The Cable Cutter That Saved Skylab." 24 September 2015. https://www.nasa.gov/image-feature/the-cable-cutter-that-saved-skylab.

———. "EMU Development, Schedules, and OFT Requirements." Memorandum from T. K. Mattingly to John W. Young, 24 January 1977.

———. "EMU PDR." Memorandum from T. K. Mattingly to John W. Young, 30 June 1977.

———. "EMU Sizing Concepts." Memorandum from T. K. Mattingly to John W. Young, undated.

———. "EVA Glove Development." Memorandum from T. K. Mattingly to Story Musgrave, 2 December 1975.

———. "EVA Insignia." http://www.collectspace.com/ubb/Forum18/HTML/000162.html.

———. "Extravehicular Mobility Unit Operational Requirements—Final." Memorandum. JSC-09973. December 1976.

———. "Gateway." 21 December 2021. https://www.nasa.gov/gateway/overview.

———. Gemini IV Mission Commentary Transcript.

———. "Gemini IX-A Technical Debriefing." Manned Spacecraft Center, 11 June 1966.

———. Gemini IX-A Voice Communications. Air-to-Ground, Ground-to-Air and On-Board Transcription.

———. "Gemini X Technical Debriefing." Manned Spacecraft Center, 26 July 1966.

———. Gemini X Voice Communications. Air-to-Ground, Ground-to-Air and On-Board Transcription.

———. "Gemini XI Technical Debriefing." Manned Spacecraft Center, 19 September 1966.

———. Gemini XI Voice Communications. Air-to-Ground, Ground-to-Air and On-Board Transcription.

———. "Gemini XII Technical Debriefing." Manned Spacecraft Center, 22 November 1966.

———. Gemini XII Voice Communications. Air-to-Ground, Ground-to-Air and On-Board Transcription.

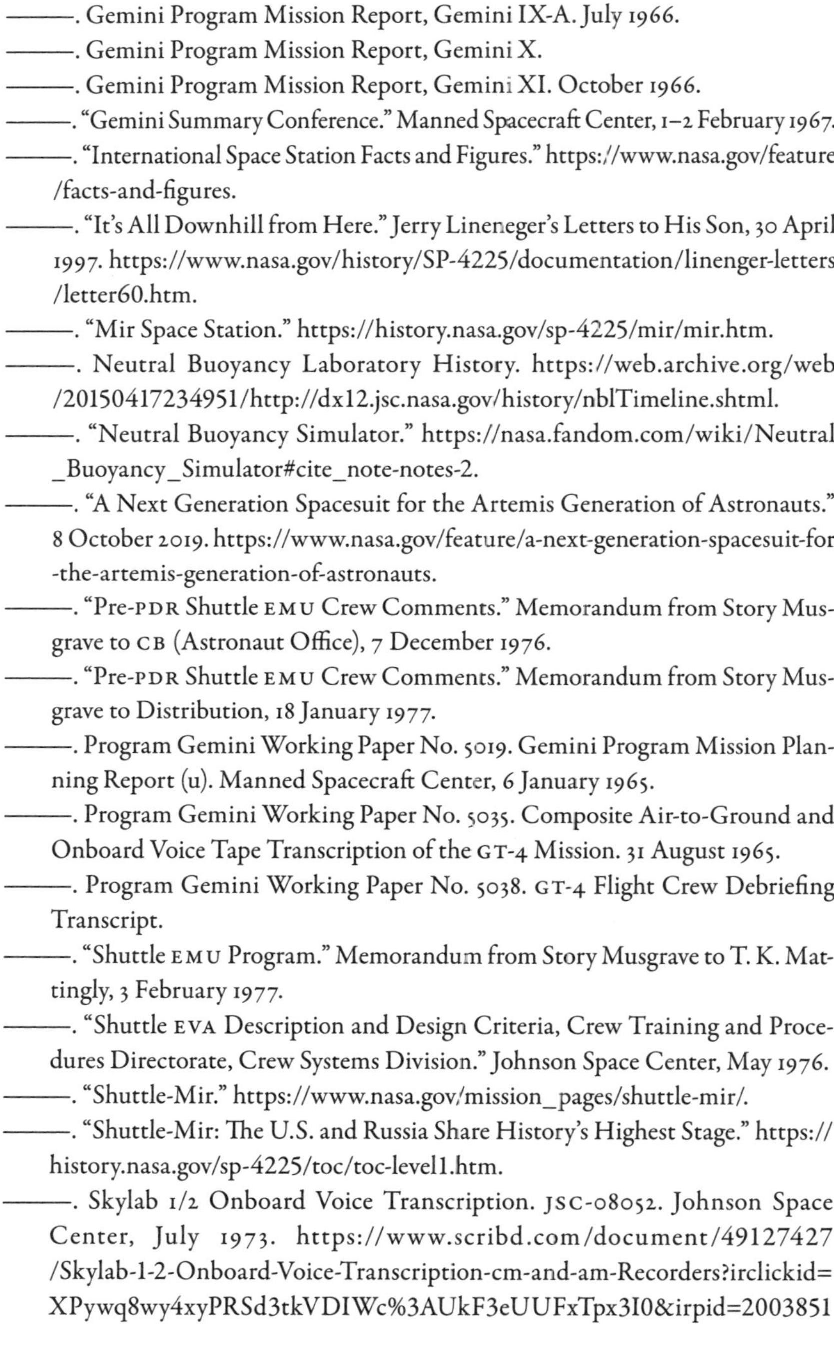

———. Gemini Program Mission Report, Gemini IX-A. July 1966.

———. Gemini Program Mission Report, Gemini X.

———. Gemini Program Mission Report, Gemini XI. October 1966.

———. "Gemini Summary Conference." Manned Spacecraft Center, 1–2 February 1967.

———. "International Space Station Facts and Figures." https://www.nasa.gov/feature/facts-and-figures.

———. "It's All Downhill from Here." Jerry Lineneger's Letters to His Son, 30 April 1997. https://www.nasa.gov/history/SP-4225/documentation/linenger-letters/letter60.htm.

———. "Mir Space Station." https://history.nasa.gov/sp-4225/mir/mir.htm.

———. Neutral Buoyancy Laboratory History. https://web.archive.org/web/20150417234951/http://dx12.jsc.nasa.gov/history/nblTimeline.shtml.

———. "Neutral Buoyancy Simulator." https://nasa.fandom.com/wiki/Neutral_Buoyancy_Simulator#cite_note-notes-2.

———. "A Next Generation Spacesuit for the Artemis Generation of Astronauts." 8 October 2019. https://www.nasa.gov/feature/a-next-generation-spacesuit-for-the-artemis-generation-of-astronauts.

———. "Pre-PDR Shuttle EMU Crew Comments." Memorandum from Story Musgrave to CB (Astronaut Office), 7 December 1976.

———. "Pre-PDR Shuttle EMU Crew Comments." Memorandum from Story Musgrave to Distribution, 18 January 1977.

———. Program Gemini Working Paper No. 5019. Gemini Program Mission Planning Report (u). Manned Spacecraft Center, 6 January 1965.

———. Program Gemini Working Paper No. 5035. Composite Air-to-Ground and Onboard Voice Tape Transcription of the GT-4 Mission. 31 August 1965.

———. Program Gemini Working Paper No. 5038. GT-4 Flight Crew Debriefing Transcript.

———. "Shuttle EMU Program." Memorandum from Story Musgrave to T. K. Mattingly, 3 February 1977.

———. "Shuttle EVA Description and Design Criteria, Crew Training and Procedures Directorate, Crew Systems Division." Johnson Space Center, May 1976.

———. "Shuttle-Mir." https://www.nasa.gov/mission_pages/shuttle-mir/.

———. "Shuttle-Mir: The U.S. and Russia Share History's Highest Stage." https://history.nasa.gov/sp-4225/toc/toc-level1.htm.

———. Skylab 1/2 Onboard Voice Transcription. JSC-08052. Johnson Space Center, July 1973. https://www.scribd.com/document/49127427/Skylab-1-2-Onboard-Voice-Transcription-cm-and-am-Recorders?irclickid=XPywq8wy4xyPRSd3tkVDIWc%3AUkF3eUUFxTpx3I0&irpid=2003851

&utm_source=impact&utm_medium=cpc&utm_campaign=affiliate_pdm_acquisition_Bing%20Rebates%20by%20Microsoft&sharedid=EdgeBingFlow&irgwc=1.

———. Skylab 1/3 Onboard Voice Transcription, Part 4 of 4. Johnson Space Center, October 1973. https://www.scribd.com/document/49133003/Skylab-1-3-Onboard-Voice-Transcription-Part-4-of-4.

———. Skylab Press Kit. NASA Release no. 73-80. Washington DC, 1 May 1973. http://www.ccas.us/CCAS_NASAPressKits/NearEarth/Skylab1-2_Press_Kit.pdf.

———. STS-1 Flight Data File—EVA Operations Book (Final), Crew Training and Procedures. Flight Operations Directorate, 9 February 1981.

———. STS-1 Press Kit. NASA Release no. 81-39. April 1981. https://historycollection.jsc.nasa.gov/JSCHistoryPortal/history/shuttle_pk/pk/Flight_001_STS-001_Press_Kit.pdf.

———. STS-5 Press Kit. NASA Release no. 82-156. October 1982. https://historycollection.jsc.nasa.gov/JSCHistoryPortal/history/shuttle_pk/pk/Flight_005_STS-005_Press_Kit.pdf.

———. STS-6 Press Kit. NASA Release no. 83-36. April 1983. https://historycollection.jsc.nasa.gov/JSCHistoryPortal/history/shuttle_pk/pk/Flight_006_STS-006_Press_Kit.pdf.

———. STS-41B MMU. YouTube, posted by lunarmodule5, retrieved 2 June 2021. https://youtu.be/Y2SW6mor0bE?si=sNG8zjm6L9oLVg0k; https://youtu.be/zBIiI1DV5Xg?si=A15pQHKqw6NdGm6T.

———. STS-49 Space Shuttle Mission Report. Johnson Space Center, July 1992.

———. "STS-80 Flight Day 11." YouTube, posted by NASA STI Program, 28 February 2013. https://www.youtube.com/watch?v=g2v2Kd7jzbs.

———. "STS-80 Flight Day 13." YouTube, posted by NASA STI Program, 28 February 2013. https://www.youtube.com/watch?v=-DkSFVi-Zdo.

———. STS-92 Report #15. 18 October 2000. https://science.ksc.nasa.gov/shuttle/missions/sts-92/news/sts-92-mcc-15.txt.

———. "STS Flight 64 Mission Highlights." Post-Flight Press Conference, 14 August 1992. YouTube, posted by NASA STI Program, 21 June 2011. https://www.youtube.com/watch?v=axo_JoMH-PU.

———. "Suited for Spacewalking: An Activity Guide for Technology Education, Mathematics, and Science." EG-1998-03-112-HQ. Washington DC, 1998.

———. *Summary of Gemini Extravehicular Activity*. Edited by Reginald M. Machell. Manned Spacecraft Center, 1967.

———. "Welcome to Shuttle/Mir—The U.S. and Russia Share History's Highest Stage." https://history.nasa.gov/SP-4225/.

Nicholas, Stephanie. "So Close Yet So Far: The Jammed Airlock Hatch of STS-80." NASA SCSC-E-0 113. Retrieved 5 January 2022. https://ntrs.nasa.gov/api/citations/20130013499/downloads/20130013499.pdf.

Paddock, Eddie. "Overview of NASA EVA Virtual Reality Training, Past, Present and Future." NASA/JSC Engineering. https://vrtech.wiki/wp-content/uploads/2021/01/NASA-EVA-Virtual-Reality-Training-Guide-Original.pdf.

Parazynski, Scott. *EVA Physiology and Medical Considerations Working in a Spacesuit*. NESC Academy, U.S. Spacesuit Knowledge Capture, 24 January 2012. Film. https://mediaex-server.larc.nasa.gov/Academy/Play/6888c89500444d84b44e8a18f31cbbf21d#!.

———. *EVA Skills Training*. NESC Academy, U.S. Spacesuit Knowledge Capture, 6 March 2012. Film. https://mediaex-server.larc.nasa.gov/Academy/Play/2c423fabc80e46bdae08b2401949e3241d#!.

———. *Real Time EVA Troubleshooting*. NESC Academy, 16 February 2021. Film. https://nescacademy.nasa.gov/video/998d7f649a644d129494d4aa24603b1c1d.

———. *TPS Inspection and Repair*. NESC Academy, U.S. Spacesuit Knowledge Capture, 23 February 2012. Film. https://nescacademy.nasa.gov/video/1c42b1687bed4910b6e6120e9a3b01681d.

PBS NOVA. "The Russian Right Stuff: The Dark Side of the Moon." YouTube, posted by Space SPAN, 27 February 1991. https://www.youtube.com/watch?v=B_ZV5l2J8MI.

"Red Star in Orbit: The Mission." YouTube, posted by Coveant, 19 May 2016. https://www.youtube.com/watch?v=E4UD2MCCRVI.

"Requirements and Potential for Enhanced EVA Information Interfaces." SAE Technical Paper Series 2003-01-2413. Thirty-Third International Conference on Environmental Systems (ICES), Vancouver BC, 7–10 July 2003. https://www.sae.org/publications/technical-papers/content/2003-01-2413/.

Ross, Amy, Richard Rhodes, and Shane McFarland. "NASA Advanced Space Suit Pressure Garment System Status and Development Priorities 2019." Internation Conference on Environmental Systems, Boston MA, 7–11 July 2019.

Scheuring, R. A., P. McCulloch., Mary Van Baalen, Charles Minard, Richard Watson, Steve Bowen, and T. Blatt. "Shoulder Injuries in US Astronauts Related to EVA Suit Design Introduction." NASA Technical Reports Server, 11 May 2012. https://ntrs.nasa.gov/api/citations/20120009404/downloads/20120009404.pdf.

Scheuring, Richard. "Musculoskeletal Injuries in US Astronauts." Johnson Space Center, 9 September 2018. https://ntrs.nasa.gov/api/citations/20190001395/downloads/20190001395.pdf.

Schweickart, Russell. Website. https://www.rustyschweickart.com/.

*Skylab Space Station*. Periscope Films. https://periscopefilm.com/skylab-news/ (site discontinued).

Soyuz TM-8 Spaceflight Mission Report. Accessed 26 August 2021. http://www.spacefacts.de/mission/english/soyuz-tm8.htm.

Steele, John W., et al. "Management of the Post-Shuttle Extravehicular Mobility Unit (EMU) Water Circuits." Hamilton Sundstrand Space Systems International, Inc. and Johnson Space Center. https://core.ac.uk/download/pdf/10566902.pdf.

Sudhakar, Rajulu, and Elizabeth Benson. "Complexity of Sizing for Space Suit Applications. NASA." Thirteenth International Conference on Human-Computer, San Diego CA, 14 July 2009. https://ntrs.nasa.gov/api/citations/20090009309/downloads/20090009309.pdf?attachment=true.

"Weightless Environment Training Facility (Hydro Lab)." Yu. A. Gagarin Research and Test Cosmonaut Training Center. http://www.gctc.su/main.php?id=130.

Whitsett, C. E., Jr., and B. McCandless II. "Skylab Experiment M509, Astronaut Maneuvering Equipment Orbital Test Results and Future Applications." Presented to the American Astronautical Society Twentieth Annual Meeting, Los Angeles CA, 20–22 August 1974.

Wikipedia. "Charles E. Whitsett." Last modified 12 April 2024. https://en.wikipedia.org/wiki/Charles_E._Whitsett.

———. "European Robotic Arm." Last modified 18 April 2024. https://en.m.wikipedia.org/wiki/European_Robotic_Arm.

———. "SuitSat." Last modified 19 May 2024. https://en.m.wikipedia.org/wiki/SuitSat.

Wiseman, Reid (@Astro.Reid). "While spacewalking I realized something: I used to think I was scared of heights but now I know I was just scared of gravity." Twitter, 10 October 2014, 4:49 a.m. https://twitter.com/astro_reid/status/520541606538387456?s=43&t=zo8aEFi5SUErxMD-TIIBjA.

# Index

**In the Outward Odyssey: A People's History of Spaceflight series**

*Into That Silent Sea: Trailblazers of the Space Era, 1961–1965*
Francis French and Colin Burgess
Foreword by Paul Haney

*In the Shadow of the Moon: A Challenging Journey to Tranquility, 1965–1969*
Francis French and Colin Burgess
Foreword by Walter Cunningham

*To a Distant Day: The Rocket Pioneers*
Chris Gainor
Foreword by Alfred Worden

*Homesteading Space: The Skylab Story*
David Hitt, Owen Garriott, and Joe Kerwin
Foreword by Homer Hickam

*Ambassadors from Earth: Pioneering Explorations with Unmanned Spacecraft*
Jay Gallentine

*Footprints in the Dust: The Epic Voyages of Apollo, 1969–1975*
Edited by Colin Burgess
Foreword by Richard F. Gordon

*Realizing Tomorrow: The Path to Private Spaceflight*
Chris Dubbs and Emeline Paat-Dahlstrom
Foreword by Charles D. Walker

*The X-15 Rocket Plane: Flying the First Wings into Space*
Michelle Evans
Foreword by Joe H. Engle

*Wheels Stop: The Tragedies and Triumphs of the Space Shuttle Program, 1986–2011*
Rick Houston
Foreword by Jerry Ross

*Bold They Rise: The Space Shuttle Early Years, 1972–1986*
David Hitt and Heather R. Smith
Foreword by Bob Crippen

*Go, Flight! The Unsung Heroes of Mission Control, 1965–1992*
Rick Houston and Milt Heflin
Foreword by John Aaron

*Infinity Beckoned: Adventuring Through the Inner Solar System, 1969–1989*
Jay Gallentine
Foreword by Bobak Ferdowsi

*Fallen Astronauts: Heroes Who Died Reaching for the Moon, Revised Edition*
Colin Burgess and Kate Doolan
with Bert Vis
Foreword by Eugene A. Cernan

*Apollo Pilot: The Memoir of Astronaut Donn Eisele*
Donn Eisele
Edited and with a foreword by Francis French
Afterword by Susie Eisele Black

*Outposts on the Frontier: A Fifty-Year History of Space Stations*
Jay Chladek
Foreword by Clayton C. Anderson

*Come Fly with Us: NASA's Payload Specialist Program*
Melvin Croft and John Youskauskas
Foreword by Don Thomas

*Shattered Dreams: The Lost and Canceled Space Missions*
Colin Burgess
Foreword by Don Thomas

*The Ultimate Engineer: The Remarkable Life of NASA's Visionary Leader George M. Low*
Richard Jurek
Foreword by Gerald D. Griffin

*Beyond Blue Skies: The Rocket Plane Programs That Led to the Space Age*
Chris Petty
Foreword by Dennis R. Jenkins

*A Long Voyage to the Moon: The Life of Naval Aviator and* Apollo 17 *Astronaut Ron Evans*
Geoffrey Bowman
Foreword by Jack Lousma

*The Light of Earth: Reflections on a Life in Space*
Al Worden with Francis French
Foreword by Dee O'Hara

*Son of Apollo: The Adventures of a Boy Whose Father Went to the Moon*
Christopher A. Roosa
Foreword by Jim Lovell

*Star Bound: A Beginner's Guide to the American Space Program, from Goddard's Rockets to Goldilocks Planets and Everything in Between*
Emily Carney and Bruce McCandless III

*Into the Void: Adventures of the Spacewalkers*
John Youskauskas and Melvin Croft
Foreword by Jerry Ross